# BIG IDEAS
# MATH.
## Course 2
## A Common Core Curriculum
### CALIFORNIA EDITION

**Ron Larson**
**Laurie Boswell**

# BIG IDEAS
# LEARNING.

Erie, Pennsylvania
*BigIdeasLearning.com*

Big Ideas Learning, LLC
1762 Norcross Road
Erie, PA 16510-3838
USA

For product information and customer support, contact Big Ideas Learning
at **1-877-552-7766** or visit us at ***BigIdeasLearning.com***.

About the Cover
The cover images on the *Big Ideas Math* series illustrate the advancements in
aviation from the hot-air balloon to spacecraft. This progression symbolizes the
launch of a student's successful journey in mathematics. The sunrise in the
background is representative of the dawn of the Common Core era in math
education, while the cradle signifies the balanced instruction that is a pillar
of the *Big Ideas Math* series.

Printed in the U.S.A.

ISBN 13: 978-1-60840-670-8
ISBN 10: 1-60840-670-9

5 6 7 8 9 10 WEB 17 16 15 14

# AUTHORS

**Ron Larson** is a professor of mathematics at Penn State Erie, The Behrend College, where he has taught since receiving his Ph.D. in mathematics from the University of Colorado. Dr. Larson is well known as the lead author of a comprehensive program for mathematics that spans middle school, high school, and college courses. His high school and Advanced Placement books are published by Houghton Mifflin Harcourt. Ron's numerous professional activities keep him in constant touch with the needs of students, teachers, and supervisors. Ron and Laurie Boswell began writing together in 1992. Since that time, they have authored over two dozen textbooks. In their collaboration, Ron is primarily responsible for the pupil edition and Laurie is primarily responsible for the teaching edition of the text.

**Laurie Boswell** is the Head of School and a mathematics teacher at the Riverside School in Lyndonville, Vermont. Dr. Boswell received her Ed.D. from the University of Vermont in 2010. She is a recipient of the Presidential Award for Excellence in Mathematics Teaching. Laurie has taught math to students at all levels, elementary through college. In addition, Laurie was a Tandy Technology Scholar, and served on the NCTM Board of Directors from 2002 to 2005. She currently serves on the board of NCSM, and is a popular national speaker. Along with Ron, Laurie has co-authored numerous math programs.

# ABOUT THE BOOK

The revised *Big Ideas Math* series uses the same research-based strategy of a balanced approach to instruction that made the first *Big Ideas Math* series so successful. This approach opens doors to abstract thought, reasoning, and inquiry as students persevere to answer the Essential Questions that introduce each section. The foundation of the program is the Common Core Standards for Mathematical Content and Standards for Mathematical Practice. Students are subtly introduced to "Habits of Mind" that help them internalize concepts for a greater depth of understanding. These habits serve students well not only in mathematics, but across all curricula throughout their academic careers.

The *Big Ideas Math* series exposes students to highly motivating and relevant problems. Woven throughout the series are the depth and rigor students need to prepare for career-readiness and other college-level courses. In addition, the *Big Ideas Math* series prepares students to meet the challenge of the new Common Core testing.

We consider the *Big Ideas Math* series to be the crowning jewel of 30 years of achievement in writing educational materials.

*Ron Larson*

*Laurie Boswell*

# TEACHER REVIEWERS

Lisa Amspacher
Milton Hershey School
Hershey, PA

Mary Ballerina
Orange County Public Schools
Orlando, FL

Lisa Bubello
School District of Palm
Beach County
Lake Worth, FL

Sam Coffman
North East School District
North East, PA

Kristen Karbon
Troy School District
Rochester Hills, MI

Laurie Mallis
Westglades Middle School
Coral Springs, FL

Dave Morris
Union City Area
School District
Union City, PA

Bonnie Pendergast
Tolleson Union High
School District
Tolleson, AZ

Valerie Sullivan
Lamoille South
Supervisory Union
Morrisville, VT

Becky Walker
Appleton Area School District
Appleton, WI

Zena Wiltshire
Dade County Public Schools
Miami, FL

# STUDENT REVIEWERS

Mike Carter
Matthew Cauley
Amelia Davis
Wisdom Dowds
John Flatley
Nick Ganger

Hannah Iadeluca
Paige Lavine
Emma Louie
David Nichols
Mikala Parnell
Jordan Pashupathi

Stephen Piglowski
Robby Quinn
Michael Rawlings
Garrett Sample
Andrew Samuels
Addie Sedelmyer
Tyler Steffy
Erin Taylor
Reid Wilson

# CONSULTANTS

**Patsy Davis**
Educational Consultant
Knoxville, Tennessee

**Bob Fulenwider**
Mathematics Consultant
Bakersfield, California

**Linda Hall**
Mathematics Assessment Consultant
Norman, Oklahoma

**Ryan Keating**
Special Education Advisor
Gilbert, Arizona

**Michael McDowell**
Project-Based Instruction Specialist
Fairfax, California

**Sean McKeighan**
Interdisciplinary Advisor
Norman, Oklahoma

**Bonnie Spence**
Differentiated Instruction Consultant
Missoula, Montana

# Common Core State Standards for Mathematical Practice

## Make sense of problems and persevere in solving them.
- Multiple representations are presented to help students move from concrete to representative and into abstract thinking
- *Essential Questions* help students focus and analyze
- *In Your Own Words* provide opportunities for students to look for meaning and entry points to a problem

## Reason abstractly and quantitatively.
- Visual problem solving models help students create a coherent representation of the problem
- Opportunities for students to decontextualize and contextualize problems are presented in every lesson

## Construct viable arguments and critique the reasoning of others.
- *Error Analysis*; *Different Words, Same Question*; and *Which One Doesn't Belong* features provide students the opportunity to construct arguments and critique the reasoning of others
- *Inductive Reasoning* activities help students make conjectures and build a logical progression of statements to explore their conjecture

## Model with mathematics.
- Real-life situations are translated into diagrams, tables, equations, and graphs to help students analyze relations and to draw conclusions
- Real-life problems are provided to help students learn to apply the mathematics that they are learning to everyday life

## Use appropriate tools strategically.
- *Graphic Organizers* support the thought process of what, when, and how to solve problems
- A variety of tool papers, such as graph paper, number lines, and manipulatives, are available as students consider how to approach a problem
- Opportunities to use the web, graphing calculators, and spreadsheets support student learning

## Attend to precision.
- *On Your Own* questions encourage students to formulate consistent and appropriate reasoning
- Cooperative learning opportunities support precise communication

## Look for and make use of structure.
- *Inductive Reasoning* activities provide students the opportunity to see patterns and structure in mathematics
- Real-world problems help students use the structure of mathematics to break down and solve more difficult problems

## Look for and express regularity in repeated reasoning.
- Opportunities are provided to help students make generalizations
- Students are continually encouraged to check for reasonableness in their solutions

Go to *BigIdeasMath.com* for more information on the Common Core State Standards for Mathematical Practice.

# Common Core State Standards for Mathematical Content for Grade 7

**Chapter Coverage for Standards**

1 - 2 - 3 - 4 - **5** - **6** - 7 - 8 - 9 - 10

## Domain  Ratios and Proportional Relationships

- Analyze proportional relationships and use them to solve real-world and mathematical problems.

**1** - **2** - 3 - 4 - 5 - 6 - 7 - 8 - 9 - 10

## Domain  The Number System

- Apply and extend previous understandings of operations with fractions to add, subtract, multiply, and divide rational numbers.

1 - 2 - **3** - **4** - 5 - 6 - 7 - 8 - 9 - 10

## Domain  Expressions and Equations

- Use properties of operations to generate equivalent expressions.
- Solve real-life and mathematical problems using numerical and algebraic expressions and equations.

1 - 2 - 3 - 4 - 5 - 6 - **7** - **8** - **9** - 10

## Domain  Geometry

- Draw, construct, and describe geometrical figures and describe the relationships between them.
- Solve real-life and mathematical problems involving angle measure, area, surface area, and volume.

1 - 2 - 3 - 4 - 5 - 6 - 7 - 8 - 9 - **10**

## Domain  Statistics and Probability

- Use random sampling to draw inferences about a population.
- Draw informal comparative inferences about two populations.
- Investigate chance processes and develop, use, and evaluate probability models.

Go to *BigIdeasMath.com* for more information on the Common Core State Standards for Mathematical Content.

# 1 Integers

*I like talking about math, and working with a partner allows me to do that.*

# Rational Numbers

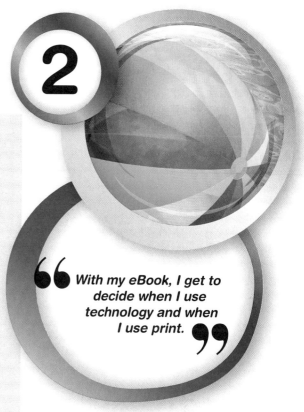

*With my eBook, I get to decide when I use technology and when I use print.*

# 3

# Expressions and Equations

66 *I like that the Essential Question helps me begin thinking about the lesson.* 99

# Inequalities

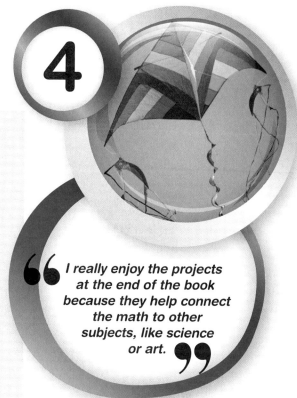

" I really enjoy the projects at the end of the book because they help connect the math to other subjects, like science or art. "

# 5 Ratios and Proportions

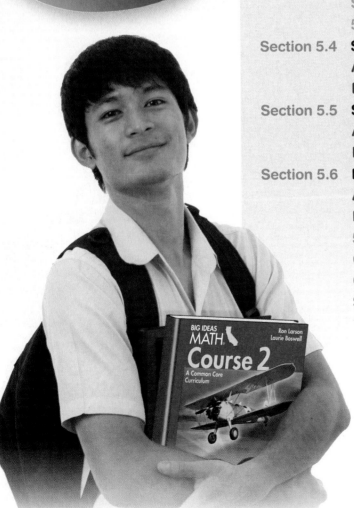

*"I like Newton and Descartes! The cartoons are funny and I like that they model the math that we are learning."*

# Percents

*I really like the Big Ideas Math website! The online resources are a huge help when I get stuck or need extra help.*

# 7
# Constructions and Scale Drawings

" *I like the real-life application exercises because they show me how I can use the math in my own life.* "

# Circles and Area

**8**

*" I like playing the games in the Game Closet! They are a fun way to practice concepts we are learning in class. "*

# Surface Area and Volume

**9**

# Probability and Statistics

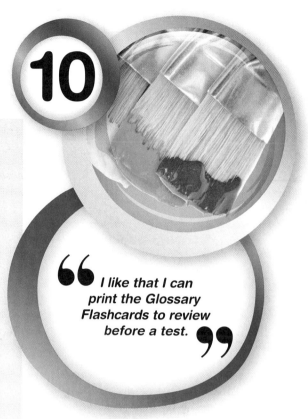

## Appendix A: My Big Ideas Projects

> *I like that I can print the Glossary Flashcards to review before a test.*

# How to Use Your Math Book

● Read the **Essential Question** in the activity.

Discuss the **Math Practice** question with your partner.

Work with a partner to decide **What Is Your Answer?**

Now you are ready to do the **Practice** problems.

● Find the **Key Vocabulary**  words, **highlighted in yellow**.

Read their definitions. Study the concepts in each **Key Idea**.
If you forget a definition, you can look it up online in the

**Multi-Language Glossary** at BigIdeasMath✓com.

● After you study each **EXAMPLE**, do the exercises in the ● **On Your Own**.

**Now You're Ready** to do the exercises that correspond to the example.

As you study, look for a **Study Tip** or a **Common Error** .

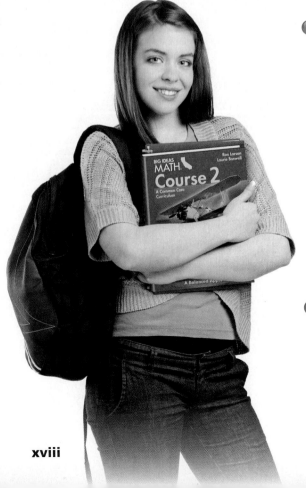

● The exercises are divided into 3 parts.

 **Vocabulary and Concept Check**

 **Practice and Problem Solving**

 **Fair Game Review**

If an exercise has a ① next to it, look back at Example 1 for help with that exercise.

More help is available at **Check It Out** Lesson Tutorials BigIdeasMath ✓com .

● To help study for your test, use the following.

**Quiz**　　**Study Help**

**Chapter Review**　　**Chapter Test**

# SCAVENGER HUNT

Use this *Scavenger Hunt* to find where things are in **Chapter 1**.

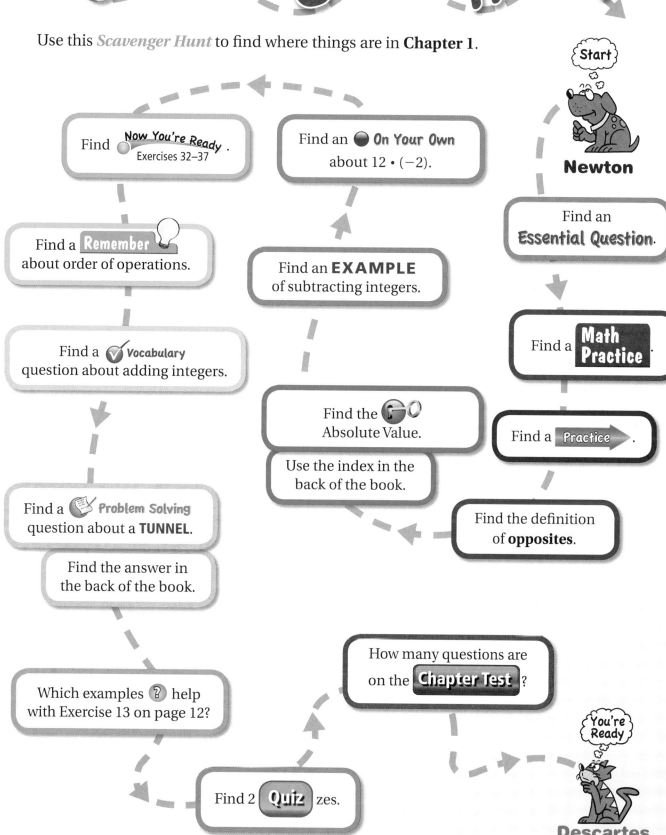

Start

**Newton**

Find an ● **On Your Own** about 12 • (−2).

Find an **Essential Question**.

Find ● **Now You're Ready**. Exercises 32–37

Find a **Remember** about order of operations.

Find a **Math Practice**.

Find an **EXAMPLE** of subtracting integers.

Find a ✓ **Vocabulary** question about adding integers.

Find a **Practice**.

Find the **Absolute Value**.

Use the index in the back of the book.

Find the definition of **opposites**.

Find a **Problem Solving** question about a **TUNNEL**.

Find the answer in the back of the book.

How many questions are on the **Chapter Test**?

Which examples ❓ help with Exercise 13 on page 12?

You're Ready

Find 2 **Quiz**zes.

**Descartes**

# 1 Integers

"Look, subtraction is not that difficult. Imagine that you have five squeaky mouse toys."

"After your friend Fluffy comes over for a visit, you notice that one of the squeaky toys is missing."

"Now, you go over to Fluffy's and retrieve the missing squeaky mouse toy. It's easy."

"Dear Sir: You asked me to 'find' the opposite of −1."

"I didn't know it was missing."

# What You Learned Before

"I liked it because it is the opposite of the freezing point on the Fahrenheit temperature scale."

Pass me my kitten mittens.

-32 Dog Dr

## ● Commutative and Associative Properties (6.EE.3)

**Example 1**  **a. Simplify the expression 6 + (14 + x).**

$$6 + (14 + x) = (6 + 14) + x \qquad \text{Associative Property of Addition}$$
$$= 20 + x \qquad \text{Add 6 and 14.}$$

**b. Simplify the expression (3.1 + x) + 7.4.**

$$(3.1 + x) + 7.4 = (x + 3.1) + 7.4 \qquad \text{Commutative Property of Addition}$$
$$= x + (3.1 + 7.4) \qquad \text{Associative Property of Addition}$$
$$= x + 10.5 \qquad \text{Add 3.1 and 7.4.}$$

**c. Simplify the expression 5(12y).**

$$5(12y) = (5 \cdot 12)y \qquad \text{Associative Property of Multiplication}$$
$$= 60y \qquad \text{Multiply 5 and 12.}$$

### Try It Yourself
**Simplify the expression. Explain each step.**

**1.** $3 + (b + 8)$       **2.** $(d + 4) + 6$       **3.** $6(5p)$

## ● Properties of Zero and One (6.EE.3)

**Example 2**  **a. Simplify the expression 6 • 0 • q.**

$$6 \cdot 0 \cdot q = (6 \cdot 0) \cdot q \qquad \text{Associative Property of Multiplication}$$
$$= 0 \cdot q = 0 \qquad \text{Multiplication Property of Zero}$$

**b. Simplify the expression 3.6 • s • 1.**

$$3.6 \cdot s \cdot 1 = 3.6 \cdot (s \cdot 1) \qquad \text{Associative Property of Multiplication}$$
$$= 3.6 \cdot s \qquad \text{Multiplication Property of One}$$
$$= 3.6s$$

### Try It Yourself
**Simplify the expression. Explain each step.**

**4.** $13 \cdot m \cdot 0$       **5.** $1 \cdot x \cdot 29$       **6.** $(n + 14) + 0$

## 1.1 Integers and Absolute Value

### Essential Question How can you use integers to represent the velocity and the speed of an object?

On these two pages, you will investigate vertical motion (up or down).

- Speed tells how fast an object is moving, but it does not tell the direction.
- Velocity tells how fast an object is moving, and it also tells the direction.
  - When velocity is positive, the object is moving up.
  - When velocity is negative, the object is moving down.

#### 1 ACTIVITY: Falling Parachute

**Work with a partner. You are gliding to the ground wearing a parachute. The table shows your height above the ground at different times.**

| Time (seconds) | 0 | 1 | 2 | 3 |
|---|---|---|---|---|
| Height (feet) | 90 | 75 | 60 | 45 |

a. Describe the pattern in the table. How many feet do you move each second? After how many seconds will you land on the ground?

b. What integer represents your speed? Give the units.

c. Do you think your velocity should be represented by a positive or negative integer? Explain your reasoning.

d. What integer represents your velocity? Give the units.

#### 2 ACTIVITY: Rising Balloons

**Work with a partner. You release a group of balloons. The table shows the height of the balloons above the ground at different times.**

| Time (seconds) | 0 | 1 | 2 | 3 |
|---|---|---|---|---|
| Height (feet) | 8 | 12 | 16 | 20 |

a. Describe the pattern in the table. How many feet do the balloons move each second? After how many seconds will the balloons be at a height of 40 feet?

b. What integer represents the speed of the balloons? Give the units.

c. Do you think the velocity of the balloons should be represented by a positive or negative integer? Explain your reasoning.

d. What integer represents the velocity of the balloons? Give the units.

**COMMON CORE**

**Integers**
In this lesson, you will
- define the absolute value of a number.
- find absolute values of numbers.
- solve real-life problems.
Preparing for Standards
7.NS.1
7.NS.2
7.NS.3

**Work with a partner. The table shows the height of a firework's parachute above the ground at different times.**

Math Practice 6

**Use Clear Definitions**

What information can you use to support your answer?

| Time (seconds) | Height (feet) |
|---|---|
| 0 | 480 |
| 1 | 360 |
| 2 | 240 |
| 3 | 120 |
| 4 | 0 |

**a.** Describe the pattern in the table. How many feet does the parachute move each second?

**b.** What integer represents the speed of the parachute? What integer represents the velocity? How are these integers similar in their relation to 0 on a number line?

## Inductive Reasoning

**4.** Copy and complete the table.

| Velocity (feet per second) | −14 | 20 | −2 | 0 | 25 | −15 |
|---|---|---|---|---|---|---|
| Speed (feet per second) | | | | | | |

**5.** Find two different velocities for which the speed is 16 feet per second.

**6.** Which number is greater: −4 or 3? Use a number line to explain your reasoning.

**7.** One object has a velocity of −4 feet per second. Another object has a velocity of 3 feet per second. Which object has the greater speed? Explain your answer.

## What Is Your Answer?

**8. IN YOUR OWN WORDS** How can you use integers to represent the velocity and the speed of an object?

**9. LOGIC** In this lesson, you will study *absolute value*. Here are some examples:

$$|-16| = 16 \qquad |16| = 16 \qquad |0| = 0 \qquad |-2| = 2$$

Which of the following is a true statement? Explain your reasoning.

$$|\text{velocity}| = \text{speed} \qquad |\text{speed}| = \text{velocity}$$

**Practice**

Use what you learned about absolute value to complete Exercises 4–11 on page 6.

**Check It Out**
Lesson Tutorials
BigIdeasMath.com

The following numbers are **integers:**

$$\ldots, -3, -2, -1, 0, 1, 2, 3, \ldots$$

**Key Vocabulary**
integer, *p. 4*
absolute value, *p. 4*

## Key Idea

**Absolute Value**

**Words** The **absolute value** of an integer is the distance between the number and 0 on a number line. The absolute value of a number $a$ is written as $|a|$.

**Numbers** $|-4| = 4$      $|4| = 4$

---

**EXAMPLE 1** **Finding Absolute Value**

**Find the absolute value of 2.**

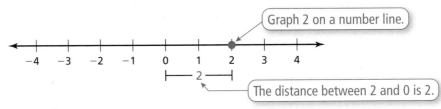

Graph 2 on a number line.

The distance between 2 and 0 is 2.

∴ So, $|2| = 2$.

---

**EXAMPLE 2** **Finding Absolute Value**

**Find the absolute value of −3.**

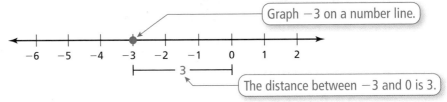

Graph −3 on a number line.

The distance between −3 and 0 is 3.

∴ So, $|-3| = 3$.

## On Your Own

**Now You're Ready**
Exercises 4–19

**Find the absolute value.**

**1.** $|7|$      **2.** $|-1|$      **3.** $|-5|$      **4.** $|14|$

## EXAMPLE 3 Comparing Values

Compare 1 and $|-4|$.

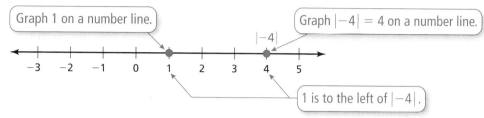

Graph 1 on a number line.

Graph $|-4| = 4$ on a number line.

$|-4|$

1 is to the left of $|-4|$.

So, $1 < |-4|$.

Now You're Ready
Exercises 20–25

### Remember

A number line can be used to compare and order integers. Numbers to the left are less than numbers to the right. Numbers to the right are greater than numbers to the left.

## On Your Own

Copy and complete the statement using <, >, or =.

5. $|-2|$ ▦ $-1$

6. $-7$ ▦ $|6|$

7. $|10|$ ▦ $11$

8. $9$ ▦ $|-9|$

## EXAMPLE 4 Real-Life Application

| Substance | Freezing Point (°C) |
|---|---|
| Butter | 35 |
| Airplane fuel | −53 |
| Honey | −3 |
| Mercury | −39 |
| Candle wax | 55 |

The *freezing point* is the temperature at which a liquid becomes a solid.

a. Which substance in the table has the lowest freezing point?

b. Is the freezing point of mercury or butter closer to the freezing point of water, 0°C?

a. Graph each freezing point.

Airplane fuel has the lowest freezing point, −53°C.

b. The freezing point of water is 0°C, so you can use absolute values.

**Mercury:** $|-39| = 39$          **Butter:** $|35| = 35$

Because 35 is less than 39, the freezing point of butter is closer to the freezing point of water.

## On Your Own

9. Is the freezing point of airplane fuel or candle wax closer to the freezing point of water? Explain your reasoning.

Vocabulary and Concept Check

1. **VOCABULARY** Which of the following numbers are integers?

$$9, 3.2, -1, \frac{1}{2}, -0.25, 15$$

2. **VOCABULARY** What is the absolute value of an integer?

3. **WHICH ONE DOESN'T BELONG?** Which expression does *not* belong with the other three? Explain your reasoning.

$|6|$  $6$  $-6$  $|-6|$

 Practice and Problem Solving

**Find the absolute value.**

4. $|9|$      5. $|-6|$      6. $|-10|$      7. $|10|$

8. $|-15|$      9. $|13|$      10. $|-7|$      11. $|-12|$

12. $|5|$      13. $|-8|$      14. $|0|$      15. $|18|$

16. $|-24|$      17. $|-45|$      18. $|60|$      19. $|-125|$

**Copy and complete the statement using <, >, or =.**

20. $2 \quad\boxed{\phantom{x}}\quad |-5|$      21. $|-4| \quad\boxed{\phantom{x}}\quad 7$      22. $-5 \quad\boxed{\phantom{x}}\quad |-9|$

23. $|-4| \quad\boxed{\phantom{x}}\quad -6$      24. $|-1| \quad\boxed{\phantom{x}}\quad |-8|$      25. $|5| \quad\boxed{\phantom{x}}\quad |-5|$

**ERROR ANALYSIS** Describe and correct the error.

26.  $|10| = -10$

27.  $|-5| < 4$

28. **SAVINGS** You deposit $50 in your savings account. One week later, you withdraw $20. Write each amount as an integer.

29. **ELEVATOR** You go down 8 floors in an elevator. Your friend goes up 5 floors in an elevator. Write each amount as an integer.

**Order the values from least to greatest.**

30. $8, |3|, -5, |-2|, -2$      31. $|-6|, -7, 8, |5|, -6$

32. $-12, |-26|, -15, |-12|, |10|$      33. $|-34|, 21, -17, |20|, |-11|$

**Simplify the expression.**

34. $|-30|$      35. $-|4|$      36. $-|-15|$

**37. PUZZLE** Use a number line.

    **a.** Graph and label the following points on a number line: $A = -3$, $E = 2$, $M = -6$, $T = 0$. What word do the letters spell?

    **b.** Graph and label the absolute value of each point in part (a). What word do the letters spell now?

**38. OPEN-ENDED** Write a negative integer whose absolute value is greater than 3.

**REASONING** Determine whether $n \geq 0$ or $n \leq 0$.

**39.** $n + \lvert -n \rvert = 2n$                **40.** $n + \lvert -n \rvert = 0$

**41. CORAL REEF** The depths of two scuba divers exploring a living coral reef are shown.

    **a.** Write an integer for the position of each diver relative to sea level.

    **b.** Which integer in part (a) is greater?

    **c.** Which integer in part (a) has the greater absolute value? Compare this absolute value with the depth of that diver.

**42. VOLCANOES** The *summit elevation* of a volcano is the elevation of the top of the volcano relative to sea level. The summit elevation of the volcano Kilauea in Hawaii is 1277 meters. The summit elevation of the underwater volcano Loihi in the Pacific Ocean is $-969$ meters. Which summit is closer to sea level?

**43. MINIATURE GOLF** The table shows golf scores, relative to *par*.

    **a.** The player with the lowest score wins. Which player wins?

    **b.** Which player is at par?

    **c.** Which player is farthest from par?

| Player | Score |
|--------|-------|
| 1 | +5 |
| 2 | 0 |
| 3 | −4 |
| 4 | −1 |
| 5 | +2 |

**True or False?** Determine whether the statement is *true* or *false*. Explain your reasoning.

**44.** If $x < 0$, then $\lvert x \rvert = -x$.

**45.** The absolute value of every integer is positive.

**Fair Game Review** What you learned in previous grades & lessons

**Add.** *(Skills Review Handbook)*

**46.** $19 + 32$     **47.** $50 + 94$     **48.** $181 + 217$     **49.** $1149 + 2021$

**50. MULTIPLE CHOICE** Which value is *not* a whole number? *(Skills Review Handbook)*

    **(A)** $-5$         **(B)** $0$         **(C)** $4$         **(D)** $113$

# 1.2 Adding Integers

## Essential Question
Is the sum of two integers *positive*, *negative*, or *zero*? How can you tell?

### 1 ACTIVITY: Adding Integers with the Same Sign

**Work with a partner. Use integer counters to find −4 + (−3).**

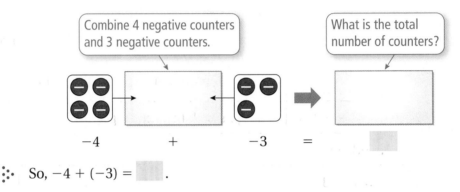

Combine 4 negative counters and 3 negative counters.

What is the total number of counters?

−4 + −3 =

∴ So, −4 + (−3) = [ ].

### 2 ACTIVITY: Adding Integers with Different Signs

**Work with a partner. Use integer counters to find −3 + 2.**

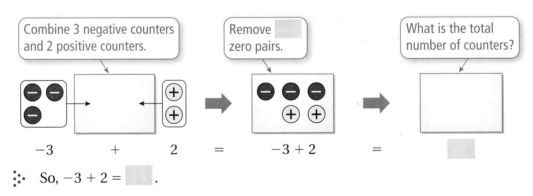

Combine 3 negative counters and 2 positive counters.

Remove [ ] zero pairs.

What is the total number of counters?

−3 + 2 = −3 + 2 =

∴ So, −3 + 2 = [ ].

**COMMON CORE**

**Integers**
In this lesson, you will
- add integers.
- show that the sum of a number and its opposite is 0.
- solve real-life problems.

Learning Standards
7.NS.1a
7.NS.1b
7.NS.1d
7.NS.3

### 3 ACTIVITY: Adding Integers with Different Signs

**Work with a partner. Use a number line to find 5 + (−3).**

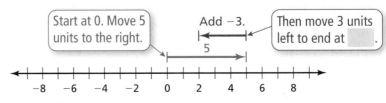

Start at 0. Move 5 units to the right.

Add −3.

Then move 3 units left to end at [ ].

5

∴ So, 5 + (−3) = [ ].

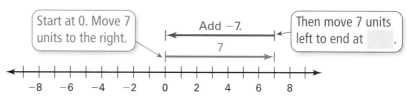

## 4 ACTIVITY: Adding Integers with Different Signs

**Math Practice 3**

**Make Conjectures**

How can the relationship between the integers help you write a rule?

**Work with a partner. Write the addition expression shown. Then find the sum. How are the integers in the expression related to 0 on a number line?**

Start at 0. Move 7 units to the right.

Add −7.

7

Then move 7 units left to end at ⬚.

−8  −6  −4  −2  0  2  4  6  8

## Inductive Reasoning

Work with a partner. Use integer counters or a number line to complete the table.

| | Exercise | Type of Sum | Sum | Sum: Positive, Negative, or Zero |
|---|---|---|---|---|
| 1 | **5.** $-4 + (-3)$ | Integers with the same sign | | |
| 2 | **6.** $-3 + 2$ | | | |
| 3 | **7.** $5 + (-3)$ | | | |
| 4 | **8.** $7 + (-7)$ | | | |
| | **9.** $2 + 4$ | | | |
| | **10.** $-6 + (-2)$ | | | |
| | **11.** $-5 + 9$ | | | |
| | **12.** $15 + (-9)$ | | | |
| | **13.** $-10 + 10$ | | | |
| | **14.** $-6 + (-6)$ | | | |
| | **15.** $13 + (-13)$ | | | |

## What Is Your Answer?

**16. IN YOUR OWN WORDS** Is the sum of two integers *positive*, *negative*, or *zero*? How can you tell?

**17. STRUCTURE** Write general rules for adding (a) two integers with the same sign, (b) two integers with different signs, and (c) two integers that vary only in sign.

**Practice**

Use what you learned about adding integers to complete Exercises 8–15 on page 12.

## 1.2 Lesson

**Key Vocabulary**
opposites, *p. 10*
additive inverse, *p. 10*

### Key Idea

**Adding Integers with the Same Sign**

**Words**   Add the absolute values of the integers. Then use the common sign.

**Numbers**   $2 + 5 = 7$       $-2 + (-5) = -7$

---

**EXAMPLE** **1**   **Adding Integers with the Same Sign**

Find $-2 + (-4)$. Use a number line to check your answer.

$-2 + (-4) = -6$     Add $|-2|$ and $|-4|$.

Use the common sign.

The sum is $-6$.

Check

---

**The Meaning of a Word**

**Opposite**

When you walk across a street, you are moving to the **opposite** side of the street.

### On Your Own

**Add.**

**1.** $7 + 13$     **2.** $-8 + (-5)$     **3.** $-20 + (-15)$

---

Two numbers that are the same distance from 0, but on opposite sides of 0, are called **opposites.** For example, $-3$ and $3$ are opposites.

### Key Ideas

**Adding Integers with Different Signs**

**Words**   Subtract the lesser absolute value from the greater absolute value. Then use the sign of the integer with the greater absolute value.

**Numbers**   $8 + (-10) = -2$   $-13 + 17 = 4$

**Additive Inverse Property**

**Words**   The sum of an integer and its **additive inverse,** or opposite, is 0.

**Numbers**   $6 + (-6) = 0$       $-25 + 25 = 0$       **Algebra**   $a + (-a) = 0$

---

Multi-Language Glossary at BigIdeasMath.com

**EXAMPLE 2 | Adding Integers with Different Signs**

**a. Find 5 + (−10).**

$$5 + (-10) = -5$$

$|-10| > |5|$. So, subtract $|5|$ from $|-10|$.

Use the sign of −10.

∴ The sum is −5.

**b. Find −3 + 7.**

$$-3 + 7 = 4$$

$|7| > |-3|$. So, subtract $|-3|$ from $|7|$.

Use the sign of 7.

∴ The sum is 4.

**c. Find −12 + 12.**

$$-12 + 12 = 0$$

The sum is 0 by the Additive Inverse Property.

−12 and 12 are opposites.

∴ The sum is 0.

**EXAMPLE 3 | Adding More Than Two Integers**

The list shows four bank account transactions in July. Find the change $C$ in the account balance.

| JULY TRANSACTIONS | |
|---|---|
| Withdrawal | -$40 |
| Deposit | $50 |
| Deposit | $75 |
| Withdrawal | -$50 |

**Study Tip**

A deposit of $50 and a withdrawal of $50 represent opposite quantities, +50 and −50, which have a sum of 0.

Find the sum of the four transactions.

$$C = -40 + 50 + 75 + (-50)$$     Write the sum.

$$= -40 + 75 + 50 + (-50)$$     Commutative Property of Addition

$$= -40 + 75 + [50 + (-50)]$$     Associative Property of Addition

$$= -40 + 75 + 0$$     Additive Inverse Property

$$= 35 + 0$$     Add −40 and 75.

$$= 35$$     Addition Property of Zero

∴ Because $C = 35$, the account balance increased $35 in July.

**On Your Own**

**Now You're Ready**
Exercises 8–23 and 28–39

Add.

**4.** $-2 + 11$      **5.** $9 + (-10)$      **6.** $-31 + 31$

**7. WHAT IF?** In Example 3, the deposit amounts are $30 and $40. Find the change $C$ in the account balance.

## 1.2 Exercises

Check It Out
Help with Homework
BigIdeasMath  com

 **Vocabulary and Concept Check**

1. **WRITING** How do you find the additive inverse of an integer?

2. **NUMBER SENSE** Is $3 + (-4)$ the same as $-4 + 3$? Explain.

**Tell whether the sum is *positive*, *negative*, or *zero* without adding. Explain your reasoning.**

3. $-8 + 20$

4. $30 + (-30)$

5. $-10 + (-18)$

**Tell whether the statement is *true* or *false*. Explain your reasoning.**

6. The sum of two negative integers is always negative.

7. An integer and its absolute value are always opposites.

 **Practice and Problem Solving**

**Add.**

8. $6 + 4$

9. $-4 + (-6)$

10. $-2 + (-3)$

11. $-5 + 12$

12. $5 + (-7)$

13. $8 + (-8)$

14. $9 + (-11)$

15. $-3 + 13$

16. $-4 + (-16)$

17. $-3 + (-1)$

18. $14 + (-5)$

19. $0 + (-11)$

20. $-10 + (-15)$

21. $-13 + 9$

22. $18 + (-18)$

23. $-25 + (-9)$

**ERROR ANALYSIS** Describe and correct the error in finding the sum.

24.

$9 + (-6) = -3$

25.

$-10 + (-10) = 0$

26. **TEMPERATURE** The temperature is $-3°F$ at 7:00 A.M. During the next 4 hours, the temperature increases 21°F. What is the temperature at 11:00 A.M.?

27. **BANKING** Your bank account has a balance of $-\$12$. You deposit $\$60$. What is your new balance?

**Tell how the Commutative and Associative Properties of Addition can help you find the sum mentally. Then find the sum.**

28. $9 + 6 + (-6)$

29. $-8 + 13 + (-13)$

30. $9 + (-17) + (-9)$

31. $7 + (-12) + (-7)$

32. $-12 + 25 + (-15)$

33. $6 + (-9) + 14$

**Add.**

34. $13 + (-21) + 16$

35. $22 + (-14) + (-35)$

36. $-13 + 27 + (-18)$

37. $-19 + 26 + 14$

38. $-32 + (-17) + 42$

39. $-41 + (-15) + (-29)$

**40.** **SCIENCE** A lithium atom has positively charged protons and negatively charged electrons. The sum of the charges represents the charge of the lithium atom. Find the charge of the atom.

Lithium Atom

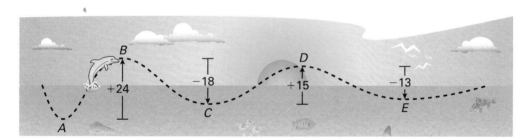

**41.** **OPEN-ENDED** Write two integers with different signs that have a sum of $-25$. Write two integers with the same sign that have a sum of $-25$.

**ALGEBRA** Evaluate the expression when $a = 4$, $b = -5$, and $c = -8$.

**42.** $a + b$

**43.** $-b + c$

**44.** $|a + b + c|$

**MENTAL MATH** Use mental math to solve the equation.

**45.** $d + 12 = 2$

**46.** $b + (-2) = 0$

**47.** $-8 + m = -15$

**48.** **PROBLEM SOLVING** Starting at point $A$, the path of a dolphin jumping out of the water is shown.

    **a.** Is the dolphin deeper at point $C$ or point $E$? Explain your reasoning.

    **b.** Is the dolphin higher at point $B$ or point $D$? Explain your reasoning.

**49.** **Puzzle** According to a legend, the Chinese Emperor Yu-Huang saw a magic square on the back of a turtle. In a *magic square*, the numbers in each row and in each column have the same sum. This sum is called the *magic sum*.

Copy and complete the magic square so that each row and each column has a magic sum of 0. Use each integer from $-4$ to 4 exactly once.

**Fair Game Review** What you learned in previous grades & lessons

**Subtract.** *(Skills Review Handbook)*

**50.** $69 - 38$

**51.** $82 - 74$

**52.** $177 - 63$

**53.** $451 - 268$

**54.** **MULTIPLE CHOICE** What is the range of the numbers below? *(Skills Review Handbook)*

    12, 8, 17, 12, 15, 18, 30

    (A) 12      (B) 15      (C) 18      (D) 22

**Essential Question** How are adding integers and subtracting integers related?

### 1 ACTIVITY: Subtracting Integers

**Work with a partner. Use integer counters to find 4 − 2.**

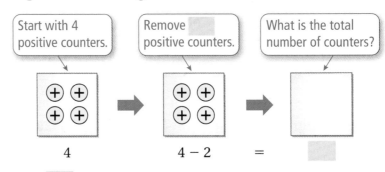

Start with 4 positive counters.

Remove [ ] positive counters.

What is the total number of counters?

4          4 − 2     =

So, 4 − 2 = [ ] .

### 2 ACTIVITY: Adding Integers

**Work with a partner. Use integer counters to find 4 + (−2).**

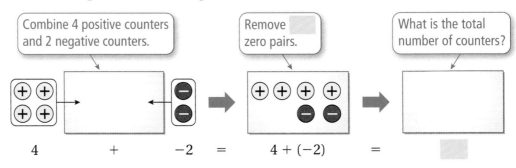

Combine 4 positive counters and 2 negative counters.

Remove [ ] zero pairs.

What is the total number of counters?

4     +     −2     =     4 + (−2)     =     [ ]

So, 4 + (−2) = [ ] .

**COMMON CORE**

**Integers**
In this lesson, you will
● subtract integers.
● solve real-life problems.
Learning Standards
7.NS.1c
7.NS.1d
7.NS.3

### 3 ACTIVITY: Subtracting Integers

**Work with a partner. Use a number line to find −3 − 1.**

Then move 1 unit left to end at [ ] .

Subtract 1.

Start at 0. Move 3 units to the left.

−3

So, −3 − 1 = [ ] .

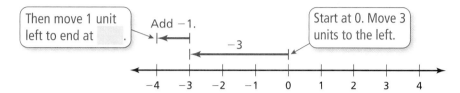

**④ ACTIVITY: Adding Integers**

**Math Practice ②**

**Make Sense of Quantities**

What integers will you use in your addition expression?

Work with a partner. Write the addition expression shown. Then find the sum.

Then move 1 unit left to end at [ ].

Add −1.

Start at 0. Move 3 units to the left.

−3

## Inductive Reasoning

Work with a partner. Use integer counters or a number line to complete the table.

| | Exercise | Operation: Add or Subtract | Answer |
|---|---|---|---|
| ① | **5.** $4 - 2$ | Subtract 2 | |
| ② | **6.** $4 + (-2)$ | | |
| ③ | **7.** $-3 - 1$ | | |
| ④ | **8.** $-3 + (-1)$ | | |
| | **9.** $3 - 8$ | | |
| | **10.** $3 + (-8)$ | | |
| | **11.** $9 - 13$ | | |
| | **12.** $9 + (-13)$ | | |
| | **13.** $-6 - (-3)$ | | |
| | **14.** $-6 + 3$ | | |
| | **15.** $-5 - (-12)$ | | |
| | **16.** $-5 + 12$ | | |

## What Is Your Answer?

**17. IN YOUR OWN WORDS** How are adding integers and subtracting integers related?

**18. STRUCTURE** Write a general rule for subtracting integers.

**19.** Use a number line to find the value of the expression $-4 + 4 - 9$. What property can you use to make your calculation easier? Explain.

**Practice**

Use what you learned about subtracting integers to complete Exercises 8–15 on page 18.

## Key Idea

**Subtracting Integers**

**Words** To subtract an integer, add its opposite.

**Numbers** $3 - 4 = 3 + (-4) = -1$

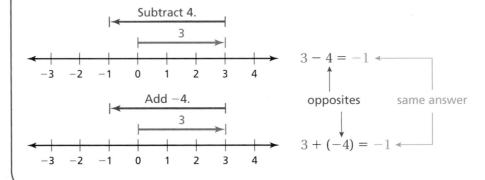

**EXAMPLE 1 Subtracting Integers**

a. **Find $3 - 12$.**

$$3 - 12 = 3 + (-12) \qquad \text{Add the opposite of 12.}$$
$$= -9 \qquad \text{Add.}$$

⋮· The difference is $-9$.

b. **Find $-8 - (-13)$.**

$$-8 - (-13) = -8 + 13 \qquad \text{Add the opposite of } -13.$$
$$= 5 \qquad \text{Add.}$$

⋮· The difference is 5.

c. **Find $5 - (-4)$.**

$$5 - (-4) = 5 + 4 \qquad \text{Add the opposite of } -4.$$
$$= 9 \qquad \text{Add.}$$

⋮· The difference is 9.

## On Your Own

**Now You're Ready**
Exercises 8–23

**Subtract.**

1. $8 - 3$
2. $9 - 17$
3. $-3 - 3$
4. $-14 - 9$
5. $9 - (-8)$
6. $-12 - (-12)$

**EXAMPLE 2** **Subtracting Integers**

Evaluate $-7 - (-12) - 14$.

$$-7 - (-12) - 14 = -7 + 12 - 14 \qquad \text{Add the opposite of } -12.$$
$$= 5 - 14 \qquad \text{Add } -7 \text{ and } 12.$$
$$= 5 + (-14) \qquad \text{Add the opposite of } 14.$$
$$= -9 \qquad \text{Add.}$$

So, $-7 - (-12) - 14 = -9$.

### On Your Own

**Now You're Ready**
Exercises 27–32

Evaluate the expression.

7. $-9 - 16 - 8$

8. $-4 - 20 - 9$

9. $0 - 9 - (-5)$

10. $-8 - (-6) - 0$

11. $15 - (-20) - 20$

12. $-14 - 9 - 36$

**EXAMPLE 3** **Real-Life Application**

Which continent has the greater range of elevations?

|  | North America | Africa |
|---|---|---|
| Highest Elevation | 6198 m | 5895 m |
| Lowest Elevation | −86 m | −155 m |

To find the range of elevations for each continent, subtract the lowest elevation from the highest elevation.

**North America**

$$\text{range} = 6198 - (-86)$$
$$= 6198 + 86$$
$$= 6284 \text{ m}$$

**Africa**

$$\text{range} = 5895 - (-155)$$
$$= 5895 + 155$$
$$= 6050 \text{ m}$$

Because 6284 is greater than 6050, North America has the greater range of elevations.

### On Your Own

13. The highest elevation in Mexico is 5700 meters, on Pico de Orizaba. The lowest elevation in Mexico is −10 meters, in Laguna Salada. Find the range of elevations in Mexico.

## Vocabulary and Concept Check

1. **WRITING** How do you subtract one integer from another?

2. **OPEN-ENDED** Write two integers that are opposites.

3. **DIFFERENT WORDS, SAME QUESTION** Which is different? Find "both" answers.

| Find the difference of 3 and $-2$. | What is 3 less than $-2$? |
|---|---|
| How much less is $-2$ than 3? | Subtract $-2$ from 3. |

**MATCHING** Match the subtraction expression with the corresponding addition expression.

4. $9 - (-5)$    5. $-9 - 5$    6. $-9 - (-5)$    7. $9 - 5$

A. $-9 + 5$    B. $9 + (-5)$    C. $-9 + (-5)$    D. $9 + 5$

 ## Practice and Problem Solving

**Subtract.**

8. $4 - 7$    9. $8 - (-5)$    10. $-6 - (-7)$    11. $-2 - 3$

12. $5 - 8$    13. $-4 - 6$    14. $-8 - (-3)$    15. $10 - 7$

16. $-8 - 13$    17. $15 - (-2)$    18. $-9 - (-13)$    19. $-7 - (-8)$

20. $-6 - (-6)$    21. $-10 - 12$    22. $32 - (-6)$    23. $0 - 20$

24. **ERROR ANALYSIS** Describe and correct the error in finding the difference $7 - (-12)$.

$$7 - (-12) = 7 + (-12) = -5$$

25. **SWIMMING POOL** The floor of the shallow end of a swimming pool is at $-3$ feet. The floor of the deep end is 9 feet deeper. Which expression can be used to find the depth of the deep end?

$$-3 + 9 \qquad -3 - 9 \qquad 9 - 3$$

26. **SHARKS** A shark is at $-80$ feet. It swims up and jumps out of the water to a height of 15 feet. Write a subtraction expression for the vertical distance the shark travels.

**Evaluate the expression.**

27. $-2 - 7 + 15$    28. $-9 + 6 - (-2)$    29. $12 - (-5) - 8$

30. $-87 - 5 - 13$    31. $-6 - (-8) + 6$    32. $-15 - 7 - (-11)$

**MENTAL MATH** Use mental math to solve the equation.

**33.** $m - 5 = 9$

**34.** $w - (-3) = 7$

**35.** $6 - c = -9$

**ALGEBRA** Evaluate the expression when $k = -3$, $m = -6$, and $n = 9$.

**36.** $4 - n$

**37.** $m - (-8)$

**38.** $-5 + k - n$

**39.** $\left| m - k \right|$

**40. PLATFORM DIVING** The figure shows a diver diving from a platform. The diver reaches a depth of 4 meters. What is the change in elevation of the diver?

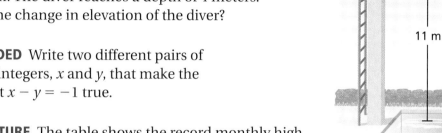

11 m

**41. OPEN-ENDED** Write two different pairs of negative integers, $x$ and $y$, that make the statement $x - y = -1$ true.

**42. TEMPERATURE** The table shows the record monthly high and low temperatures for a city in Alaska.

| | Jan | Feb | Mar | Apr | May | Jun | Jul | Aug | Sep | Oct | Nov | Dec |
|---|---|---|---|---|---|---|---|---|---|---|---|---|
| **High (°F)** | 56 | 57 | 56 | 72 | 82 | 92 | 84 | 85 | 73 | 64 | 62 | 53 |
| **Low (°F)** | −35 | −38 | −24 | −15 | 1 | 29 | 34 | 31 | 19 | −6 | −21 | −36 |

   **a.** Find the range of temperatures for each month.

   **b.** What are the all-time high and all-time low temperatures?

   **c.** What is the range of the temperatures in part (b)?

**REASONING** Tell whether the difference between the two integers is *always*, *sometimes*, or *never* positive. Explain your reasoning.

**43.** two positive integers

**44.** two negative integers

**45.** a positive integer and a negative integer

**46.** a negative integer and a positive integer

 **Number Sense** For what values of $a$ and $b$ is the statement true?

**47.** $\left| a - b \right| = \left| b - a \right|$

**48.** $\left| a + b \right| = \left| a \right| + \left| b \right|$

**49.** $\left| a - b \right| = \left| a \right| - \left| b \right|$

 **Fair Game Review** What you learned in previous grades & lessons

**Add.** *(Section 1.2)*

**50.** $-5 + (-5) + (-5) + (-5)$

**51.** $-9 + (-9) + (-9) + (-9) + (-9)$

**Multiply.** *(Skills Review Handbook)*

**52.** $8 \times 5$

**53.** $6 \times 78$

**54.** $36 \times 41$

**55.** $82 \times 29$

**56. MULTIPLE CHOICE** Which value of $n$ makes the value of the expression $4n + 3$ a composite number? *(Skills Review Handbook)*

   Ⓐ 1        Ⓑ 2        Ⓒ 3        Ⓓ 4

You can use an **idea and examples chart** to organize information about a concept. Here is an example of an idea and examples chart for absolute value.

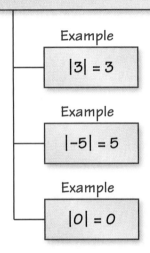

Absolute Value: the distance between a number and 0 on the number line

Example

$|3| = 3$

Example

$|-5| = 5$

Example

$|0| = 0$

## On Your Own

**Make idea and examples charts to help you study these topics.**

1. integers

2. adding integers

   **a.** with the same sign

   **b.** with different signs

3. Additive Inverse Property

4. subtracting integers

**After you complete this chapter, make idea and examples charts for the following topics.**

5. multiplying integers

   **a.** with the same sign

   **b.** with different signs

6. dividing integers

   **a.** with the same sign

   **b.** with different signs

Biscuits: Doggy treats

Cheese

Chicken

Liver

7 birthdays a year. Is that fair?

"I made an idea and examples chart to give my owner ideas for my birthday next week."

**Copy and complete the statement using <, >, or =.** *(Section 1.1)*

1. $\left| -8 \right|$ ▢ $3$

2. $7$ ▢ $\left| -7 \right|$

**Order the values from least to greatest.** *(Section 1.1)*

3. $-4, \left| -5 \right|, \left| -4 \right|, 3, -6$

4. $12, -8, \left| -15 \right|, -10, \left| -9 \right|$

**Evaluate the expression.** *(Section 1.2 and Section 1.3)*

5. $-3 + (-8)$

6. $-4 + 16$

7. $3 - 9$

8. $-5 - (-5)$

**Evaluate the expression when $a = -2$, $b = -8$, and $c = 5$.** *(Section 1.2 and Section 1.3)*

9. $4 - a - c$

10. $\left| b - c \right|$

11. **EXPLORING** Two climbers explore a cave. *(Section 1.1)*

   a. Write an integer for the position of each climber relative to the surface.

   b. Which integer in part (a) is greater?

   c. Which integer in part (a) has the greater absolute value?

10 ft    7 ft

12. **SCHOOL CARNIVAL** The table shows the income and expenses for a school carnival. The school's goal was to raise $1100. Did the school reach its goal? Explain. *(Section 1.2)*

| Games | Concessions | Donations | Flyers | Decorations |
|-------|-------------|-----------|--------|-------------|
| $650 | $530 | $52 | −$28 | −$75 |

13. **TEMPERATURE** Temperatures in the Gobi Desert reach −40°F in the winter and 90°F in the summer. Find the range of the temperatures. *(Section 1.3)*

# 1.4 Multiplying Integers

**Essential Question** Is the product of two integers *positive*, *negative*, or *zero*? How can you tell?

## 1 ACTIVITY: Multiplying Integers with the Same Sign

**Work with a partner. Use repeated addition to find 3 • 2.**

Recall that multiplication is repeated addition. 3 • 2 means to add 3 groups of 2.

Now you can write

3 • 2 = ▢ + ▢ + ▢

= ▢ .

So, 3 • 2 = ▢ .

## 2 ACTIVITY: Multiplying Integers with Different Signs

**Work with a partner. Use repeated addition to find 3 • (−2).**

3 • (−2) means to add 3 groups of −2.

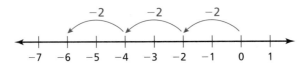

Now you can write

3 • (−2) = ▢ + ▢ + ▢

= ▢ .

So, 3 • (−2) = ▢ .

## 3 ACTIVITY: Multiplying Integers with Different Signs

**Work with a partner. Use a table to find −3 • 2.**

Describe the pattern of the products in the table. Then complete the table.

| 2 | • | 2 | = | 4 |
|---|---|---|---|---|
| 1 | • | 2 | = | 2 |
| 0 | • | 2 | = | 0 |
| −1 | • | 2 | = | ▢ |
| −2 | • | 2 | = | ▢ |
| −3 | • | 2 | = | ▢ |

So, −3 • 2 = ▢ .

**COMMON CORE**

**Integers**

In this lesson, you will

• multiply integers.
• solve real-life problems.

Learning Standards
7.NS.2a
7.NS.2c
7.NS.3

**Work with a partner. Use a table to find −3 • (−2).**

Describe the pattern of the products in the table. Then complete the table.

**Math Practice 7**

**Look for Patterns**
How can you use the pattern to complete the table?

| −3 | • | 3 | = | −9 |
| −3 | • | 2 | = | −6 |
| −3 | • | 1 | = | −3 |
| −3 | • | 0 | = | |
| −3 | • | −1 | = | |
| −3 | • | −2 | = | |

So, −3 • (−2) = .

## Inductive Reasoning

Work with a partner. Complete the table.

| Exercise | Type of Product | Product | Product: Positive or Negative |
|---|---|---|---|
| **5.** 3 • 2 | Integers with the same sign | | |
| **6.** 3 • (−2) | | | |
| **7.** −3 • 2 | | | |
| **8.** −3 • (−2) | | | |
| **9.** 6 • 3 | | | |
| **10.** 2 • (−5) | | | |
| **11.** −6 • 5 | | | |
| **12.** −5 • (−3) | | | |

## What Is Your Answer?

**13.** Write two integers whose product is 0.

**14.** **IN YOUR OWN WORDS** Is the product of two integers *positive*, *negative*, or *zero*? How can you tell?

**15.** **STRUCTURE** Write general rules for multiplying (a) two integers with the same sign and (b) two integers with different signs.

**Practice** Use what you learned about multiplying integers to complete Exercises 8–15 on page 26.

**Key Ideas**

**Multiplying Integers with the Same Sign**

**Words** The product of two integers with the same sign is positive.

**Numbers** $2 \cdot 3 = 6$      $-2 \cdot (-3) = 6$

**Multiplying Integers with Different Signs**

**Words** The product of two integers with different signs is negative.

**Numbers** $2 \cdot (-3) = -6$     $-2 \cdot 3 = -6$

---

**EXAMPLE 1   Multiplying Integers with the Same Sign**

Find $-5 \cdot (-6)$.

The integers have the same sign.

$$-5 \cdot (-6) = 30$$

The product is positive.

∴ The product is 30.

---

**EXAMPLE 2   Multiplying Integers with Different Signs**

**Multiply.**

**a.** $3(-4)$          **b.** $-7 \cdot 4$

The integers have different signs.

$$3(-4) = -12 \qquad\qquad -7 \cdot 4 = -28$$

The product is negative.

∴ The product is $-12$.          ∴ The product is $-28$.

---

**On Your Own**

**Multiply.**

*Now You're Ready*
Exercises 8–23

**1.** $5 \cdot 5$                    **2.** $4(11)$

**3.** $-1(-9)$                    **4.** $-7 \cdot (-8)$

**5.** $12 \cdot (-2)$              **6.** $4(-6)$

**7.** $-10(-6)(0)$                **8.** $-7 \cdot (-5) \cdot (-4)$

---

EXAMPLE 3 Using Exponents

a. **Evaluate $(-2)^2$.**

$$(-2)^2 = (-2) \cdot (-2)$$   Write $(-2)^2$ as repeated multiplication.

$$= 4$$   Multiply.

b. **Evaluate $-5^2$.**

$$-5^2 = -(5 \cdot 5)$$   Write $5^2$ as repeated multiplication.

$$= -25$$   Multiply.

c. **Evaluate $(-4)^3$.**

$$(-4)^3 = (-4) \cdot (-4) \cdot (-4)$$   Write $(-4)^3$ as repeated multiplication.

$$= 16 \cdot (-4)$$   Multiply.

$$= -64$$   Multiply.

**Study Tip**

Place parentheses around a negative number to raise it to a power.

## On Your Own

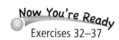

**Evaluate the expression.**

**9.** $(-3)^2$    **10.** $(-2)^3$    **11.** $-7^2$    **12.** $-6^3$

EXAMPLE 4 Real-Life Application

**The bar graph shows the number of taxis a company has in service. The number of taxis decreases by the same amount each year for 4 years. Find the total change in the number of taxis.**

**Taxis in Service**

50 fewer taxis

The bar graph shows that the number of taxis in service decreases by 50 each year. Use a model to solve the problem.

total change   =   change per year   •   number of years

$$= -50 \cdot 4$$

Use $-50$ for the change per year because the number *decreases* each year.

$$= -200$$

∴ The total change in the number of taxis is $-200$.

## On Your Own

**13.** A manatee population decreases by 15 manatees each year for 3 years. Find the total change in the manatee population.

## 1.4 Exercises

Check It Out
Help with Homework
BigIdeasMath.com

 ## Vocabulary and Concept Check

1. **WRITING** What can you conclude about the signs of two integers whose product is (a) positive and (b) negative?

2. **OPEN-ENDED** Write two integers whose product is negative.

**Tell whether the product is *positive* or *negative* without multiplying. Explain your reasoning.**

3. $4(-8)$ 

4. $-5(-7)$ 

5. $-3 \cdot 12$

**Tell whether the statement is *true* or *false*. Explain your reasoning.**

6. The product of three positive integers is positive.

7. The product of three negative integers is positive.

 ## Practice and Problem Solving

**Multiply.**

① ② 8. $6 \cdot 4$ 

9. $7(-3)$ 

10. $-2(8)$ 

11. $-3(-4)$

12. $-6 \cdot 7$ 

13. $3 \cdot 9$ 

14. $8 \cdot (-5)$ 

15. $-1 \cdot (-12)$

16. $-5(10)$ 

17. $-13(0)$ 

18. $-9 \cdot 9$ 

19. $15(-2)$

20. $-10 \cdot 11$ 

21. $-6 \cdot (-13)$ 

22. $7(-14)$ 

23. $-11 \cdot (-11)$

24. **JOGGING** You burn 10 calories each minute you jog. What integer represents the change in your calories after you jog for 20 minutes?

25. **WETLANDS** About 60,000 acres of wetlands are lost each year in the United States. What integer represents the change in wetlands after 4 years?

**Multiply.**

26. $3 \cdot (-8) \cdot (-2)$ 

27. $6(-9)(-1)$ 

28. $-3(-5)(-4)$

29. $(-5)(-7)(-20)$ 

30. $-6 \cdot 3 \cdot (-2)$ 

31. $3 \cdot (-12) \cdot 0$

**Evaluate the expression.**

 32. $(-4)^2$ 

33. $(-1)^3$ 

34. $-8^2$

35. $-6^2$ 

36. $-5^2 \cdot 4$ 

37. $-2 \cdot (-3)^3$

**ERROR ANALYSIS** Describe and correct the error in evaluating the expression.

38.

$$\times \quad -2(-7) = -14$$

39.

$$\times \quad -10^2 = 100$$

**ALGEBRA** Evaluate the expression when $a = -2$, $b = 3$, and $c = -8$.

**40.** $ab$

**41.** $\left| a^2 c \right|$

**42.** $-ab^3 - ac$

**NUMBER SENSE** Find the next two numbers in the pattern.

**43.** $-12, 60, -300, 1500, \ldots$

**44.** $7, -28, 112, -448, \ldots$

**45. GYM CLASS** You lose four points each time you attend gym class without sneakers. You forget your sneakers three times. What integer represents the change in your points?

**46. MODELING** The height of an airplane during a landing is given by $22{,}000 + (-480t)$, where $t$ is the time in minutes.

   **a.** Copy and complete the table.

   **b.** Estimate how many minutes it takes the plane to land. Explain your reasoning.

| Time (minutes) | 5 | 10 | 15 | 20 |
|---|---|---|---|---|
| Height (feet) | | | | |

**47. INLINE SKATES** In June, the price of a pair of inline skates is \$165. The price changes each of the next 3 months.

   **a.** Copy and complete the table.

| Month | Price of Skates |
|---|---|
| June | $165 \hspace{2em} = \$165$ |
| July | $165 + (-12) = \$\underline{\hspace{2em}}$ |
| August | $165 + 2(-12) = \$\underline{\hspace{2em}}$ |
| September | $165 + 3(-12) = \$\underline{\hspace{2em}}$ |

   **b.** Describe the change in the price of the inline skates for each month.

   **c.** The table at the right shows the amount of money you save each month to buy the inline skates. Do you have enough money saved to buy the inline skates in August? September? Explain your reasoning.

| Amount Saved | |
|---|---|
| June | \$35 |
| July | \$55 |
| August | \$45 |
| September | \$18 |

**48.** **Reasoning** Two integers, $a$ and $b$, have a product of 24. What is the least possible sum of $a$ and $b$?

## Fair Game Review  What you learned in previous grades & lessons

**Divide.** *(Skills Review Handbook)*

**49.** $27 \div 9$

**50.** $48 \div 6$

**51.** $56 \div 4$

**52.** $153 \div 9$

**53. MULTIPLE CHOICE** What is the prime factorization of 84? *(Skills Review Handbook)*

    (A) $2^2 \times 3^2$      (B) $2^3 \times 7$      (C) $3^3 \times 7$      (D) $2^2 \times 3 \times 7$

## 1.5 Dividing Integers

### Essential Question Is the quotient of two integers *positive*, *negative*, or *zero*? How can you tell?

---

#### 1 ACTIVITY: Dividing Integers with Different Signs

**Work with a partner. Use integer counters to find** $-15 \div 3$.

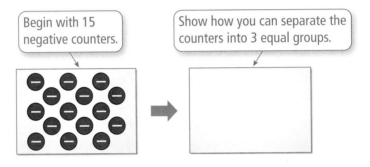

Begin with 15 negative counters.

Show how you can separate the counters into 3 equal groups.

⋮ Because there are [ ] negative counters in each group, $-15 \div 3 = $ [ ].

---

#### 2 ACTIVITY: Rewriting a Product as a Quotient

**Work with a partner. Rewrite the product** $3 \cdot 4 = 12$ **as a quotient in two different ways.**

*First Way*

12 is equal to 3 groups of [ ].

⋮ So, $12 \div 3 = $ [ ].

*Second Way*

12 is equal to 4 groups of [ ].

⋮ So, $12 \div 4 = $ [ ].

---

**COMMON CORE**

**Integers**
In this lesson, you will
- divide integers.
- solve real-life problems.

Learning Standards
7.NS.2b
7.NS.3

#### 3 ACTIVITY: Dividing Integers with Different Signs

**Work with a partner. Rewrite the product** $-3 \cdot (-4) = 12$ **as a quotient in two different ways. What can you conclude?**

*First Way*

$12 \div \left( \text{[ ]} \right) = $ [ ]

*Second Way*

$12 \div \left( \text{[ ]} \right) = $ [ ]

⋮ In each case, when you divide a [ ] [ ] integer by a [ ] [ ] integer, you get a [ ] integer.

**4** **ACTIVITY: Dividing Negative Integers**

## Math Practice 8

**Maintain Oversight**

How do you know what the sign will be when you divide two integers?

**Work with a partner. Rewrite the product $3 \cdot (-4) = -12$ as a quotient in two different ways. What can you conclude?**

*First Way*

$$-12 \div \left( \boxed{\phantom{xx}} \right) = \boxed{\phantom{xx}}$$

*Second Way*

$$-12 \div \left( \boxed{\phantom{xx}} \right) = \boxed{\phantom{xx}}$$

⋮⋮ When you divide a ▭ integer by a ▭ integer, you get a ▭ integer. When you divide a ▭ integer by a ▭ integer, you get a ▭ integer.

## Inductive Reasoning

Work with a partner. Complete the table.

| Exercise | Type of Quotient | Quotient | Quotient: Positive, Negative, or Zero |
|---|---|---|---|
| **5.** $-15 \div 3$ | Integers with different signs | | |
| **6.** $12 \div 4$ | | | |
| **7.** $12 \div (-3)$ | | | |
| **8.** $-12 \div (-4)$ | | | |
| **9.** $-6 \div 2$ | | | |
| **10.** $-21 \div (-7)$ | | | |
| **11.** $10 \div (-2)$ | | | |
| **12.** $12 \div (-6)$ | | | |
| **13.** $0 \div (-15)$ | | | |
| **14.** $0 \div 4$ | | | |

## What Is Your Answer?

**15. IN YOUR OWN WORDS** Is the quotient of two integers *positive*, *negative*, or *zero*? How can you tell?

**16. STRUCTURE** Write general rules for dividing (a) two integers with the same sign and (b) two integers with different signs.

**Practice**
Use what you learned about dividing integers to complete Exercises 8–15 on page 32.

**Section 1.5**   Dividing Integers   **29**

**Check It Out**
Lesson Tutorials
BigIdeasMath **✓**com

## 🔑 Key Ideas

**Remember**

Division by 0 is undefined.

### Dividing Integers with the Same Sign

**Words** The quotient of two integers with the same sign is positive.

**Numbers** $8 \div 2 = 4$     $-8 \div (-2) = 4$

### Dividing Integers with Different Signs

**Words** The quotient of two integers with different signs is negative.

**Numbers** $8 \div (-2) = -4$     $-8 \div 2 = -4$

---

**EXAMPLE 1** **Dividing Integers with the Same Sign**

**Find $-18 \div (-6)$.**

The integers have the same sign.

$-18 \div (-6) = 3$

The quotient is positive.

∴ The quotient is 3.

---

**EXAMPLE 2** **Dividing Integers with Different Signs**

**Divide.**

**a.** $75 \div (-25)$          **b.** $\dfrac{-54}{6}$

The integers have different signs.

$75 \div (-25) = -3$          $\dfrac{-54}{6} = -9$

The quotient is negative.

∴ The quotient is $-3$.          ∴ The quotient is $-9$.

---

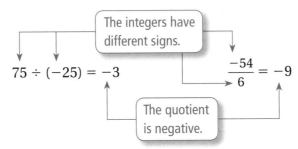

## ⬤ On Your Own

**Now You're Ready**
Exercises 8–23

**Divide.**

**1.** $14 \div 2$          **2.** $-32 \div (-4)$          **3.** $-40 \div (-8)$

**4.** $0 \div (-6)$          **5.** $\dfrac{-49}{7}$          **6.** $\dfrac{21}{-3}$

---

Evaluate $10 - x^2 \div y$ when $x = 8$ and $y = -4$.

$$\begin{aligned}10 - x^2 \div y &= 10 - 8^2 \div (-4) & \text{Substitute 8 for } x \text{ and } -4 \text{ for } y.\\ &= 10 - 8 \cdot 8 \div (-4) & \text{Write } 8^2 \text{ as repeated multiplication.}\\ &= 10 - 64 \div (-4) & \text{Multiply 8 and 8.}\\ &= 10 - (-16) & \text{Divide 64 by } -4.\\ &= 26 & \text{Subtract.}\end{aligned}$$

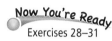 **Remember**

Use order of operations when evaluating an expression.

● **On Your Own**

**Now You're Ready**
Exercises 28–31

Evaluate the expression when $a = -18$ and $b = -6$.

**7.** $a \div b$

**8.** $\dfrac{a + 6}{3}$

**9.** $\dfrac{b^2}{a} + 4$

---

EXAMPLE ④ **Real-Life Application**

You measure the height of the tide using the support beams of a pier. Your measurements are shown in the picture. What is the mean hourly change in the height?

59 inches at 2 P.M.→
8 inches at 8 P.M.→

Use a model to solve the problem.

$$\text{mean hourly change} = \frac{\text{final height} \quad - \quad \text{initial height}}{\text{elapsed time}}$$

$$\begin{aligned} &= \frac{8 - 59}{6} & \text{Substitute. The elapsed time from 2 P.M. to 8 P.M. is 6 hours.}\\ &= \frac{-51}{6} & \text{Subtract.}\\ &= -8.5 & \text{Divide.}\end{aligned}$$

∴ The mean change in the height of the tide is $-8.5$ inches per hour.

● **On Your Own**

**10.** The height of the tide at the Bay of Fundy in New Brunswick decreases 36 feet in 6 hours. What is the mean hourly change in the height?

Vocabulary and Concept Check

1. **WRITING** What can you tell about two integers when their quotient is positive? negative? zero?

2. **VOCABULARY** A quotient is undefined. What does this mean?

3. **OPEN-ENDED** Write two integers whose quotient is negative.

4. **WHICH ONE DOESN'T BELONG?** Which expression does *not* belong with the other three? Explain your reasoning.

$$\frac{10}{-5} \qquad \frac{-10}{5} \qquad \frac{-10}{-5} \qquad -\left(\frac{10}{5}\right)$$

**Tell whether the quotient is *positive* or *negative* without dividing.**

5. $-12 \div 4$

6. $\dfrac{-6}{-2}$

7. $15 \div (-3)$

 Practice and Problem Solving

**Divide, if possible.**

 8. $4 \div (-2)$

9. $21 \div (-7)$

10. $-20 \div 4$

11. $-18 \div (-3)$

12. $\dfrac{-14}{7}$

13. $\dfrac{0}{6}$

14. $\dfrac{-15}{-5}$

15. $\dfrac{54}{-9}$

16. $-33 \div 11$

17. $-49 \div (-7)$

18. $0 \div (-2)$

19. $60 \div (-6)$

20. $\dfrac{-56}{14}$

21. $\dfrac{18}{0}$

22. $\dfrac{65}{-5}$

23. $\dfrac{-84}{-7}$

**ERROR ANALYSIS** Describe and correct the error in finding the quotient.

24. $\times \quad \dfrac{-63}{-9} = -7$

25. $\times \quad 0 \div (-5) = -5$

26. **ALLIGATORS** An alligator population in a nature preserve in the Everglades decreases by 60 alligators over 5 years. What is the mean yearly change in the alligator population?

27. **READING** You read 105 pages of a novel over 7 days. What is the mean number of pages you read each day?

**ALGEBRA** Evaluate the expression when $x = 10$, $y = -2$, and $z = -5$.

28. $x \div y$

29. $\dfrac{10y^2}{z}$

30. $\left| \dfrac{xz}{-y} \right|$

31. $\dfrac{-x^2 + 6z}{y}$

**Find the mean of the integers.**

**32.** $3, -10, -2, 13, 11$

**33.** $-26, 39, -10, -16, 12, 31$

**Evaluate the expression.**

**34.** $-8 - 14 \div 2 + 5$

**35.** $24 \div (-4) + (-2) \cdot (-5)$

**36. PATTERN** Find the next two numbers in the pattern $-128, 64, -32, 16, \ldots$. Explain your reasoning.

**37. SNOWBOARDING** A snowboarder descends a 1200-foot hill in 3 minutes. What is the mean change in elevation per minute?

**38. GOLF** The table shows a golfer's score for each round of a tournament.

    **a.** What was the golfer's total score?

    **b.** What was the golfer's mean score per round?

| Scorecard | |
|---|---|
| Round 1 | $-2$ |
| Round 2 | $-6$ |
| Round 3 | $-7$ |
| Round 4 | $-3$ |

**39. TUNNEL** The Detroit-Windsor Tunnel is an underwater highway that connects the cities of Detroit, Michigan, and Windsor, Ontario. How many times deeper is the roadway than the bottom of the ship?

**40. AMUSEMENT PARK** The regular admission price for an amusement park is $72. For a group of 15 or more, the admission price is reduced by $25. How many people need to be in a group to save $500?

**41.**  **Number Sense** Write five different integers that have a mean of $-10$. Explain how you found your answer.

---

**Fair Game Review** What you learned in previous grades & lessons

**Graph the values on a number line. Then order the values from least to greatest.** *(Section 1.1)*

**42.** $-6, 4, \left|2\right|, -1, \left|-10\right|$

**43.** $3, \left|0\right|, \left|-4\right|, -3, -8$

**44.** $\left|5\right|, -2, -5, \left|-2\right|, -7$

**45. MULTIPLE CHOICE** What is the value of $4 \cdot 3 + (12 \div 2)^2$? *(Skills Review Handbook)*

    **Ⓐ** 15     **Ⓑ** 48     **Ⓒ** 156     **Ⓓ** 324

**Evaluate the expression.** *(Section 1.4 and Section 1.5)*

**1.** $-7(6)$

**2.** $-1(-10)$

**3.** $\dfrac{-72}{-9}$

**4.** $-24 \div 3$

**5.** $-3 \cdot 4 \cdot (-6)$

**6.** $(-3)^3$

**Evaluate the expression when $a = 4$, $b = -6$, and $c = -12$.** *(Section 1.4 and Section 1.5)*

**7.** $c^2$

**8.** $bc$

**9.** $\dfrac{ab}{c}$

**10.** $\dfrac{|c - b|}{a}$

**11. SPEECH** In speech class, you lose 3 points for every 30 seconds you go over the time limit. Your speech is 90 seconds over the time limit. What integer represents the change in your points? *(Section 1.4)*

**12. MOUNTAIN CLIMBING** On a mountain, the temperature decreases by 18°F every 5000 feet. What integer represents the change in temperature at 20,000 feet? *(Section 1.4)*

**13. GAMING** You play a video game for 15 minutes. You lose 165 points. What is the mean change in points per minute? *(Section 1.5)*

**14. DIVING** You dive 21 feet from the surface of a lake in 7 seconds. *(Section 1.4 and Section 1.5)*

  **a.** What is the mean change in your position in feet per second?

  **b.** You continue diving. What is your position relative to the surface after 5 more seconds?

**15. HIBERNATION** A female grizzly bear weighs 500 pounds. After hibernating for 6 months, she weighs only 200 pounds. What is the mean change in weight per month? *(Section 1.5)*

# 1 Chapter Review

Check It Out
Vocabulary Help
BigIdeasMath ✓com

## Review Key Vocabulary

integer, *p. 4*
absolute value, *p. 4*

opposites, *p. 10*
additive inverse, *p. 10*

## Review Examples and Exercises

### 1.1 Integers and Absolute Value *(pp. 2–7)*

**Find the absolute value of −2.**

Graph −2 on a number line.

The distance between −2 and 0 is 2.

∴ So, $|-2| = 2$.

#### Exercises

**Find the absolute value.**

1. $|3|$    2. $|-9|$    3. $|-17|$    4. $|8|$

5. **ELEVATION** The elevation of Death Valley, California, is −282 feet. The Mississippi River in Illinois has an elevation of 279 feet. Which is closer to sea level?

### 1.2 Adding Integers *(pp. 8–13)*

**Find 6 + (−14).**

$$6 + (-14) = -8 \qquad |-14| > |6|. \text{ So, subtract } |6| \text{ from } |-14|.$$

Use the sign of −14.

∴ The sum is −8.

#### Exercises

**Add.**

6. $-16 + (-11)$    7. $-15 + 5$    8. $100 + (-75)$    9. $-32 + (-2)$

**Subtracting Integers** *(pp. 14–19)*

**Subtract.**

**a.** $7 - 19 = 7 + (-19)$      Add the opposite of 19.

         $= -12$      Add.

   :•:   The difference is $-12$.

**b.** $-6 - (-10) = -6 + 10$      Add the opposite of $-10$.

           $= 4$      Add.

   :•:   The difference is 4.

### Exercises

**Subtract.**

**10.** $8 - 18$      **11.** $-16 - (-5)$      **12.** $-18 - 7$      **13.** $-12 - (-27)$

**14. GAME SHOW** Your score on a game show is $-300$. You answer the final question incorrectly, so you lose 400 points. What is your final score?

**Multiplying Integers** *(pp. 22–27)*

**a. Find $-7 \cdot (-9)$.**

The integers have the same sign.

$$-7 \cdot (-9) = 63$$

The product is positive.

   :•:   The product is 63.

**b. Find $-6(14)$.**

The integers have different signs.

$$-6(14) = -84$$

The product is negative.

   :•:   The product is $-84$.

### Exercises

**Multiply.**

**15.** $-8 \cdot 6$      **16.** $10(-7)$      **17.** $-3 \cdot (-6)$      **18.** $-12(5)$

## 1.5 Dividing Integers (pp. 28–33)

**a. Find $30 \div (-10)$.**

> The integers have different signs.

$$30 \div (-10) = -3$$

> The quotient is negative.

⋮ The quotient is −3.

**b. Find $\dfrac{-72}{-9}$.**

> The integers have the same sign.

$$\frac{-72}{-9} = 8$$

> The quotient is positive.

⋮ The quotient is 8.

### Exercises

**Divide.**

**19.** $-18 \div 9$

**20.** $\dfrac{-42}{-6}$

**21.** $\dfrac{-30}{6}$

**22.** $84 \div (-7)$

**Evaluate the expression when $x = 3$, $y = -4$, and $z = -6$.**

**23.** $z \div x$

**24.** $\dfrac{xy}{z}$

**25.** $\dfrac{z - 2x}{y}$

**Find the mean of the integers.**

**26.** $-3, -8, 12, -15, 9$

**27.** $-54, -32, -70, -25, -65, -42$

**28. PROFITS** The table shows the weekly profits of a fruit vendor. What is the mean profit for these weeks?

| Week | 1 | 2 | 3 | 4 |
|---|---|---|---|---|
| Profit | −$125 | −$86 | $54 | −$35 |

**29. RETURNS** You return several shirts to a store. The receipt shows that the amount placed back on your credit card is −$30.60. Each shirt is −$6.12. How many shirts did you return?

**Check It Out**
Test Practice
BigIdeasMath ✓ com

**Find the absolute value.**

**1.** $\left| -9 \right|$

**2.** $\left| 64 \right|$

**3.** $\left| -22 \right|$

**Copy and complete the statement using <, >, or =.**

**4.** $4 \quad \boxed{\phantom{x}} \quad \left| -8 \right|$

**5.** $\left| -7 \right| \quad \boxed{\phantom{x}} \quad -12$

**6.** $-7 \quad \boxed{\phantom{x}} \quad \left| 3 \right|$

**Evaluate the expression.**

**7.** $-6 + (-11)$

**8.** $2 - (-9)$

**9.** $-9 \cdot 2$

**10.** $-72 \div (-3)$

**Evaluate the expression when $x = 5$, $y = -3$, and $z = -2$.**

**11.** $\dfrac{y + z}{x}$

**12.** $\dfrac{x - 5z}{y}$

**Find the mean of the integers.**

**13.** $11, -7, -14, 10, -5$

**14.** $-32, -41, -39, -27, -33, -44$

**15. NASCAR** A driver receives $-25$ points for each rule violation. What integer represents the change in points after 4 rule violations?

**16. GOLF** The table shows your scores, relative to *par*, for nine holes of golf. What is your total score for the nine holes?

| Hole | 1 | 2 | 3 | 4 | 5 | 6 | 7 | 8 | 9 | Total |
|------|----|----|----|----|----|----|----|----|----|-------|
| Score | +1 | -2 | -1 | 0 | -1 | +3 | -1 | -3 | +1 | ? |

**17. VISITORS** In a recent 10-year period, the change in the number of visitors to U.S. national parks was about $-11{,}150{,}000$ visitors.

a. What was the mean yearly change in the number of visitors?

b. During the seventh year, the change in the number of visitors was about $10{,}800{,}000$. Explain how the change for the 10-year period can be negative.

1. A football team gains 2 yards on the first play, loses 5 yards on the second play, loses 3 yards on the third play, and gains 4 yards on the fourth play. What is the team's overall gain or loss for all four plays? *(7.NS.1b)*

   **A.** a gain of 14 yards     **C.** a loss of 2 yards

   **B.** a gain of 2 yards     **D.** a loss of 14 yards

2. Which expression is *not* equal to the number 0? *(7.NS.1a)*

   **F.** $5 - 5$     **H.** $6 - (-6)$

   **G.** $-7 + 7$     **I.** $-8 - (-8)$

**Test-Taking Strategy**

**Solve Directly or Eliminate Choices**

You ripped out $(-1)^2 + (-2)(-3)$ whiskers. How many did you rip out?
Ⓐ −5   Ⓑ 5   Ⓒ −7   Ⓓ 7

*Yeow, why the biggest number?*

"You can eliminate A and C. Then, solve directly to determine that the correct answer is D."

3. What is the value of the expression below when $a = -2$, $b = 3$, and $c = -5$? *(7.NS.3)*

   $$\left| a^2 - 2ac + 5b \right|$$

   **A.** $-9$          **C.** $1$

   **B.** $-1$          **D.** $9$

4. What is the value of the expression below? *(7.NS.1c)*

   $$17 - (-8)$$

5. Sam was evaluating an expression in the box below.

   $$(-2)^3 \cdot 3 - (-5) = 8 \cdot 3 - (-5)$$
   $$= 24 + 5$$
   $$= 29$$

   What should Sam do to correct the error that he made? *(7.NS.3)*

   **F.** Subtract 5 from 24 instead of adding.

   **G.** Rewrite $(-2)^3$ as $-8$.

   **H.** Subtract $-5$ from 3 before multiplying by $(-2)^3$.

   **I.** Multiply $-2$ by 3 before raising the quantity to the third power.

**6.** What is the value of the expression below when $x = 6$, $y = -4$, and $z = -2$? *(7.NS.3)*

$$\frac{x - 2y}{-z}$$

**A.** $-7$

**B.** $-1$

**C.** $1$

**D.** $7$

**7.** What is the missing number in the sequence below? *(7.NS.1c)*

$$39, 24, 9, \underline{\quad}, -21$$

**8.** You are playing a game using the spinner shown. You start with a score of 0 and spin the spinner four times. When you spin blue or green, you add the number to your score. When you spin red or orange, you subtract the number from your score. Which sequence of colors represents the greatest score? *(7.NS.3)*

**F.** red, green, green, red

**G.** orange, orange, green, blue

**H.** red, blue, orange, green

**I.** blue, red, blue, red

**9.** Which expression represents a negative integer? *(7.NS.3)*

**A.** $5 - (-6)$

**B.** $(-3)^3$

**C.** $-12 \div (-6)$

**D.** $(-2)(-4)$

**10.** Which expression has the greatest value when $x = -2$ and $y = -3$? *(7.NS.3)*

**F.** $-xy$

**G.** $xy$

**H.** $x - y$

**I.** $-x - y$

**11.** What is the value of the expression below?  *(7.NS.3)*

$$-5 \cdot (-4)^2 - (-3)$$

   **A.** $-83$                                    **C.** $77$

   **B.** $-77$                                    **D.** $83$

**12.** Which property does the equation below represent?  *(7.NS.1d)*

$$-80 + 30 + (-30) = -80 + [30 + (-30)]$$

   **F.** Commutative Property of Addition

   **G.** Associative Property of Addition

   **H.** Additive Inverse Property

   **I.** Addition Property of Zero

**13.** What is the mean of the data set in the box below?  *(7.NS.3)*

$$-8, -6, -2, 0, -6, -8, 4, -7, -8, 1$$

   **A.** $-8$                                    **C.** $-6$

   **B.** $-7$                                    **D.** $-4$

**14.** Consider the number line shown below.  *(7.NS.1b, 7.NS.1c)*

   *Part A*   Use the number line to explain how to add $-2$ and $-3$.

   *Part B*   Use the number line to explain how to subtract 5 from 2.

**15.** What is the value of the expression below?  *(7.NS.3)*

$$\frac{-3 - 2^2}{-1}$$

   **F.** $-25$                                    **H.** $7$

   **G.** $-1$                                    **I.** $25$

# 2 Rational Numbers

"On the count of 5, I'm going to give you half of my dog biscuits."

"1, 2, 3, 4, 4½, 4¾, 4⅞,..."

"I entered a contest for dog biscuits."

"I was notified that the number of biscuits I won was in the three-digit range."

# What You Learned Before

"Let's play a game. The goal is to say a positive rational number that is less than the other pet's number... You go first."

*This feels like a setup.*

## Writing Decimals and Fractions (4.NF.6)

**Example 1** Write 0.37 as a fraction.

$$0.37 = \frac{37}{100}$$

**Example 2** Write $\frac{2}{5}$ as a decimal.

$$\frac{2}{5} = \frac{2 \cdot 2}{5 \cdot 2} = \frac{4}{10} = 0.4$$

### Try It Yourself
Write the decimal as a fraction or the fraction as a decimal.

**1.** 0.51

**2.** 0.731

**3.** $\frac{3}{5}$

**4.** $\frac{7}{8}$

## Adding and Subtracting Fractions (5.NF.1)

**Example 3** Find $\frac{1}{3} + \frac{1}{5}$.

$$\frac{1}{3} + \frac{1}{5} = \frac{1 \cdot 5}{3 \cdot 5} + \frac{1 \cdot 3}{5 \cdot 3}$$
$$= \frac{5}{15} + \frac{3}{15}$$
$$= \frac{8}{15}$$

**Example 4** Find $\frac{1}{4} - \frac{2}{9}$.

$$\frac{1}{4} - \frac{2}{9} = \frac{1 \cdot 9}{4 \cdot 9} - \frac{2 \cdot 4}{9 \cdot 4}$$
$$= \frac{9}{36} - \frac{8}{36}$$
$$= \frac{1}{36}$$

## Multiplying and Dividing Fractions (5.NF.4, 6.NS.1)

**Example 5** Find $\frac{5}{6} \cdot \frac{3}{4}$.

$$\frac{5}{6} \cdot \frac{3}{4} = \frac{5 \cdot \overset{1}{\cancel{3}}}{\underset{2}{\cancel{6}} \cdot 4}$$
$$= \frac{5}{8}$$

**Example 6** Find $\frac{2}{3} \div \frac{9}{10}$.

$$\frac{2}{3} \div \frac{9}{10} = \frac{2}{3} \cdot \frac{10}{9} \quad \longleftarrow \text{Multiply by the reciprocal of the divisor.}$$
$$= \frac{2 \cdot 10}{3 \cdot 9}$$
$$= \frac{20}{27}$$

### Try It Yourself
Evaluate the expression.

**5.** $\frac{1}{4} + \frac{13}{20}$

**6.** $\frac{14}{15} - \frac{1}{3}$

**7.** $\frac{3}{7} \cdot \frac{9}{10}$

**8.** $\frac{4}{5} \div \frac{16}{17}$

## 2.1 Rational Numbers

### Essential Question How can you use a number line to order rational numbers?

#### The Meaning of a Word ● Rational

The word **rational** comes from the word *ratio*. Recall that you can write a ratio using fraction notation.

If you sleep for 8 hours in a day, then the ratio of your sleeping time to the total hours in a day can be written as

$$\frac{8\ h}{24\ h}.$$

A **rational number** is a number that can be written as the ratio of two integers.

$$2 = \frac{2}{1} \qquad -3 = \frac{-3}{1} \qquad -\frac{1}{2} = \frac{-1}{2} \qquad 0.25 = \frac{1}{4}$$

### 1 ACTIVITY: Ordering Rational Numbers

**Work in groups of five. Order the numbers from least to greatest.**

- Use masking tape and a marker to make a number line on the floor similar to the one shown.

- Write the numbers on pieces of paper. Then each person should choose one.

- Stand on the location of your number on the number line.

- Use your positions to order the numbers from least to greatest.

**a.** $-0.5, 1.25, -\dfrac{1}{3}, 0.5, -\dfrac{5}{3}$

**b.** $-\dfrac{7}{4}, 1.1, \dfrac{1}{2}, -\dfrac{1}{10}, -1.3$

**c.** $-1.4, -\dfrac{3}{5}, \dfrac{9}{2}, \dfrac{1}{4}, 0.9$

**d.** $\dfrac{5}{4}, 0.75, -\dfrac{5}{4}, -0.8, -1.1$

**COMMON CORE**

Rational Numbers

In this lesson, you will
- understand that a rational number is an integer divided by an integer.
- convert rational numbers to decimals.

Learning Standards
7.NS.2b
7.NS.2d

**Preparation:**

- Cut index cards to make 40 playing cards.
- Write each number in the table on a card.

**To Play:**

- Play with a partner.
- Deal 20 cards to each player facedown.
- Each player turns one card faceup. The player with the greater number wins. The winner collects both cards and places them at the bottom of his or her cards.
- Suppose there is a tie. Each player lays three cards facedown, then a new card faceup. The player with the greater of these new cards wins. The winner collects all ten cards and places them at the bottom of his or her cards.
- Continue playing until one player has all the cards. This player wins the game.

| $-\dfrac{3}{2}$ | $\dfrac{3}{10}$ | $-\dfrac{3}{4}$ | $-0.6$ | $1.25$ | $-0.15$ | $\dfrac{5}{4}$ | $\dfrac{3}{5}$ | $-1.6$ | $-0.3$ |
|---|---|---|---|---|---|---|---|---|---|
| $\dfrac{3}{20}$ | $\dfrac{8}{5}$ | $-1.2$ | $\dfrac{19}{10}$ | $0.75$ | $-1.5$ | $-\dfrac{6}{5}$ | $-\dfrac{3}{5}$ | $1.2$ | $0.3$ |
| $1.5$ | $1.9$ | $-0.75$ | $-0.4$ | $\dfrac{3}{4}$ | $-\dfrac{5}{4}$ | $-1.9$ | $\dfrac{2}{5}$ | $-\dfrac{3}{20}$ | $-\dfrac{19}{10}$ |
| $\dfrac{6}{5}$ | $-\dfrac{3}{10}$ | $1.6$ | $-\dfrac{2}{5}$ | $0.6$ | $0.15$ | $\dfrac{3}{2}$ | $-1.25$ | $0.4$ | $-\dfrac{8}{5}$ |

## What Is Your Answer?

3. **IN YOUR OWN WORDS** How can you use a number line to order rational numbers? Give an example.

**The numbers are in order from least to greatest. Fill in the blank spaces with rational numbers.**

4. $-\dfrac{1}{2},$ ___ $, \dfrac{1}{3},$ ___ $, \dfrac{7}{5},$ ___

5. $-\dfrac{5}{2},$ ___ $, -1.9,$ ___ $, -\dfrac{2}{3},$ ___

6. $-\dfrac{1}{3},$ ___ $, -0.1,$ ___ $, \dfrac{4}{5},$ ___

7. $-3.4,$ ___ $, -1.5,$ ___ $, 2.2,$ ___

**Practice** ➤ Use what you learned about ordering rational numbers to complete Exercises 28–30 on page 48.

Check It Out
Lesson Tutorials
BigIdeasMath.com

**Key Vocabulary**
rational number,
   p. 46
terminating decimal,
   p. 46
repeating decimal,
   p. 46

## Key Idea

**Rational Numbers**

A **rational number** is a number that can be written as $\frac{a}{b}$ where $a$ and $b$ are integers and $b \neq 0$.

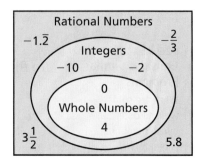

Because you can divide any integer by any nonzero integer, you can use long division to write fractions and mixed numbers as decimals. These decimals are also rational numbers and will either *terminate* or *repeat*.

A **terminating decimal** is a decimal that ends.

$$1.5, \; -0.25, \; 10.625$$

A **repeating decimal** is a decimal that has a pattern that repeats.

$$-1.333\ldots = -1.\overline{3}$$
$$0.151515\ldots = 0.\overline{15}$$

> Use *bar notation* to show which of the digits repeat.

## EXAMPLE 1 Writing Rational Numbers as Decimals

**a.** Write $-2\frac{1}{4}$ as a decimal.

Notice that $-2\frac{1}{4} = -\frac{9}{4}$.

Divide 9 by 4.

$$\begin{array}{r} 2.25 \\ 4\overline{)9.00} \\ -8 \phantom{.00} \\ \hline 1\,0 \phantom{0} \\ -8 \phantom{0} \\ \hline 20 \\ -20 \\ \hline 0 \end{array}$$

The remainder is 0. So, it is a terminating decimal.

∴ So, $-2\frac{1}{4} = -2.25$.

**b.** Write $\frac{5}{11}$ as a decimal.

Divide 5 by 11.

$$\begin{array}{r} 0.4545 \\ 11\overline{)5.0000} \\ -4\,4 \phantom{.000} \\ \hline 60 \phantom{00} \\ -55 \phantom{00} \\ \hline 50 \phantom{0} \\ -44 \phantom{0} \\ \hline 60 \\ -55 \\ \hline 5 \end{array}$$

The remainder repeats. So, it is a repeating decimal.

∴ So, $\frac{5}{11} = 0.\overline{45}$.

## On Your Own

*Now You're Ready*
Exercises 11–18

Write the rational number as a decimal.

**1.** $-\frac{6}{5}$   **2.** $-7\frac{3}{8}$   **3.** $-\frac{3}{11}$   **4.** $1\frac{5}{27}$

Multi-Language Glossary at BigIdeasMath.com

EXAMPLE 2 **Writing a Decimal as a Fraction**

Write −0.26 as a fraction in simplest form.

**Study Tip**

If $p$ and $q$ are integers, then $-\dfrac{p}{q} = \dfrac{-p}{q} = \dfrac{p}{-q}$.

$$-0.26 = -\frac{26}{100}$$

Write the digits after the decimal point in the numerator.

The last digit is in the hundredths place. So, use 100 in the denominator.

$$= -\frac{13}{50}$$ Simplify.

**On Your Own**

*Now You're Ready*
Exercises 20–27

**Write the decimal as a fraction or a mixed number in simplest form.**

**5.** −0.7　　**6.** 0.125　　**7.** −3.1　　**8.** −10.25

---

**EXAMPLE 3** **Ordering Rational Numbers**

| Creature | Elevation (kilometers) |
|----------|------------------------|
| Anglerfish | $-\dfrac{13}{10}$ |
| Squid | $-2\dfrac{1}{5}$ |
| Shark | $-\dfrac{2}{11}$ |
| Whale | −0.8 |

The table shows the elevations of four sea creatures relative to sea level. Which of the sea creatures are deeper than the whale? Explain.

Write each rational number as a decimal.

$$-\frac{13}{10} = -1.3$$

$$-2\frac{1}{5} = -2.2$$

$$-\frac{2}{11} = -0.\overline{18}$$

Then graph each decimal on a number line.

Both −2.2 and −1.3 are less than −0.8. So, the squid and the anglerfish are deeper than the whale.

**On Your Own**

*Now You're Ready*
Exercises 28–33

**9. WHAT IF?** The elevation of a dolphin is $-\dfrac{1}{10}$ kilometer. Which of the sea creatures in Example 3 are deeper than the dolphin? Explain.

## 2.1 Exercises

### ✓ Vocabulary and Concept Check

1. **VOCABULARY** Is the quotient of two integers always a rational number? Explain.

2. **WRITING** Are all terminating and repeating decimals rational numbers? Explain.

**Tell whether the number belongs to each of the following number sets:**
*rational numbers, integers, whole numbers.*

3. $-5$     4. $-2.1\overline{6}$     5. $12$     6. $0$

**Tell whether the decimal is *terminating* or *repeating*.**

7. $-0.4848\ldots$     8. $-0.151$     9. $72.72$     10. $-5.2\overline{36}$

### Practice and Problem Solving

**Write the rational number as a decimal.**

① 11. $\dfrac{7}{8}$     12. $\dfrac{1}{11}$     13. $-\dfrac{7}{9}$     14. $-\dfrac{17}{40}$

15. $1\dfrac{5}{6}$     16. $-2\dfrac{17}{18}$     17. $-5\dfrac{7}{12}$     18. $8\dfrac{15}{22}$

19. **ERROR ANALYSIS** Describe and correct the error in writing the rational number as a decimal.

$$\boxed{\times} \quad -\dfrac{7}{11} = -0.6\overline{3}$$

**Write the decimal as a fraction or a mixed number in simplest form.**

② 20. $-0.9$     21. $0.45$     22. $-0.258$     23. $-0.312$

24. $-2.32$     25. $-1.64$     26. $6.012$     27. $-12.405$

**Order the numbers from least to greatest.**

③ 28. $-\dfrac{3}{4}, 0.5, \dfrac{2}{3}, -\dfrac{7}{3}, 1.2$     29. $\dfrac{9}{5}, -2.5, -1.1, -\dfrac{4}{5}, 0.8$     30. $-1.4, -\dfrac{8}{5}, 0.6, -0.9, \dfrac{1}{4}$

31. $2.1, -\dfrac{6}{10}, -\dfrac{9}{4}, -0.75, \dfrac{5}{3}$     32. $-\dfrac{7}{2}, -2.8, -\dfrac{5}{4}, \dfrac{4}{3}, 1.3$     33. $-\dfrac{11}{5}, -2.4, 1.6, \dfrac{15}{10}, -2.25$

34. **COINS** You lose one quarter, two dimes, and two nickels.

    a. Write the amount as a decimal.

    b. Write the amount as a fraction in simplest form.

35. **HIBERNATION** A box turtle hibernates in sand at $-1\dfrac{5}{8}$ feet. A spotted turtle hibernates at $-1\dfrac{16}{25}$ feet. Which turtle is deeper?

**Copy and complete the statement using <, >, or =.**

**36.** $-2.2$ [ ] $-2.42$

**37.** $-1.82$ [ ] $-1.81$

**38.** $\dfrac{15}{8}$ [ ] $1\dfrac{7}{8}$

**39.** $-4\dfrac{6}{10}$ [ ] $-4.65$

**40.** $-5\dfrac{3}{11}$ [ ] $-5.\overline{2}$

**41.** $-2\dfrac{13}{16}$ [ ] $-2\dfrac{11}{14}$

**42. OPEN-ENDED** Find one terminating decimal and one repeating decimal between $-\dfrac{1}{2}$ and $-\dfrac{1}{3}$.

| Player | Hits | At Bats |
|--------|------|---------|
| Eva | 42 | 90 |
| Michelle | 38 | 80 |

**43. SOFTBALL** In softball, a batting average is the number of hits divided by the number of times at bat. Does Eva or Michelle have the higher batting average?

**44. PROBLEM SOLVING** You miss 3 out of 10 questions on a science quiz and 4 out of 15 questions on a math quiz. Which quiz has a higher percent of correct answers?

**45. SKATING** Is the half pipe deeper than the skating pool? Explain.

**46. ENVIRONMENT** The table shows the changes from the average water level of a pond over several weeks. Order the numbers from least to greatest.

| Week | 1 | 2 | 3 | 4 |
|------|---|---|---|---|
| Change (inches) | $-\dfrac{7}{5}$ | $-1\dfrac{5}{11}$ | $-1.45$ | $-1\dfrac{91}{200}$ |

**47.** **Critical Thinking** Given: $a$ and $b$ are integers.

**a.** When is $-\dfrac{1}{a}$ positive?

**b.** When is $\dfrac{1}{ab}$ positive?

 **Fair Game Review** What you learned in previous grades & lessons

**Add or subtract.** *(Skills Review Handbook)*

**48.** $\dfrac{3}{5} + \dfrac{2}{7}$

**49.** $\dfrac{9}{10} - \dfrac{2}{3}$

**50.** $8.79 - 4.07$

**51.** $11.81 + 9.34$

**52. MULTIPLE CHOICE** In one year, a company has a profit of $-\$2$ million. In the next year, the company has a profit of $\$7$ million. How much more profit did the company make the second year? *(Section 1.3)*

(A) $2 million   (B) $5 million   (C) $7 million   (D) $9 million

## 2.2 Adding Rational Numbers

### Essential Question
How can you use what you know about adding integers to add rational numbers?

---

**1 ACTIVITY: Adding Rational Numbers**

**Work with a partner. Use a number line to find the sum.**

**a.** $2.7 + (-3.4)$

Start at 0. Move 2.7 units to the right.

Add $-3.4$.

Then move 3.4 units left to end at [ ].

2.7

So, $2.7 + (-3.4) =$ [ ].

**b.** $1.3 + (-1.5)$

**c.** $-2.1 + 0.8$

**d.** $-1\frac{1}{4} + \frac{3}{4}$

**e.** $\frac{3}{10} + \left(-\frac{3}{10}\right)$

---

**2 ACTIVITY: Adding Rational Numbers**

**Work with a partner. Use a number line to find the sum.**

**a.** $-1\frac{2}{5} + \left(-\frac{4}{5}\right)$

Then move $\frac{4}{5}$ unit left to end at [ ].

Add $-\frac{4}{5}$.

Start at 0. Move $1\frac{2}{5}$ units to the left.

$-1\frac{2}{5}$

So, $-1\frac{2}{5} + \left(-\frac{4}{5}\right) =$ [ ].

**b.** $-\frac{7}{10} + \left(-1\frac{7}{10}\right)$

**c.** $-1\frac{2}{3} + \left(-1\frac{1}{3}\right)$

**d.** $-0.4 + (-1.9)$

**e.** $-2.3 + (-0.6)$

COMMON CORE

**Rational Numbers**

In this lesson, you will
- add rational numbers.
- solve real-life problems.

Learning Standards
7.NS.1a
7.NS.1b
7.NS.1d
7.NS.3

**ACTIVITY: Writing Expressions**

Work with a partner. Write the addition expression shown. Then find the sum.

**a.**

**b.**

**c.**

# What Is Your Answer?

4. **IN YOUR OWN WORDS** How can you use what you know about adding integers to add rational numbers?

**PUZZLE** Find a path through the table so that the numbers add up to the sum. You can move horizontally or vertically.

5. Sum: $\frac{3}{4}$

Start →

| $\frac{1}{2}$ | $\frac{2}{3}$ | $-\frac{5}{7}$ |
|---|---|---|
| $-\frac{1}{8}$ | $-\frac{3}{4}$ | $\frac{1}{3}$ |

← End

6. Sum: $-0.07$

Start →

| 2.43 | 1.75 | $-0.98$ |
|---|---|---|
| $-1.09$ | 3.47 | $-4.88$ |

← End

 Use what you learned about adding rational numbers to complete Exercises 4–6 on page 54.

**Key Idea**

**Adding Rational Numbers**

**Words** To add rational numbers, use the same rules for signs as you used for integers.

**Numbers** $-\dfrac{1}{3} + \dfrac{1}{6} = \dfrac{-2}{6} + \dfrac{1}{6} = \dfrac{-2+1}{6} = \dfrac{-1}{6} = -\dfrac{1}{6}$

**EXAMPLE** **1** **Adding Rational Numbers**

**Find** $-\dfrac{8}{3} + \dfrac{5}{6}.$    **Estimate** $-3 + 1 = -2$

**Study Tip**

In Example 1, notice how $-\dfrac{8}{3}$ is written as $-\dfrac{8}{3} = \dfrac{-8}{3} = \dfrac{-16}{6}.$

$-\dfrac{8}{3} + \dfrac{5}{6} = \dfrac{-16}{6} + \dfrac{5}{6}$    Rewrite using the LCD (least common denominator).

$= \dfrac{-16 + 5}{6}$    Write the sum of the numerators over the common denominator.

$= \dfrac{-11}{6}$    Add.

$= -1\dfrac{5}{6}$    Write the improper fraction as a mixed number.

∴ The sum is $-1\dfrac{5}{6}.$    **Reasonable?** $-1\dfrac{5}{6} \approx -2$ ✓

**EXAMPLE** **2** **Adding Rational Numbers**

**Find** $-4.05 + 7.62.$

$-4.05 + 7.62 = 3.57$    $|7.62| > |-4.05|$. So, subtract $|-4.05|$ from $|7.62|$.

Use the sign of 7.62.

∴ The sum is 3.57.

● **On Your Own**

*Now You're Ready*
Exercises 4–12

**Add.**

**1.** $-\dfrac{7}{8} + \dfrac{1}{4}$    **2.** $-6\dfrac{1}{3} + \dfrac{20}{3}$    **3.** $2 + \left(-\dfrac{7}{2}\right)$

**4.** $-12.5 + 15.3$    **5.** $-8.15 + (-4.3)$    **6.** $0.65 + (-2.75)$

## EXAMPLE 3 Evaluating Expressions

Evaluate $2x + y$ when $x = \dfrac{1}{4}$ and $y = -\dfrac{3}{2}$.

$$2x + y = 2\left(\dfrac{1}{4}\right) + \left(-\dfrac{3}{2}\right)$$     Substitute $\dfrac{1}{4}$ for $x$ and $-\dfrac{3}{2}$ for $y$.

$$= \dfrac{1}{2} + \left(\dfrac{-3}{2}\right)$$     Multiply.

$$= \dfrac{1 + (-3)}{2}$$     Write the sum of the numerators over the common denominator.

$$= -1$$     Simplify.

## EXAMPLE 4 Real-Life Application

| Year | Profit (billions of dollars) |
|------|------------------------------|
| 2008 | −1.7 |
| 2009 | −4.75 |
| 2010 | 1.7 |
| 2011 | 0.85 |
| 2012 | 3.6 |

**The table shows the annual profits (in billions of dollars) of a financial company from 2008 to 2012. Positive numbers represent *gains*, and negative numbers represent *losses*. Which statement describes the profit over the five-year period?**

   (A) gain of $0.3 billion      (B) gain of $30 million

   (C) loss of $3 million      (D) loss of $300 million

To determine whether there was a gain or a loss, find the sum of the profits.

five-year profit $= -1.7 + (-4.75) + 1.7 + 0.85 + 3.6$    Write the sum.

$$= -1.7 + 1.7 + (-4.75) + 0.85 + 3.6$$    Comm. Prop. of Add.

$$= 0 + (-4.75) + 0.85 + 3.6$$    Additive Inv. Prop.

$$= -4.75 + 0.85 + 3.6$$    Add. Prop. of Zero

$$= -3.9 + 3.6$$    Add $-4.75$ and $0.85$.

$$= -0.3$$    Add $-3.9$ and $3.6$.

The five-year profit is $-$0.3 billion. So, the company has a five-year loss of $0.3 billion, or $300 million.

   ∴ The correct answer is (D).

### On Your Own

**Now You're Ready**
Exercises 15–17

Evaluate the expression when $a = \dfrac{1}{2}$ and $b = -\dfrac{5}{2}$.

**7.** $b + 4a$          **8.** $|a + b|$

**9. WHAT IF?** In Example 4, the 2013 profit is $1.07 billion. State the company's gain or loss over the six-year period in millions of dollars.

Check It Out
Help with Homework
BigIdeasMath✓com

## ✓ Vocabulary and Concept Check

1. **WRITING** Explain how to find the sum $-8.46 + 5.31$.

2. **OPEN-ENDED** Write an addition expression using fractions that equals $-\dfrac{1}{2}$.

3. **DIFFERENT WORDS, SAME QUESTION** Which is different? Find "both" answers.

| | |
|---|---|
| Add $-4.5$ and $3.5$. | What is the distance between $-4.5$ and $3.5$? |
| What is $-4.5$ increased by $3.5$? | Find the sum of $-4.5$ and $3.5$. |

## Practice and Problem Solving

**Add. Write fractions in simplest form.**

 ① ②

4. $\dfrac{11}{12} + \left(-\dfrac{7}{12}\right)$
   5. $-1\dfrac{1}{5} + \left(-\dfrac{3}{5}\right)$
   6. $-4.2 + 3.3$

7. $-\dfrac{9}{14} + \dfrac{2}{7}$
   8. $4 + \left(-1\dfrac{2}{3}\right)$
   9. $\dfrac{15}{4} + \left(-4\dfrac{1}{3}\right)$

10. $-3.1 + (-0.35)$
    11. $12.48 + (-10.636)$
    12. $20.25 + (-15.711)$

**ERROR ANALYSIS** Describe and correct the error in finding the sum.

13.
$$\times \quad \begin{array}{r} -3.7 \\ + \ (-0.25) \\ \hline -0.62 \end{array}$$

14.
$$\times \quad -\dfrac{5}{8} + \dfrac{1}{8} = \dfrac{-5+1}{8} = \dfrac{-6}{8} = -\dfrac{3}{4}$$

**Evaluate the expression when $x = \dfrac{1}{3}$ and $y = -\dfrac{7}{4}$.**

③ 15. $x + y$
    16. $3x + y$
    17. $-x + |y|$

18. **BANKING** Your bank account balance is −$20.85. You deposit $15.50. What is your new balance?

19. **HOT DOGS** You eat $\dfrac{3}{10}$ of a pack of hot dogs.

    Your friend eats $\dfrac{1}{5}$ of the pack of hot dogs.

    What fraction of the pack of hot dogs do

    you and your friend eat?

**Add. Write fractions in simplest form.**

**20.** $6 + \left(-4\dfrac{3}{4}\right) + \left(-2\dfrac{1}{8}\right)$

**21.** $-5\dfrac{2}{3} + 3\dfrac{1}{4} + \left(-7\dfrac{1}{3}\right)$

**22.** $10.9 + (-15.6) + 2.1$

| June | July | August |
|------|------|--------|
| $-2\dfrac{1}{8}$ | $1\dfrac{1}{4}$ | $-\dfrac{9}{16}$ |

**23. NUMBER SENSE** When is the sum of two negative mixed numbers an integer?

**24. WRITING** You are adding two rational numbers with different signs. How can you tell if the sum will be *positive*, *negative*, or *zero*?

**25. RESERVOIR** The table at the left shows the water level (in inches) of a reservoir for three months compared to the yearly average. Is the water level for the three-month period greater than or less than the yearly average? Explain.

**26. BREAK EVEN** The table at the right shows the annual profits (in thousands of dollars) of a county fair from 2008 to 2012. What must the 2012 profit be (in hundreds of dollars) to break even over the five-year period?

| Year | Profit (thousands of dollars) |
|------|-------------------------------|
| 2008 | 2.5 |
| 2009 | 1.75 |
| 2010 | $-3.3$ |
| 2011 | $-1.4$ |
| 2012 | ? |

**27. REASONING** Is $|a + b| = |a| + |b|$ for all rational numbers $a$ and $b$? Explain.

**28.** **Repeated Reasoning** Evaluate the expression.

$$\dfrac{19}{20} + \left(\dfrac{-18}{20}\right) + \dfrac{17}{20} + \left(\dfrac{-16}{20}\right) + \cdots + \left(\dfrac{-4}{20}\right) + \dfrac{3}{20} + \left(\dfrac{-2}{20}\right) + \dfrac{1}{20}$$

 **Fair Game Review** *What you learned in previous grades & lessons*

**Identify the property. Then simplify.** *(Skills Review Handbook)*

**29.** $8 + (-3) + 2 = 8 + 2 + (-3)$

**30.** $2 \cdot (4.5 \cdot 9) = (2 \cdot 4.5) \cdot 9$

**31.** $\dfrac{1}{4} + \left(\dfrac{3}{4} + \dfrac{1}{8}\right) = \left(\dfrac{1}{4} + \dfrac{3}{4}\right) + \dfrac{1}{8}$

**32.** $\dfrac{3}{7} \cdot \dfrac{4}{5} \cdot \dfrac{14}{27} = \dfrac{3}{7} \cdot \dfrac{14}{27} \cdot \dfrac{4}{5}$

**33. MULTIPLE CHOICE** The regular price of a photo album is $18. You have a coupon for 15% off. How much is the discount? *(Skills Review Handbook)*

    **Ⓐ** $2.70      **Ⓑ** $3      **Ⓒ** $15      **Ⓓ** $15.30

You can use a **process diagram** to show the steps involved in a procedure. Here is an example of a process diagram for adding rational numbers.

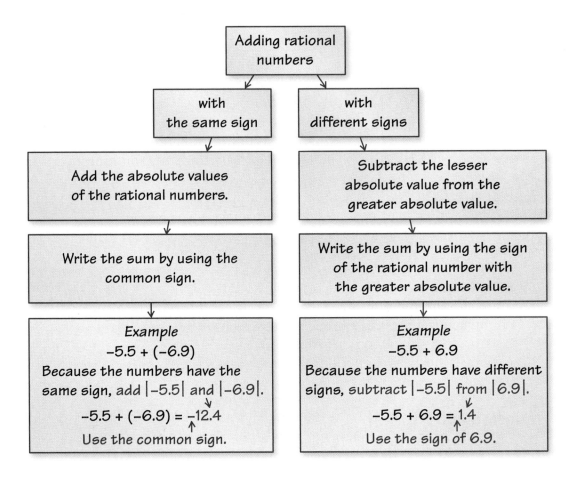

## On Your Own

**Make a process diagram with examples to help you study the topic.**

1. writing rational numbers as decimals

**After you complete this chapter, make process diagrams with examples for the following topics.**

2. subtracting rational numbers

3. multiplying rational numbers

4. dividing rational numbers

"Does this **process diagram** accurately show how a cat claws furniture?"

**Write the rational number as a decimal.**    *(Section 2.1)*

**1.** $-\dfrac{3}{20}$

**2.** $-\dfrac{11}{6}$

**Write the decimal as a fraction or a mixed number in simplest form.**    *(Section 2.1)*

**3.** $-0.325$

**4.** $-1.28$

**Order the numbers from least to greatest.**    *(Section 2.1)*

**5.** $-\dfrac{1}{3},\, -0.2,\, \dfrac{5}{3},\, 0.4,\, 1.3$

**6.** $-\dfrac{4}{3},\, -1.2,\, 0.3,\, \dfrac{4}{9},\, -0.8$

**Add. Write fractions in simplest form.**    *(Section 2.2)*

**7.** $-\dfrac{4}{5} + \left(-\dfrac{3}{8}\right)$

**8.** $-\dfrac{13}{6} + \dfrac{7}{12}$

**9.** $-5.8 + 2.6$

**10.** $-4.28 + (-2.56)$

**Evaluate the expression when $x = \dfrac{3}{4}$ and $y = -\dfrac{1}{2}$.**    *(Section 2.2)*

**11.** $x + y$

**12.** $2x + y$

**13.** $x + |y|$

**14.** $|-x + y|$

**15. STOCK** The value of Stock A changes $-\$3.68$, and the value of Stock B changes $-\$3.72$. Which stock has the greater loss? Explain.    *(Section 2.1)*

**16. LEMONADE** You drink $\dfrac{2}{7}$ of a pitcher of lemonade. Your friend drinks $\dfrac{3}{14}$ of the pitcher. What fraction of the pitcher do you and your friend drink?    *(Section 2.2)*

**17. FOOTBALL** The table shows the statistics of a running back in a football game. Did he gain more than 50 yards total? Explain.    *(Section 2.2)*

| Quarter | 1 | 2 | 3 | 4 | Total |
|---------|---|---|---|---|-------|
| Yards | $-8\tfrac{1}{2}$ | 23 | $42\tfrac{1}{2}$ | $-2\tfrac{1}{4}$ | ? |

## 2.3 Subtracting Rational Numbers

**Essential Question** How can you use what you know about subtracting integers to subtract rational numbers?

---

### 1 ACTIVITY: Subtracting Rational Numbers

**Work with a partner. Use a number line to find the difference.**

a. $-1\dfrac{1}{2} - \dfrac{1}{2}$

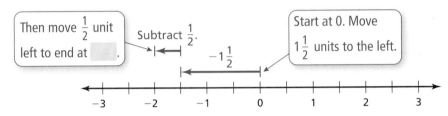

> Then move $\dfrac{1}{2}$ unit left to end at ▢.

> Subtract $\dfrac{1}{2}$.

> $-1\dfrac{1}{2}$

> Start at 0. Move $1\dfrac{1}{2}$ units to the left.

⋮ So, $-1\dfrac{1}{2} - \dfrac{1}{2} = $ ▢ .

b. $\dfrac{6}{10} - 1\dfrac{3}{10}$

c. $-1\dfrac{1}{4} - 1\dfrac{3}{4}$

d. $-1.9 - 0.8$

e. $0.2 - 0.7$

---

### 2 ACTIVITY: Finding Distances on a Number Line

**Work with a partner.**

a. Plot $-3$ and 2 on the number line. Then find $-3 - 2$ and $2 - (-3)$. What do you notice about your results?

b. Plot $\dfrac{3}{4}$ and 1 on the number line. Then find $\dfrac{3}{4} - 1$ and $1 - \dfrac{3}{4}$. What do you notice about your results?

c. Choose any two points $a$ and $b$ on a number line. Find the values of $a - b$ and $b - a$. What do the absolute values of these differences represent? Is this true for any pair of rational numbers? Explain.

**COMMON CORE**

**Rational Numbers**

In this lesson, you will

- subtract rational numbers.
- solve real-life problems.

Learning Standards
7.NS.1c
7.NS.1d
7.NS.3

**Work with a partner. The table shows the balance in a checkbook.**

- Black numbers are amounts added to the account.
- Red numbers are amounts taken from the account.

| Date | Check # | Transaction | Amount | Balance |
|---|---|---|---|---|
| -- | -- | Previous balance | -- | 100.00 |
| 1/02/2013 | 124 | Groceries | 34.57 | |
| 1/07/2013 | | Check deposit | 875.50 | |
| 1/11/2013 | | ATM withdrawal | 40.00 | |
| 1/14/2013 | 125 | Electric company | 78.43 | |
| 1/17/2013 | | Music store | 10.55 | |
| 1/18/2013 | 126 | Shoes | 47.21 | |
| 1/22/2013 | | Check deposit | 125.00 | |
| 1/24/2013 | | Interest | 2.12 | |
| 1/25/2013 | 127 | Cell phone | 59.99 | |
| 1/26/2013 | 128 | Clothes | 65.54 | |
| 1/30/2013 | 129 | Cable company | 75.00 | |

**Math Practice 4**

**Interpret Results**

What does your answer represent? Does your answer make sense?

You can find the balance in the **second row** two different ways.

$$100.00 - 34.57 = 65.43 \qquad \text{Subtract 34.57 from 100.00.}$$
$$100.00 + (-34.57) = 65.43 \qquad \text{Add } -34.57 \text{ to 100.00.}$$

**a.** Copy the table. Then complete the balance column.

**b.** How did you find the balance in the twelfth row?

**c.** Use a different way to find the balance in part (b).

## What Is Your Answer?

**4. IN YOUR OWN WORDS** How can you use what you know about subtracting integers to subtract rational numbers?

**5.** Give two real-life examples of subtracting rational numbers that are not integers.

Use what you learned about subtracting rational numbers to complete Exercises 3–5 on page 62.

**Key Idea**

**Subtracting Rational Numbers**

**Words** To subtract rational numbers, use the same rules for signs as you used for integers.

**Numbers** $\dfrac{2}{5} - \left(-\dfrac{1}{5}\right) = \dfrac{2}{5} + \dfrac{1}{5} = \dfrac{2+1}{5} = \dfrac{3}{5}$

**EXAMPLE 1** **Subtracting Rational Numbers**

Find $-4\dfrac{1}{7} - \left(-\dfrac{6}{7}\right)$.  **Estimate** $-4 - (-1) = -3$

$-4\dfrac{1}{7} - \left(-\dfrac{6}{7}\right) = -4\dfrac{1}{7} + \dfrac{6}{7}$  Add the opposite of $-\dfrac{6}{7}$.

$= -\dfrac{29}{7} + \dfrac{6}{7}$  Write the mixed number as an improper fraction.

$= \dfrac{-29 + 6}{7}$  Write the sum of the numerators over the common denominator.

$= \dfrac{-23}{7}$  Add.

$= -3\dfrac{2}{7}$  Write the improper fraction as a mixed number.

∴ The difference is $-3\dfrac{2}{7}$.  **Reasonable?** $-3\dfrac{2}{7} \approx -3$ ✓

**EXAMPLE 2** **Subtracting Rational Numbers**

Find $12.8 - 21.6$.

$12.8 - 21.6 = 12.8 + (-21.6)$  Add the opposite of 21.6.

$= -8.8$  $|-21.6| > |12.8|$. So, subtract $|12.8|$ from $|-21.6|$.

∴ The difference is $-8.8$.  Use the sign of $-21.6$.

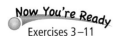 **On Your Own**

Now You're Ready
Exercises 3–11

**1.** $\dfrac{1}{3} - \left(-\dfrac{1}{3}\right)$  **2.** $-3\dfrac{1}{3} - \dfrac{5}{6}$  **3.** $4\dfrac{1}{2} - 5\dfrac{1}{4}$

**4.** $-8.4 - 6.7$  **5.** $-20.5 - (-20.5)$  **6.** $0.41 - (-0.07)$

The distance between any two numbers on a number line is the absolute value of the difference of the numbers.

**EXAMPLE 3**  **Finding Distances Between Numbers on a Number Line**

**Find the distance between the two numbers on the number line.**

To find the distance between the numbers, first find the difference of the numbers.

$$-2\frac{2}{3} - 2\frac{1}{3} = -2\frac{2}{3} + \left(-2\frac{1}{3}\right) \quad \text{Add the opposite of } 2\frac{1}{3}.$$

$$= -\frac{8}{3} + \left(-\frac{7}{3}\right) \quad \text{Write the mixed numbers as improper fractions.}$$

$$= \frac{-15}{3} \quad \text{Add.}$$

$$= -5 \quad \text{Simplify.}$$

∴ Because $|-5| = 5$, the distance between $-2\frac{2}{3}$ and $2\frac{1}{3}$ is 5.

**EXAMPLE 4**  **Real-Life Application**

Clearance: 11 ft 8 in.

**In the water, the bottom of a boat is 2.1 feet below the surface, and the top of the boat is 8.7 feet above it. Towed on a trailer, the bottom of the boat is 1.3 feet above the ground. Can the boat and trailer pass under the bridge?**

**Step 1:** Find the height $h$ of the boat.

$$h = 8.7 - (-2.1) \quad \text{Subtract the lowest point from the highest point.}$$

$$= 8.7 + 2.1 \quad \text{Add the opposite of } -2.1.$$

$$= 10.8 \quad \text{Add.}$$

**Step 2:** Find the height $t$ of the boat and trailer.

$$t = 10.8 + 1.3 \quad \text{Add the trailer height to the boat height.}$$

$$= 12.1 \quad \text{Add.}$$

∴ Because 12.1 feet is greater than 11 feet 8 inches, the boat and trailer cannot pass under the bridge.

**On Your Own**

Now You're Ready
Exercises 13–15

**7.** Find the distance between $-7.5$ and $-15.3$ on a number line.

**8. WHAT IF?** In Example 4, the clearance is 12 feet 1 inch. Can the boat and trailer pass under the bridge?

Vocabulary and Concept Check

1. **WRITING** Explain how to find the difference $-\dfrac{4}{5} - \dfrac{3}{5}$.

2. **WHICH ONE DOESN'T BELONG?** Which expression does *not* belong with the other three? Explain your reasoning.

$$-\dfrac{5}{8} - \dfrac{3}{4} \qquad -\dfrac{3}{4} + \dfrac{5}{8} \qquad -\dfrac{5}{8} + \left(-\dfrac{3}{4}\right) \qquad -\dfrac{3}{4} - \dfrac{5}{8}$$

 Practice and Problem Solving

**Subtract. Write fractions in simplest form.**

**① ②** 3. $\dfrac{5}{8} - \left(-\dfrac{7}{8}\right)$

4. $-1\dfrac{1}{3} - 1\dfrac{2}{3}$

5. $-1 - 2.5$

6. $-5 - \dfrac{5}{3}$

7. $-8\dfrac{3}{8} - 10\dfrac{1}{6}$

8. $-\dfrac{1}{2} - \left(-\dfrac{5}{9}\right)$

9. $5.5 - 8.1$

10. $-7.34 - (-5.51)$

11. $6.673 - (-8.29)$

12. **ERROR ANALYSIS** Describe and correct the error in finding the difference.

$$\cancel{\times} \quad \dfrac{3}{4} - \dfrac{9}{2} = \dfrac{3 - 9}{4 - 2} = \dfrac{-6}{2} = -3$$

**Find the distance between the two numbers on a number line.**

**③** 13. $-2\dfrac{1}{2}, -5\dfrac{3}{4}$

14. $-2.2, 8.4$

15. $-7, -3\dfrac{2}{3}$

16. **SPORTS DRINK** Your sports drink bottle is $\dfrac{5}{6}$ full. After practice, the bottle is $\dfrac{3}{8}$ full. Write the difference of the amounts after practice and before practice.

17. **SUBMARINE** The figure shows the depths of a submarine.

   a. Find the vertical distance traveled by the submarine.

   b. Find the mean hourly vertical distance traveled by the submarine.

−314.9 ft (now)

−725.6 ft (3 hours ago)

**Evaluate.**

18. $2\dfrac{1}{6} - \left(-\dfrac{8}{3}\right) + \left(-4\dfrac{7}{9}\right)$

19. $6.59 + (-7.8) - (-2.41)$

20. $-\dfrac{12}{5} + \left|-\dfrac{13}{6}\right| + \left(-3\dfrac{2}{3}\right)$

**21. REASONING** When is the difference of two decimals an integer? Explain.

**22. RECIPE** A cook has $2\frac{2}{3}$ cups of flour. A recipe calls for $2\frac{3}{4}$ cups of flour. Does the cook have enough flour? If not, how much more flour is needed?

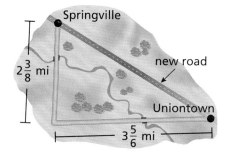

**23. ROADWAY** A new road that connects Uniontown to Springville is $4\frac{1}{3}$ miles long. What is the change in distance when using the new road instead of the dirt roads?

**RAINFALL** In Exercises 24–26, the bar graph shows the differences in a city's rainfall from the historical average.

**24.** What is the difference in rainfall between the wettest and the driest months?

**25.** Find the sum of the differences for the year.

**26.** What does the sum in Exercise 25 tell you about the rainfall for the year?

**27. OPEN-ENDED** Write two different pairs of negative decimals, $x$ and $y$, that make the statement $x - y = 0.6$ true.

**REASONING** Tell whether the difference between the two numbers is *always*, *sometimes*, or *never* positive. Explain your reasoning.

**28.** two negative fractions

**29.** a positive decimal and a negative decimal

**30.** **Structure** Fill in the blanks to make the solution correct.

$$5. \boxed{\phantom{x}}\, 4 - \left(\boxed{\phantom{x}}\,.8\,\boxed{\phantom{x}}\right) = -3.61$$

 **Fair Game Review** What you learned in previous grades & lessons

**Evaluate.** *(Skills Review Handbook)*

**31.** $5.2 \times 6.9$

**32.** $7.2 \div 2.4$

**33.** $2\frac{2}{3} \times 3\frac{1}{4}$

**34.** $9\frac{4}{5} \div 3\frac{1}{2}$

**35. MULTIPLE CHOICE** A sports store has 116 soccer balls. Over 6 months, it sells 8 soccer balls per month. How many soccer balls are in inventory at the end of the 6 months? *(Section 1.3 and Section 1.4)*

Ⓐ −48 Ⓑ 48 Ⓒ 68 Ⓓ 108

## 2.4 Multiplying and Dividing Rational Numbers

**Essential Question** Why is the product of two negative rational numbers positive?

In Section 1.4, you used a table to see that the product of two negative integers is a positive integer. In this activity, you will find that same result another way.

### 1 ACTIVITY: Showing $(-1)(-1) = 1$

**Work with a partner. How can you show that $(-1)(-1) = 1$?**

To begin, assume that $(-1)(-1) = 1$ is a true statement. From the Additive Inverse Property, you know that $1 + (-1) = 0$. So, substitute $(-1)(-1)$ for 1 to get $(-1)(-1) + (-1) = 0$. If you can show that $(-1)(-1) + (-1) = 0$ is true, then you have shown that $(-1)(-1) = 1$.

Justify each step.

$$(-1)(-1) + (-1) = (-1)(-1) + 1(-1)$$

$$= (-1)[(-1) + 1]$$

$$= (-1)0$$

$$= 0$$

So, $(-1)(-1) = 1$.

### 2 ACTIVITY: Multiplying by $-1$

**Work with a partner.**

a. Graph each number below on three different number lines. Then multiply each number by $-1$ and graph the product on the appropriate number line.

$$2 \qquad 8 \qquad -1$$

b. How does multiplying by $-1$ change the location of the points in part (a)? What is the relationship between the number and the product?

c. Graph each number below on three different number lines. Where do you think the points will be after multiplying by $-1$? Plot the points. Explain your reasoning.

$$\frac{1}{2} \qquad 2.5 \qquad -\frac{5}{2}$$

d. What is the relationship between a rational number $-a$ and the product $-1(a)$? Explain your reasoning.

**COMMON CORE**

**Rational Numbers**
In this lesson, you will
• multiply and divide rational numbers.
• solve real-life problems.
Learning Standards
7.NS.2a
7.NS.2b
7.NS.2c
7.NS.3

**ACTIVITY: Understanding the Product of Rational Numbers**

**Work with a partner. Let $a$ and $b$ be positive rational numbers.**

**a.** Because $a$ and $b$ are positive, what do you know about $-a$ and $-b$?

**b.** Justify each step.

$$(-a)(-b) = (-1)(a)(-1)(b)$$

$$= (-1)(-1)(a)(b)$$

$$= (1)(a)(b)$$

$$= ab$$

**c.** Because $a$ and $b$ are positive, what do you know about the product $ab$?

**d.** What does this tell you about products of rational numbers? Explain.

**4** **ACTIVITY: Writing a Story**

**Work with a partner. Write a story that uses addition, subtraction, multiplication, or division of rational numbers.**

- At least one of the numbers in the story has to be negative and *not* an integer.
- Draw pictures to help illustrate what is happening in the story.
- Include the solution of the problem in the story.

**Math Practice 6**

**Specify Units**

What units are in your story?

If you are having trouble thinking of a story, here are some common uses of negative numbers:

- A profit of $-\$15$ is a loss of $15.
- An elevation of $-100$ feet is a depth of 100 feet below sea level.
- A gain of $-5$ yards in football is a loss of 5 yards.
- A score of $-4$ in golf is 4 strokes under par.

## What Is Your Answer?

**5.** **IN YOUR OWN WORDS** Why is the product of two negative rational numbers positive?

**6.** **PRECISION** Show that $(-2)(-3) = 6$.

**7.** How can you show that the product of a negative rational number and a positive rational number is negative?

**Practice**

Use what you learned about multiplying rational numbers to complete Exercises 7–9 on page 68.

## 2.4 Lesson

### Key Idea

**Multiplying and Dividing Rational Numbers**

**Words**  To multiply or divide rational numbers, use the same rules for signs as you used for integers.

**Remember**

The *reciprocal* of $\frac{a}{b}$ is $\frac{b}{a}$.

**Numbers**

$$-\frac{2}{7} \cdot \frac{1}{3} = \frac{-2 \cdot 1}{7 \cdot 3} = \frac{-2}{21} = -\frac{2}{21}$$

$$-\frac{1}{2} \div \frac{4}{9} = \frac{-1}{2} \cdot \frac{9}{4} = \frac{-1 \cdot 9}{2 \cdot 4} = \frac{-9}{8} = -\frac{9}{8}$$

---

**EXAMPLE** **1**  **Dividing Rational Numbers**

**Find** $-5\frac{1}{5} \div 2\frac{1}{3}$.

**Estimate** $-5 \div 2 = -2\frac{1}{2}$

$$-5\frac{1}{5} \div 2\frac{1}{3} = -\frac{26}{5} \div \frac{7}{3}$$  Write mixed numbers as improper fractions.

$$= \frac{-26}{5} \cdot \frac{3}{7}$$  Multiply by the reciprocal of $\frac{7}{3}$.

$$= \frac{-26 \cdot 3}{5 \cdot 7}$$  Multiply the numerators and the denominators.

$$= \frac{-78}{35}, \text{ or } -2\frac{8}{35}$$  Simplify.

The quotient is $-2\frac{8}{35}$.

**Reasonable?** $-2\frac{8}{35} \approx -2\frac{1}{2}$  ✔

---

**EXAMPLE** **2**  **Multiplying Rational Numbers**

**Find** $-2.5 \cdot 3.6$.

$$
\begin{array}{r}
-2.5 \\
\times 3.6 \\
\hline
1\,5\,0 \\
7\,5\,0 \\
\hline
-9.0\,0
\end{array}
$$

The decimals have different signs.

The product is negative.

The product is $-9$.

EXAMPLE 3  Multiplying More Than Two Rational Numbers

Find $-\dfrac{1}{7} \cdot \left[\dfrac{4}{5} \cdot (-7)\right]$.

You can use properties of multiplication to make the product easier to find.

$$-\dfrac{1}{7} \cdot \left[\dfrac{4}{5} \cdot (-7)\right] = -\dfrac{1}{7} \cdot \left(-7 \cdot \dfrac{4}{5}\right)$$    Commutative Property of Multiplication

$$= -\dfrac{1}{7} \cdot (-7) \cdot \dfrac{4}{5}$$    Associative Property of Multiplication

$$= 1 \cdot \dfrac{4}{5}$$    Multiplicative Inverse Property

$$= \dfrac{4}{5}$$    Multiplication Property of One

∴ The product is $\dfrac{4}{5}$.

## On Your Own

Now You're Ready
Exercises 10–30

**Multiply or divide. Write fractions in simplest form.**

1. $-\dfrac{6}{5} \div \left(-\dfrac{1}{2}\right)$

2. $\dfrac{1}{3} \div \left(-2\dfrac{2}{3}\right)$

3. $1.8(-5.1)$

4. $-6.3(-0.6)$

5. $-\dfrac{2}{3} \cdot 7\dfrac{7}{8} \cdot \dfrac{3}{2}$

6. $-7.2 \cdot 0.1 \cdot (-100)$

---

EXAMPLE 4  Real-Life Application

| Account Positions | ⟳ | | |
|---|---|---|---|
| **Stock** | **Original Value** | **Current Value** | **Change** |
| A | 600.54 | 420.15 | −180.39 |
| B | 391.10 | 518.38 | 127.28 |
| C | 380.22 | 99.70 | −280.52 |

An investor owns Stocks A, B, and C. What is the mean change in the value of the stocks?

$$\text{mean} = \dfrac{-180.39 + 127.28 + (-280.52)}{3} = \dfrac{-333.63}{3} = -111.21$$

∴ The mean change in the value of the stocks is −$111.21.

## On Your Own

7. **WHAT IF?** The change in the value of Stock D is $568.23. What is the mean change in the value of the four stocks?

Vocabulary and Concept Check

1. **WRITING** How is multiplying and dividing rational numbers similar to multiplying and dividing integers?

2. **NUMBER SENSE** Find the reciprocal of $-\frac{2}{5}$.

**Tell whether the expression is *positive* or *negative* without evaluating.**

3. $-\frac{3}{10} \times \left(-\frac{8}{15}\right)$    4. $1\frac{1}{2} \div \left(-\frac{1}{4}\right)$    5. $-6.2 \times 8.18$    6. $\frac{-8.16}{-2.72}$

 Practice and Problem Solving

**Multiply.**

7. $-1\left(\frac{4}{5}\right)$    8. $-1\left(-3\frac{1}{2}\right)$    9. $-0.25(-1)$

**Divide. Write fractions in simplest form.**

① 10. $-\frac{7}{10} \div \frac{2}{5}$    11. $\frac{1}{4} \div \left(-\frac{3}{8}\right)$    12. $-\frac{8}{9} \div \left(-\frac{8}{9}\right)$    13. $-\frac{1}{5} \div 20$

14. $-2\frac{4}{5} \div (-7)$    15. $-10\frac{2}{7} \div \left(-4\frac{4}{11}\right)$    16. $-9 \div 7.2$    17. $8 \div 2.2$

18. $-3.45 \div (-15)$    19. $-0.18 \div 0.03$    20. $8.722 \div (-3.56)$    21. $12.42 \div (-4.8)$

**Multiply. Write fractions in simplest form.**

② ③ 22. $-\frac{1}{4} \times \left(-\frac{4}{3}\right)$    23. $\frac{5}{6}\left(-\frac{8}{15}\right)$    24. $-2\left(-1\frac{1}{4}\right)$

25. $-3\frac{1}{3} \cdot \left(-2\frac{7}{10}\right)$    26. $0.4 \times (-0.03)$    27. $-0.05 \times (-0.5)$

28. $-8(0.09)(-0.5)$    29. $\frac{5}{6} \cdot \left(-4\frac{1}{2}\right) \cdot \left(-2\frac{1}{5}\right)$    30. $\left(-1\frac{2}{3}\right)^3$

**ERROR ANALYSIS  Describe and correct the error.**

31.

$-2.2 \times 3.7 = 8.14$

32.

$-\frac{1}{4} \div \frac{3}{2} = -\frac{4}{1} \times \frac{3}{2} = -\frac{12}{2} = -6$

33. **HOUR HAND** The hour hand of a clock moves $-30°$ every hour. How many degrees does it move in $2\frac{1}{5}$ hours?

34. **SUNFLOWER SEEDS** How many 0.75-pound packages can you make with 6 pounds of sunflower seeds?

**Evaluate.**

**35.** $-4.2 + 8.1 \times (-1.9)$      **36.** $2.85 - 6.2 \div 2^2$      **37.** $-3.64 \cdot \left| -5.3 \right| - 1.5^3$

**38.** $1\dfrac{5}{9} \div \left(-\dfrac{2}{3}\right) + \left(-2\dfrac{3}{5}\right)$      **39.** $-3\dfrac{3}{4} \times \dfrac{5}{6} - 2\dfrac{1}{3}$      **40.** $\left(-\dfrac{2}{3}\right)^2 - \dfrac{3}{4}\left(2\dfrac{1}{3}\right)$

**41. OPEN-ENDED** Write two fractions whose product is $-\dfrac{3}{5}$.

**42. FENCING** A farmer needs to enclose two adjacent rectangular pastures. How much fencing does the farmer need?

$30\dfrac{2}{9}$ yd

$50\dfrac{5}{8}$ yd

**43. GASOLINE** A 14.5-gallon gasoline tank is $\dfrac{3}{4}$ full. How many gallons will it take to fill the tank?

**44. PRECISION** A section of a boardwalk is made using 15 boards. Each board is $9\dfrac{1}{4}$ inches wide. The total width of the section is 144 inches. The spacing between each board is equal. What is the width of the spacing between each board?

**45. RUNNING** The table shows the changes in the times (in seconds) of four teammates. What is the mean change?

| Teammate | Change |
|----------|--------|
| 1 | $-2.43$ |
| 2 | $-1.85$ |
| 3 | $0.61$ |
| 4 | $-1.45$ |

**46.**  The daily changes in the barometric pressure for four days are $-0.05$, $0.09$, $-0.04$, and $-0.08$ inches.

   **a.** What is the mean change?

   **b.** The mean change after five days is $-0.01$ inch. What is the change on the fifth day? Explain.

## Fair Game Review *What you learned in previous grades & lessons*

**Add or subtract.** *(Section 2.2 and Section 2.3)*

**47.** $-6.2 + 4.7$      **48.** $-8.1 - (-2.7)$      **49.** $\dfrac{9}{5} - \left(-2\dfrac{7}{10}\right)$      **50.** $-4\dfrac{5}{6} + \left(-3\dfrac{4}{9}\right)$

**51. MULTIPLE CHOICE** What are the coordinates of the point in Quadrant IV? *(Skills Review Handbook)*

   **(A)** $(-4, 1)$      **(B)** $(-3, -3)$

   **(C)** $(0, -2)$      **(D)** $(3, -3)$

**Subtract. Write fractions in simplest form.** *(Section 2.3)*

**1.** $\dfrac{2}{7} - \left(\dfrac{6}{7}\right)$

**2.** $\dfrac{12}{7} - \left(-\dfrac{2}{9}\right)$

**3.** $9.1 - 12.9$

**4.** $5.647 - (-9.24)$

**Find the distance between the two numbers on the number line.** *(Section 2.3)*

**5.**

**6.**

**Divide. Write fractions in simplest form.** *(Section 2.4)*

**7.** $\dfrac{2}{3} \div \left(-\dfrac{5}{6}\right)$

**8.** $-8\dfrac{5}{9} \div \left(-1\dfrac{4}{7}\right)$

**9.** $-8.4 \div 2.1$

**10.** $32.436 \div (-4.24)$

**Multiply. Write fractions in simplest form.** *(Section 2.4)*

**11.** $\dfrac{5}{8} \times \left(-\dfrac{4}{15}\right)$

**12.** $-2\dfrac{3}{8} \times \dfrac{8}{5}$

**13.** $-9.4 \times (-4.7)$

**14.** $-100(-0.6)(0.01)$

**15. PARASAILING** A parasail is at 200.6 feet above the water. After 5 minutes, the parasail is at 120.8 feet above the water. What is the change in height of the parasail? *(Section 2.3)*

**16. TEMPERATURE** Use the thermometer shown. How much did the temperature drop from 5:00 P.M. to 10:00 P.M.? *(Section 2.3)*

**17. LATE FEES** You were overcharged $4.52 on your cell phone bill 3 months in a row. The cell phone company says that it will add −$4.52 to your next bill for each month you were overcharged. On the next bill, you see an adjustment of −13.28. Is this amount correct? Explain. *(Section 2.4)*

**18. CASHEWS** How many $1\dfrac{1}{4}$-pound packages can you make with $7\dfrac{1}{2}$ pounds of cashews? *(Section 2.4)*

Check It Out
Vocabulary Help
BigIdeasMath ✓com

## Review Key Vocabulary

rational number, *p. 46*                    repeating decimal, *p. 46*
terminating decimal, *p. 46*

## Review Examples and Exercises

**2.1** **Rational Numbers** *(pp. 44–49)*

a.  **Write $4\frac{3}{5}$ as a decimal.**

Notice that $4\frac{3}{5} = \frac{23}{5}$.

$$\begin{array}{r} 4.6 \\ 5\overline{)23.0} \\ -20 \phantom{.0}\\ \hline 3\,0 \\ -3\,0 \\ \hline 0 \end{array}$$

Divide 23 by 5.

The remainder is 0. So, it is a terminating decimal.

∴ So, $4\frac{3}{5} = 4.6$.

b.  **Write −0.14 as a fraction in simplest form.**

$$-0.14 = -\frac{14}{100}$$

Write the digits after the decimal point in the numerator.

The last digit is in the hundredths place. So, use 100 in the denominator.

$$= -\frac{7}{50} \qquad \text{Simplify.}$$

### Exercises

**Write the rational number as a decimal.**

**1.** $-\frac{8}{15}$     **2.** $\frac{5}{8}$     **3.** $-\frac{13}{6}$     **4.** $1\frac{7}{16}$

**Write the decimal as a fraction or a mixed number in simplest form.**

**5.** $-0.6$     **6.** $-0.35$     **7.** $-5.8$     **8.** $24.23$

**Adding Rational Numbers**  *(pp. 50–55)*

Find $-\dfrac{7}{2} + \dfrac{5}{4}$.

$$-\dfrac{7}{2} + \dfrac{5}{4} = \dfrac{-14}{4} + \dfrac{5}{4}$$  Rewrite using the LCD (least common denominator).

$$= \dfrac{-14 + 5}{4}$$  Write the sum of the numerators over the common denominator.

$$= \dfrac{-9}{4}$$  Add.

$$= -2\dfrac{1}{4}$$  Write the improper fraction as a mixed number.

∴ The sum is $-2\dfrac{1}{4}$.

## Exercises

**Add. Write fractions in simplest form.**

**9.** $\dfrac{9}{10} + \left(-\dfrac{4}{5}\right)$

**10.** $-4\dfrac{5}{9} + \dfrac{8}{9}$

**11.** $-1.6 + (-2.4)$

**Subtracting Rational Numbers**  *(pp. 58–63)*

Find $-4\dfrac{2}{5} - \left(-\dfrac{3}{5}\right)$.

$$-4\dfrac{2}{5} - \left(-\dfrac{3}{5}\right) = -4\dfrac{2}{5} + \dfrac{3}{5}$$  Add the opposite of $-\dfrac{3}{5}$.

$$= -\dfrac{22}{5} + \dfrac{3}{5}$$  Write the mixed number as an improper fraction.

$$= \dfrac{-22 + 3}{5}$$  Write the sum of the numerators over the common denominator.

$$= \dfrac{-19}{5}, \text{ or } -3\dfrac{4}{5}$$  Simplify.

∴ The difference is $-3\dfrac{4}{5}$.

## Exercises

**Subtract. Write fractions in simplest form.**

**12.** $-\dfrac{5}{12} - \dfrac{3}{10}$

**13.** $3\dfrac{3}{4} - \dfrac{7}{8}$

**14.** $3.8 - (-7.45)$

**15.** **TURTLE** A turtle is $20\dfrac{5}{6}$ inches below the surface of a pond. It dives to a depth of $32\dfrac{1}{4}$ inches. What is the change in the turtle's position?

**Multiplying and Dividing Rational Numbers** *(pp. 64–69)*

a. Find $-4\frac{1}{6} \div 1\frac{1}{3}$.

$$-4\frac{1}{6} \div 1\frac{1}{3} = -\frac{25}{6} \div \frac{4}{3}$$   Write mixed numbers as improper fractions.

$$= -\frac{25}{6} \cdot \frac{3}{4}$$   Multiply by the reciprocal of $\frac{4}{3}$.

$$= \frac{-25 \cdot 3}{6 \cdot 4}$$   Multiply the numerators and the denominators.

$$= \frac{-25}{8}, \text{ or } -3\frac{1}{8}$$   Simplify.

∴ The quotient is $-3\frac{1}{8}$.

b. Find $-1.6 \cdot 2.4$.

$$\begin{array}{r} -1.6 \\ \times\ 2.4 \\ \hline 64 \\ 320 \\ \hline -3.84 \end{array}$$

← The decimals have different signs.

← The product is negative.

∴ The product is $-3.84$.

### Exercises

**Divide. Write fractions in simplest form.**

16. $\frac{9}{10} \div \left(-\frac{6}{5}\right)$   17. $-\frac{4}{11} \div \frac{2}{7}$   18. $6.4 \div (-3.2)$   19. $-15.4 \div (-2.5)$

**Multiply. Write fractions in simplest form.**

20. $-\frac{4}{9}\left(-\frac{7}{9}\right)$   21. $\frac{8}{15}\left(-\frac{2}{3}\right)$   22. $-5.9(-9.7)$

23. $4.5(-5.26)$   24. $-\frac{2}{3} \cdot \left(2\frac{1}{2}\right) \cdot (-3)$   25. $-1.6 \cdot (0.5) \cdot (-20)$

26. **SUNKEN SHIP** The elevation of a sunken ship is $-120$ feet. Your elevation is $\frac{5}{8}$ of the ship's elevation. What is your elevation?

Check It Out
Test Practice
BigIdeasMath ✓com

**Write the rational number as a decimal.**

**1.** $\dfrac{7}{40}$      **2.** $-\dfrac{1}{9}$      **3.** $-\dfrac{21}{16}$      **4.** $\dfrac{36}{5}$

**Write the decimal as a fraction or a mixed number in simplest form.**

**5.** $-0.122$      **6.** $0.33$      **7.** $-4.45$      **8.** $-7.09$

**Add or subtract. Write fractions in simplest form.**

**9.** $-\dfrac{4}{9} + \left(-\dfrac{23}{18}\right)$      **10.** $\dfrac{17}{12} - \left(-\dfrac{1}{8}\right)$      **11.** $9.2 + (-2.8)$      **12.** $2.86 - 12.1$

**Multiply or divide. Write fractions in simplest form.**

**13.** $3\dfrac{9}{10} \times \left(-\dfrac{8}{3}\right)$      **14.** $-1\dfrac{5}{6} \div 4\dfrac{1}{6}$

**15.** $-4.4 \times (-6.02)$      **16.** $-5 \div 1.5$

**17.** $-\dfrac{3}{5} \cdot \left(2\dfrac{2}{7}\right) \cdot \left(-3\dfrac{3}{4}\right)$      **18.** $-6 \cdot (-0.05) \cdot (-0.4)$

**19. ALMONDS** How many 2.25-pound containers can you make with 24.75 pounds of almonds?

**20. FISH** The elevation of a fish is $-27$ feet.

    **a.** The fish decreases its elevation by 32 feet, and then increases its elevation by 14 feet. What is its new elevation?

    **b.** Your elevation is $\dfrac{2}{5}$ of the fish's new elevation. What is your elevation?

**21. RAINFALL** The table shows the rainfall (in inches) for three months compared to the yearly average. Is the total rainfall for the three-month period greater than or less than the yearly average? Explain.

| November | December | January |
|----------|----------|---------|
| $-0.86$  | $2.56$   | $-1.24$ |

**22. BANK ACCOUNTS** Bank Account A has $750.92, and Bank Account B has $675.44. Account A changes by $-\$216.38$, and Account B changes by $-\$168.49$. Which account has the greater balance? Explain.

1. When José and Sean were each 5 years old, José was $1\frac{1}{2}$ inches taller than Sean. José grew at an average rate of $2\frac{3}{4}$ inches per year from the time that he was 5 years old until the time he was 13 years old. José was 63 inches tall when he was 13 years old. How tall was Sean when he was 5 years old? *(7.NS.3)*

   **A.** $39\frac{1}{2}$ in.          **C.** $44\frac{3}{4}$ in.

   **B.** $42\frac{1}{2}$ in.          **D.** $47\frac{3}{4}$ in.

2. Which expression represents a positive integer? *(7.NS.2a)*

   **F.** $-6^2$          **H.** $(-5)^2$

   **G.** $(-3)^3$          **I.** $-2^3$

3. What is the missing number in the sequence below? *(7.NS.2a)*

   $$\frac{9}{16}, \ -\frac{9}{8}, \ \frac{9}{4}, \ -\frac{9}{2}, \ 9, \ \underline{\hspace{1cm}}$$

4. What is the value of the expression below? *(7.NS.1c)*

   $$\left| -2 - (-2.5) \right|$$

   **A.** $-4.5$          **C.** $0.5$

   **B.** $-0.5$          **D.** $4.5$

5. What is the distance between the two numbers on the number line? *(7.NS.1c)*

   **F.** $-2\frac{1}{8}$          **H.** $1\frac{3}{8}$

   **G.** $-1\frac{3}{8}$          **I.** $2\frac{1}{8}$

Test-Taking Strategy
Estimate the Answer

One-fourth of the 36 cats in our town are tabbies. How many are not tabbies?
Ⓐ 9   Ⓑ 18   Ⓒ 27   Ⓓ 36

IC.

"Using estimation you can see that there are about 10 tabbies. So about 30 are not tabbies."

**6.** Sandra was evaluating an expression in the box below.

$$-4\frac{3}{4} \div 2\frac{1}{5} = -\frac{19}{4} \div \frac{11}{5}$$

$$= \frac{-4}{19} \cdot \frac{5}{11}$$

$$= \frac{-4 \cdot 5}{19 \cdot 11}$$

$$= \frac{-20}{209}$$

What should Sandra do to correct the error that she made?   *(7.NS.3)*

**A.** Rewrite $-\frac{19}{4}$ as $-\frac{4}{19}$ and multiply by $\frac{11}{5}$.

**B.** Rewrite $\frac{11}{5}$ as $\frac{5}{11}$ and multiply by $-\frac{19}{4}$.

**C.** Rewrite $\frac{11}{5}$ as $-\frac{5}{11}$ and multiply by $-\frac{19}{4}$.

**D.** Rewrite $-4\frac{3}{4}$ as $-\frac{13}{4}$ and multiply by $\frac{5}{11}$.

**7.** What is the value of the expression below when $q = -2$, $r = -12$, and $s = 8$?   *(7.NS.3)*

$$\frac{-q^2 - r}{s}$$

**F.** $-2$                    **H.** 1

**G.** $-1$                    **I.** 2

**8.** You are stacking wooden blocks with the dimensions shown below. How many blocks do you need to stack to build a block tower that is $7\frac{1}{2}$ inches tall?   *(7.NS.3)*

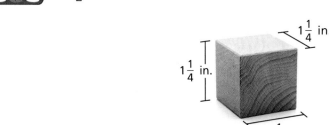

9. What is the area of a triangle with a base length of $2\frac{1}{2}$ inches and a height of 2 inches? *(7.NS.2c)*

   **A.** $2\frac{1}{4}$ in.$^2$            **C.** $4\frac{1}{2}$ in.$^2$

   **B.** $2\frac{1}{2}$ in.$^2$            **D.** 5 in.$^2$

10. What is the value of the expression below? *(7.NS.3)*
    $$\frac{-4^2 - (-2)^3}{4}$$

    **F.** $-6$            **H.** 2

    **G.** $-2$            **I.** 6

11. Four points are graphed on the number line below. *(7.NS.3)*

   *Part A*  Choose the two points whose values have the greatest sum. Approximate this sum. Explain your reasoning.

   *Part B*  Choose the two points whose values have the greatest difference. Approximate this difference. Explain your reasoning.

   *Part C*  Choose the two points whose values have the greatest product. Approximate this product. Explain your reasoning.

   *Part D*  Choose the two points whose values have the greatest quotient. Approximate this quotient. Explain your reasoning.

12. What number belongs in the box to make the equation true? *(7.NS.3)*
    $$\frac{-0.4}{\Box} + 0.8 = -1.2$$

    **A.** $-1$            **C.** 0.2

    **B.** $-0.2$            **D.** 1

# 3 Expressions and Equations

"I can't find my algebra tiles, so I am painting some of my dog biscuits."

"Now I will be able to solve the equation 2x + (−2) = 2."

"Descartes, if you solve for

in the equation, what do you get?"

# What You Learned Before

"Hey, Descartes ... True or False: The expressions are equivalent."

## Evaluating Expressions (7.NS.3)

**Example 1** Evaluate $6x + 2y$ when $x = -3$ and $y = 5$.

$$6x + 2y = 6(-3) + 2(5)$$      Substitute $-3$ for $x$ and 5 for $y$.

$$= -18 + 10$$      Using order of operations, multiply 6 and $-3$, and 2 and 5.

$$= -8$$      Add $-18$ and 10.

**Example 2** Evaluate $6x^2 - 3(y + 2) + 8$ when $x = -2$ and $y = 4$.

$$6x^2 - 3(y + 2) + 8 = 6(-2)^2 - 3(4 + 2) + 8$$    Substitute $-2$ for $x$ and 4 for $y$.

$$= 6(-2)^2 - 3(6) + 8$$    Using order of operations, evaluate within the parentheses.

$$= 6(4) - 3(6) + 8$$    Using order of operations, evaluate the exponent.

$$= 24 - 18 + 8$$    Using order of operations, multiply 6 and 4, and 3 and 6.

$$= 14$$    Subtract 18 from 24. Add the result to 8.

### Try It Yourself
Evaluate the expression when $x = -\dfrac{1}{4}$ and $y = 3$.

**1.** $2xy$      **2.** $12x - 3y$      **3.** $-4x - y + 4$      **4.** $8x - y^2 - 3$

## Writing Algebraic Expressions (6.EE.2a)

**Example 3** Write the phrase as an algebraic expression.

  **a.** the sum of twice a number $m$ and four      **b.** eight less than three times a number $x$

                    $2m + 4$                                  $3x - 8$

### Try It Yourself
Write the phrase as an algebraic expression.

  **5.** five more than three times a number $q$      **6.** nine less than a number $n$

  **7.** the product of a number $p$ and six      **8.** the quotient of eight and a number $h$

  **9.** four more than three times a number $t$      **10.** two less than seven times a number $c$

**Essential Question** How can you simplify an algebraic expression?

### 1 ACTIVITY: Simplifying Algebraic Expressions

**Work with a partner.**

**a.** Evaluate each algebraic expression when $x = 0$ and when $x = 1$. Use the results to match each expression in the left table with its equivalent expression in the right table.

| | Expression | Value When $x = 0$ | $x = 1$ |
|---|---|---|---|
| A. | $3x + 2 - x + 4$ | | |
| B. | $5(x - 3) + 2$ | | |
| C. | $x + 3 - (2x + 1)$ | | |
| D. | $-4x + 2 - x + 3x$ | | |
| E. | $-(1 - x) + 3$ | | |
| F. | $2x + x - 3x + 4$ | | |
| G. | $4 - 3 + 2(x - 1)$ | | |
| H. | $2(1 - x + 4)$ | | |
| I. | $5 - (4 - x + 2x)$ | | |
| J. | $5x - (2x + 4 - x)$ | | |

| | Expression | Value When $x = 0$ | $x = 1$ |
|---|---|---|---|
| a. | $4$ | | |
| b. | $-x + 1$ | | |
| c. | $4x - 4$ | | |
| d. | $2x + 6$ | | |
| e. | $5x - 13$ | | |
| f. | $-2x + 10$ | | |
| g. | $x + 2$ | | |
| h. | $2x - 1$ | | |
| i. | $-2x + 2$ | | |
| j. | $-x + 2$ | | |

**COMMON CORE**

**Algebraic Expressions**

In this lesson, you will

● apply properties of operations to simplify algebraic expressions.

● solve real-life problems.

Learning Standards
7.EE.1
7.EE.2

**b.** Compare each expression in the left table with its equivalent expression in the right table. In general, how do you think you obtain the equivalent expression in the right column?

**Work with a partner. Use your results from Activity 1 to write a lesson on simplifying an algebraic expression.**

**Math Practice 6**

**Communicate Precisely**

What can you do to make sure that you are communicating exactly what is needed in the Key Idea?

Describe steps you can use to simplify an expression.

**Simplifying an Algebraic Expression**

**Key Idea** Use the following steps to simplify an algebraic expression.
1.
2.
3.

Write 3 examples. Use expressions from Activity 1.

**Examples**
a.

b.

c.

Write 3 exercises. Use expressions different from the ones in Activity 1.

**Exercises**
Simplify the expression.
1.
2.
3.

## What Is Your Answer?

3. **IN YOUR OWN WORDS** How can you simplify an algebraic expression? Give an example that demonstrates your procedure.

4. **REASONING** Why would you want to simplify an algebraic expression? Discuss several reasons.

**Practice**

Use what you learned about simplifying algebraic expressions to complete Exercises 12–14 on page 84.

Check It Out
Lesson Tutorials
BigIdeasMath ✓com

**Key Vocabulary**
like terms, *p. 82*
simplest form, *p. 82*

Parts of an algebraic expression are called *terms*. **Like terms** are terms that have the same variables raised to the same exponents. Constant terms are also like terms. To identify terms and like terms in an expression, first write the expression as a sum of its terms.

**EXAMPLE 1** **Identifying Terms and Like Terms**

**Identify the terms and like terms in each expression.**

**a.** $9x - 2 + 7 - x$
Rewrite as a sum of terms.

$$9x + (-2) + 7 + (-x)$$

**Terms:** $9x$, $-2$, $7$, $-x$

**Like terms:** $9x$ and $-x$, $-2$ and $7$

**b.** $z^2 + 5z - 3z^2 + z$
Rewrite as a sum of terms.

$$z^2 + 5z + (-3z^2) + z$$

**Terms:** $z^2$, $5z$, $-3z^2$, $z$

**Like terms:** $z^2$ and $-3z^2$, $5z$ and $z$

An algebraic expression is in **simplest form** when it has no like terms and no parentheses. To *combine* like terms that have variables, use the Distributive Property to add or subtract the coefficients.

**EXAMPLE 2** **Simplifying an Algebraic Expression**

**Study Tip**
To subtract a variable term, add the term with the opposite coefficient.

**Simplify $\frac{3}{4}y + 12 - \frac{1}{2}y - 6$.**

$$\frac{3}{4}y + 12 - \frac{1}{2}y - 6 = \frac{3}{4}y + 12 + \left(-\frac{1}{2}y\right) + (-6) \quad \text{Rewrite as a sum.}$$

$$= \frac{3}{4}y + \left(-\frac{1}{2}y\right) + 12 + (-6) \quad \text{Commutative Property of Addition}$$

$$= \left[\frac{3}{4} + \left(-\frac{1}{2}\right)\right]y + 12 + (-6) \quad \text{Distributive Property}$$

$$= \frac{1}{4}y + 6 \quad \text{Combine like terms.}$$

⬤ **On Your Own**

**Now You're Ready**
Exercises 5–10
and 12–17

**Identify the terms and like terms in the expression.**

**1.** $y + 10 - \frac{3}{2}y$     **2.** $2r^2 + 7r - r^2 - 9$     **3.** $7 + 4p - 5 + p + 2q$

**Simplify the expression.**

**4.** $14 - 3z + 8 + z$     **5.** $2.5x + 4.3x - 5$     **6.** $\frac{3}{8}b - \frac{3}{4}b$

🔊 Multi-Language Glossary at BigIdeasMath✓com

EXAMPLE 3 · Simplifying an Algebraic Expression

Simplify $-\frac{1}{2}(6n + 4) + 2n$.

$$-\frac{1}{2}(6n + 4) + 2n = -\frac{1}{2}(6n) + \left(-\frac{1}{2}\right)(4) + 2n \quad \text{Distributive Property}$$

$$= -3n + (-2) + 2n \quad \text{Multiply.}$$

$$= -3n + 2n + (-2) \quad \text{Commutative Property of Addition}$$

$$= (-3 + 2)n + (-2) \quad \text{Distributive Property}$$

$$= -n - 2 \quad \text{Simplify.}$$

## On Your Own

Now You're Ready
Exercises 18–20

**Simplify the expression.**

**7.** $3(q + 1) - 4$

**8.** $-2(g + 4) + 7g$

**9.** $7 - 4\left(\frac{3}{4}x - \frac{1}{4}\right)$

EXAMPLE 4 · Real-Life Application

**Each person in a group buys a ticket, a medium drink, and a large popcorn. Write an expression in simplest form that represents the amount of money the group spends at the movies. Interpret the expression.**

**Royal Cinemas**

| Evening Tickets | $7.50 |
|---|---|
| **REFRESHMENTS** | |
| **Drinks** | |
| Small | $1.75 |
| Medium | $2.75 |
| Large | $3.50 |
| **Popcorn** | |
| Small | $3.00 |
| Large | $4.00 |

**Words**    Each ticket is $7.50, each medium drink is $2.75, and each large popcorn is $4.

**Variable**    The same number of each item is purchased. So, $x$ can represent the number of tickets, the number of medium drinks, and the number of large popcorns.

**Expression**    $7.50\,x$    +    $2.75\,x$    +    $4\,x$

$$7.50x + 2.75x + 4x = (7.50 + 2.75 + 4)x \quad \text{Distributive Property}$$

$$= 14.25x \quad \text{Add coefficients.}$$

∴ The expression $14.25x$ indicates that the total cost per person is $14.25.

## On Your Own

**10.** **WHAT IF?** Each person buys a ticket, a large drink, and a small popcorn. How does the expression change? Explain.

**Vocabulary and Concept Check**

1. **WRITING** Explain how to identify the terms of $3y - 4 - 5y$.

2. **WRITING** Describe how to combine like terms in the expression $3n + 4n - 2$.

3. **VOCABULARY** Is the expression $3x + 2x - 4$ in simplest form? Explain.

4. **REASONING** Which algebraic expression is in simplest form? Explain.

$$5x - 4 + 6y$$  $$4x + 8 - x$$

$$3(7 + y)$$  $$12n - n$$

 **Practice and Problem Solving**

**Identify the terms and like terms in the expression.**

**5.** $t + 8 + 3t$

**6.** $3z + 4 + 2 + 4z$

**7.** $2n - n - 4 + 7n$

**8.** $-x - 9x^2 + 12x^2 + 7$

**9.** $1.4y + 5 - 4.2 - 5y^2 + z$

**10.** $\frac{1}{2}s - 4 + \frac{3}{4}s + \frac{1}{8} - s^3$

11. **ERROR ANALYSIS** Describe and correct the error in identifying the like terms in the expression.

 $3x - 5 + 2x^2 + 9x = 3x + 2x^2 + 9x - 5$

Like Terms: $3x$, $2x^2$, and $9x$

**Simplify the expression.**

**12.** $12g + 9g$

**13.** $11x + 9 - 7$

**14.** $8s - 11s + 6$

**15.** $4.2v - 5 - 6.5v$

**16.** $8 + 4a + 6.2 - 9a$

**17.** $\frac{2}{5}y - 4 + 7 - \frac{9}{10}y$

**18.** $4(b - 6) + 19$

**19.** $4p - 5(p + 6)$

**20.** $-\frac{2}{3}(12c - 9) + 14c$

21. **HIKING** On a hike, each hiker carries the items shown. Write an expression in simplest form that represents the weight carried by $x$ hikers. Interpret the expression.

4.6 lb

3.4 lb

2.2 lb

22. **STRUCTURE** Evaluate the expression $-8x + 5 - 2x - 4 + 5x$ when $x = 2$ before and after simplifying. Which method do you prefer? Explain.

23. **REASONING** Are the expressions $8x^2 + 3(x^2 + y)$ and $7x^2 + 7y + 4x^2 - 4y$ equivalent? Explain your reasoning.

24. **CRITICAL THINKING** Which solution shows a correct way of simplifying $6 - 4(2 - 5x)$? Explain the errors made in the other solutions.

     **(A)**   $6 - 4(2 - 5x) = 6 - 4(-3x) = 6 + 12x$       **(B)**   $6 - 4(2 - 5x) = 6 - 8 + 20x = -2 + 20x$

     **(C)**   $6 - 4(2 - 5x) = 2(2 - 5x) = 4 - 10x$       **(D)**   $6 - 4(2 - 5x) = 6 - 8 - 20x = -2 - 20x$

25. **BANNER** Write an expression in simplest form that represents the area of the banner.

3 ft

$(3 + x)$ ft

26. **CAR WASH** Write an expression in simplest form that represents the earnings for washing and waxing $x$ cars and $y$ trucks.

|      | Car  | Truck |
|------|------|-------|
| Wash | $8  | $10   |
| Wax  | $12 | $15   |

**MODELING** Draw a diagram that shows how the expression can represent the area of a figure. Then simplify the expression.

27. $5(2 + x + 3)$

28. $(4 + 1)(x + 2x)$

29. **Critical Thinking** You apply gold foil to a piece of red poster board to make the design shown.

    **a.** Write an expression in simplest form that represents the area of the gold foil.

    **b.** Find the area of the gold foil when $x = 3$.

    **c.** The pattern at the right is called "St. George's Cross." Find a country that uses this pattern as its flag.

$x$ in.

$x$ in.

12 in.

20 in.

 **Fair Game Review** *What you learned in previous grades & lessons*

**Order the lengths from least to greatest.** *(Skills Review Handbook)*

30. 15 in., 14.8 in., 15.8 in., 14.5 in., 15.3 in.

31. 0.65 m, 0.6 m, 0.52 m, 0.55 m, 0.545 m

32. **MULTIPLE CHOICE** A bird's nest is 12 feet above the ground. A mole's den is 12 inches below the ground. What is the difference in height of these two positions? *(Section 1.3)*

    **(A)** 24 in.       **(B)** 11 ft       **(C)** 13 ft       **(D)** 24 ft

**Essential Question** How can you use algebra tiles to add or subtract algebraic expressions?

**Key:** [ + ] = variable   [ − ] = −variable   [ + ][ − ] = zero pair

[+] = 1   [−] = −1   [+][−] = zero pair

---

### 1 ACTIVITY: Writing Algebraic Expressions

**Work with a partner. Write an algebraic expression shown by the algebra tiles.**

a. [ + ] [+] [+] [+]

b. [ + ] [−] [−]
   [ + ]

c. [ + ] [+] [+] [+] [+] [+]
   [ + ] [−] [−]

d. [ + ] [+] [+] [+]
   [ + ] [−] [−] [−] [−] [−]
   [ + ] [−] [−]

---

### 2 ACTIVITY: Adding Algebraic Expressions

**Work with a partner. Write the sum of two algebraic expressions modeled by the algebra tiles. Then use the algebra tiles to simplify the expression.**

a. ( [ + ] [+] [+] ) + ( [ + ] [+] [+] [+] [+] )

b. ( [ + ] [−] [−] [−] [−] [−] ) + ( [ + ] [−] [−] )

c. ( [ + ] [+] [+] [+] [+] [+] ) + ( [ + ] [−] [−] [−] )
                                    ( [ + ]              )

d. ( [ + ] [−] [−] [−] [−] [−] ) + ( [ + ] [+] [+] [+] )
   ( [ + ] [−] [−] [−] [−]     )   ( [ + ] [+] [+]     )
                                   ( [ + ]              )

---

**COMMON CORE**

**Linear Expressions**

In this lesson, you will

- apply properties of operations to add and subtract linear expressions.
- solve real-life problems.

Learning Standards
7.EE.1
7.EE.2

**ACTIVITY: Subtracting Algebraic Expressions**

**Math Practice 2**

**Use Expressions**
What do the tiles represent? How does this help you write an expression?

Work with a partner. Write the difference of two algebraic expressions modeled by the algebra tiles. Then use the algebra tiles to simplify the expression.

**a.** $\left( \boxed{+} \; + \; + \; + \right) - \left( \boxed{+} \; + \right)$

**b.** $\left( \boxed{+} \; - \; - \; - \; - \right) - \left( \boxed{+} \; - \; - \; - \right)$

**c.** $\left( \begin{array}{l} \boxed{+} \; + \; + \; + \; + \; + \\ \boxed{+} \end{array} \right) - \left( \boxed{+} \; - \right)$

**d.** $\left( \begin{array}{l} \boxed{+} \; - \; - \; - \; - \; - \\ \boxed{+} \; - \; - \\ \boxed{+} \end{array} \right) - \left( \begin{array}{l} \boxed{+} \; + \; + \; + \\ \boxed{+} \end{array} \right)$

---

**4** **ACTIVITY: Adding and Subtracting Algebraic Expressions**

Work with a partner. Use algebra tiles to model the sum or difference. Then use the algebra tiles to simplify the expression.

**a.** $(2x + 1) + (x - 1)$

**b.** $(2x - 6) + (3x + 2)$

**c.** $(2x + 4) - (x + 2)$

**d.** $(4x + 3) - (2x - 1)$

## What Is Your Answer?

**5. IN YOUR OWN WORDS** How can you use algebra tiles to add or subtract algebraic expressions?

**6.** Write the difference of two algebraic expressions modeled by the algebra tiles. Then use the algebra tiles to simplify the expression.

$$\left( \boxed{-} \; + \; + \; + \right) - \left( \begin{array}{l} \boxed{-} \; - \; - \\ \boxed{-} \end{array} \right)$$

**Practice**

Use what you learned about adding and subtracting algebraic expressions to complete Exercises 6 and 7 on page 90.

**Key Vocabulary** 🔊
linear expression,
p. 88

A **linear expression** is an algebraic expression in which the exponent of the variable is 1.

| Linear Expressions | $-4x$ | $3x + 5$ | $5 - \dfrac{1}{6}x$ |
|---|---|---|---|
| Nonlinear Expressions | $x^2$ | $-7x^3 + x$ | $x^5 + 1$ |

You can use a vertical or a horizontal method to add linear expressions.

**EXAMPLE** ① **Adding Linear Expressions**

Find each sum.

**a.** $(x - 2) + (3x + 8)$

**Vertical method:** Align like terms vertically and add.

$$\begin{array}{r} x - 2 \\ + \; 3x + 8 \\ \hline 4x + 6 \end{array}$$

**b.** $(-4y + 3) + (11y - 5)$

**Horizontal method:** Use properties of operations to group like terms and simplify.

$$(-4y + 3) + (11y - 5) = -4y + 3 + 11y - 5 \qquad \text{Rewrite the sum.}$$

$$= -4y + 11y + 3 - 5 \qquad \text{Commutative Property of Addition}$$

$$= (-4y + 11y) + (3 - 5) \qquad \text{Group like terms.}$$

$$= 7y - 2 \qquad \text{Combine like terms.}$$

**EXAMPLE** ② **Adding Linear Expressions**

Find $2(-7.5z + 3) + (5z - 2)$.

$$2(-7.5z + 3) + (5z - 2) = -15z + 6 + 5z - 2 \qquad \text{Distributive Property}$$

$$= -15z + 5z + 6 - 2 \qquad \text{Commutative Property of Addition}$$

$$= -10z + 4 \qquad \text{Combine like terms.}$$

⚫ **On Your Own**

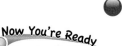

Now You're Ready
Exercises 8–16

**Find the sum.**

1. $(x + 3) + (2x - 1)$

2. $(-8z + 4) + (8z - 7)$

3. $(4 - n) + 2(-5n + 3)$

4. $\dfrac{1}{2}(w - 6) + \dfrac{1}{4}(w + 12)$

To subtract one linear expression from another, add the opposite of each term in the expression. You can use a vertical or a horizontal method.

**EXAMPLE** ③ **Subtracting Linear Expressions**

**Find each difference.**

**a.** $(5x + 6) - (-x + 6)$        **b.** $(7y + 5) - 2(4y - 3)$

**Study Tip**

To find the opposite of a linear expression, you can multiply the expression by $-1$.

**a.** **Vertical method:** Align like terms vertically and subtract.

$$\begin{array}{r} (5x + 6) \\ - (-x + 6) \end{array}$$   **Add the opposite.**   $$\begin{array}{r} 5x + 6 \\ + \phantom{5}x - 6 \\ \hline 6x \end{array}$$

**b.** **Horizontal method:** Use properties of operations to group like terms and simplify.

$$
\begin{aligned}
(7y + 5) - 2(4y - 3) &= 7y + 5 - 8y + 6 && \text{Distributive Property} \\
&= 7y - 8y + 5 + 6 && \text{Commutative Property of Addition} \\
&= (7y - 8y) + (5 + 6) && \text{Group like terms.} \\
&= -y + 11 && \text{Combine like terms.}
\end{aligned}
$$

**EXAMPLE** ④ **Real-Life Application**

**The original price of a cowboy hat is $d$ dollars. You use a coupon and buy the hat for $(d - 2)$ dollars. You decorate the hat and sell it for $(2d - 4)$ dollars. Write an expression that represents your earnings from buying and selling the hat. Interpret the expression.**

$$
\begin{aligned}
\text{earnings} &= \text{selling price} - \text{purchase price} && \text{Use a model.} \\
&= (2d - 4) - (d - 2) && \text{Write the difference.} \\
&= (2d - 4) + (-d + 2) && \text{Add the opposite.} \\
&= 2d - d - 4 + 2 && \text{Group like terms.} \\
&= d - 2 && \text{Combine like terms.}
\end{aligned}
$$

⋰ You earn $(d - 2)$ dollars. You also paid $(d - 2)$ dollars, so you doubled your money by selling the hat for twice as much as you paid for it.

**On Your Own**

**Now You're Ready**
Exercises 19–24

**Find the difference.**

**5.** $(m - 3) - (-m + 12)$        **6.** $-2(c + 2.5) - 5(1.2c + 4)$

**7.** **WHAT IF?** In Example 4, you sell the hat for $(d + 2)$ dollars. How much do you earn from buying and selling the hat?

**Vocabulary and Concept Check**

**VOCABULARY** Determine whether the algebraic expression is a linear expression. Explain.

**1.** $x^2 + x + 1$  **2.** $-2x - 8$  **3.** $x - x^4$

**4. WRITING** Describe two methods for adding or subtracting linear expressions.

**5. DIFFERENT WORDS, SAME QUESTION** Which is different? Find "both" answers.

> Subtract $x$ from $3x - 1$.

> Find $3x - 1$ decreased by $x$.

> What is $x$ more than $3x - 1$?

> What is the difference of $3x - 1$ and $x$?

 **Practice and Problem Solving**

Write the sum or difference of two algebraic expressions modeled by the algebra tiles. Then use the algebra tiles to simplify the expression.

**6.** $\left( \boxed{+} \; \boxed{-}\boxed{-}\boxed{-} \atop \boxed{+} \; \boxed{-}\boxed{-}\boxed{-} \right) + \left( \boxed{+} \; \boxed{+}\boxed{+}\boxed{+}\boxed{+}\boxed{+} \right)$

**7.** $\left( \boxed{+} \; \boxed{+}\boxed{+}\boxed{+}\boxed{+}\boxed{+} \atop \boxed{+} \; \boxed{+}\boxed{+} \right) - \left( \boxed{+} \; \boxed{-}\boxed{-}\boxed{-}\boxed{-} \atop \boxed{+} \right)$

**Find the sum.**

 **① ②**  **8.** $(n + 8) + (n - 12)$  **9.** $(7 - b) + (3b + 2)$  **10.** $(2w - 9) + (-4w - 5)$

**11.** $(2x - 6) + 4(x - 3)$  **12.** $5(-3.4k - 7) + (3k + 21)$  **13.** $(1 - 5q) + 2(2.5q + 8)$

**14.** $3(2 - 0.9h) + (-1.3h - 4)$  **15.** $\frac{1}{3}(9 - 6m) + \frac{1}{4}(12m - 8)$  **16.** $-\frac{1}{2}(7z + 4) + \frac{1}{5}(5z - 15)$

**17. BANKING** You start a new job. After $w$ weeks, you have $(10w + 120)$ dollars in your savings account and $(45w + 25)$ dollars in your checking account. Write an expression that represents the total in both accounts.

**18. FIREFLIES** While catching fireflies, you and a friend decide to have a competition. After $m$ minutes, you have $(3m + 13)$ fireflies and your friend has $(4m + 6)$ fireflies.

**a.** Write an expression that represents the number of fireflies you and your friend caught together.

**b.** The competition ends after 5 minutes. Who has more fireflies?

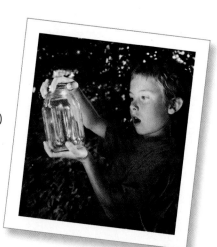

**Find the difference.**

③ **19.** $(-2g + 7) - (g + 11)$      **20.** $(6d + 5) - (2 - 3d)$      **21.** $(4 - 5y) - 2(3.5y - 8)$

**22.** $(2n - 9) - 5(-2.4n + 4)$      **23.** $\frac{1}{8}(-8c + 16) - \frac{1}{3}(6 + 3c)$      **24.** $\frac{3}{4}(3x + 6) - \frac{1}{4}(5x - 24)$

**25. ERROR ANALYSIS** Describe and correct the error in finding the difference.

$$
\begin{aligned}
(4m + 9) - 3(2m - 5) &= 4m + 9 - 6m - 15 \\
&= 4m - 6m + 9 - 15 \\
&= -2m - 6
\end{aligned}
$$

**26. STRUCTURE** Refer to the expressions in Exercise 18.

     **a.** How many fireflies are caught each minute during the competition?

     **b.** How many fireflies are caught before the competition starts?

**27. LOGIC** Your friend says the sum of two linear expressions is always a linear expression. Is your friend correct? Explain.

**28. GEOMETRY** The expression $17n + 11$ represents the perimeter (in feet) of the triangle. Write an expression that represents the measure of the third side.

**29. TAXI** Taxi Express charges $2.60 plus $3.65 per mile, and Cab Cruiser charges $2.75 plus $3.90 per mile. Write an expression that represents how much more Cab Cruiser charges than Taxi Express.

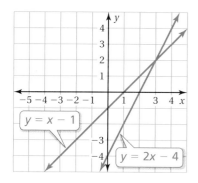

**30. MODELING** A rectangular room is 10 feet longer than it is wide. One-foot-by-one-foot tiles cover the entire floor. Write an expression that represents the number of tiles along the outside of the room.

**31. Reasoning** Write an expression in simplest form that represents the vertical distance between the two lines shown. What is the distance when $x = 3$? when $x = -3$?

**Fair Game Review**   What you learned in previous grades & lessons

Evaluate the expression when $x = -\dfrac{4}{5}$ and $y = \dfrac{1}{3}$.  *(Section 2.2)*

**32.** $x + y$      **33.** $2x + 6y$      **34.** $-x + 4y$

**35. MULTIPLE CHOICE** What is the surface area of a cube that has a side length of 5 feet? *(Skills Review Handbook)*

     (A) $25$ ft$^2$      (B) $75$ ft$^2$      (C) $125$ ft$^2$      (D) $150$ ft$^2$

Check It Out
Lesson Tutorials
BigIdeasMath.com

**Key Vocabulary**
factoring an
expression, p. 92

When **factoring an expression**, you write the expression as a product of factors. You can use the Distributive Property to factor expressions.

## EXAMPLE 1 Factoring Out the GCF

**Factor $24x - 18$ using the GCF.**

Find the GCF of $24x$ and $18$ by writing their prime factorizations.

$$24x = 2 \cdot 2 \cdot 2 \cdot 3 \cdot x$$
$$18 = 2 \cdot 3 \cdot 3$$

Circle the common prime factors.

So, the GCF of $24x$ and $18$ is $2 \cdot 3 = 6$. Use the GCF to factor the expression.

$$24x - 18 = 6(4x) - 6(3)$$
Rewrite using GCF.

$$= 6(4x - 3)$$
Distributive Property

So, $24x - 18 = 6(4x - 3)$.

You can also use the Distributive Property to factor out any rational number from an expression.

## EXAMPLE 2 Factoring Out a Fraction

**Factor $\frac{1}{2}$ out of $\frac{1}{2}x + \frac{3}{2}$.**

Write each term as a product of $\frac{1}{2}$ and another factor.

$$\frac{1}{2}x = \frac{1}{2} \cdot x$$
Think: $\frac{1}{2}x$ is $\frac{1}{2}$ times what?

$$\frac{3}{2} = \frac{1}{2} \cdot 3$$
Think: $\frac{3}{2}$ is $\frac{1}{2}$ times what?

**COMMON CORE**

**Linear Expressions**
In this extension, you will
• factor linear expressions.
Learning Standard
7.EE.1

Use the Distributive Property to factor out $\frac{1}{2}$.

$$\frac{1}{2}x + \frac{3}{2} = \frac{1}{2} \cdot x + \frac{1}{2} \cdot 3$$
Rewrite the expression.

$$= \frac{1}{2}(x + 3)$$
Distributive Property

So, $\frac{1}{2}x + \frac{3}{2} = \frac{1}{2}(x + 3)$.

Multi-Language Glossary at BigIdeasMath.com

**EXAMPLE 3** **Factoring Out a Negative Number**

Factor −2 out of −4p + 10.

Write each term as a product of −2 and another factor.

$$-4p = -2 \cdot 2p$$ Think: −4p is −2 times what?

$$10 = -2 \cdot (-5)$$ Think: 10 is −2 times what?

Use the Distributive Property to factor out −2.

$$-4p + 10 = -2 \cdot 2p + (-2) \cdot (-5)$$ Rewrite the expression.

$$= -2[2p + (-5)]$$ Distributive Property

$$= -2(2p - 5)$$ Simplify.

So, $-4p + 10 = -2(2p - 5)$.

**Math Practice 7**

**View as Components**

How does rewriting each term as a product help you see the common factor?

## Practice

**Factor the expression using the GCF.**

1. $9 + 21$   2. $32 - 48$   3. $8x + 2$   4. $3y - 24$

5. $20z - 8$   6. $15w + 65$   7. $36a + 16b$   8. $21m - 49n$

**Factor out the coefficient of the variable.**

9. $\frac{1}{3}b - \frac{1}{3}$   10. $\frac{3}{8}d + \frac{3}{4}$   11. $2.2x + 4.4$   12. $4h - 3$

13. Factor $-\frac{1}{2}$ out of $-\frac{1}{2}x + 6$.   14. Factor $-\frac{1}{4}$ out of $-\frac{1}{2}x - \frac{5}{4}y$.

15. **WRESTLING** A square wrestling mat has a perimeter of $(12x - 32)$ feet. Write an expression that represents the side length of the mat (in feet).

16. **MAKING A DIAGRAM** A table is 6 feet long and 3 feet wide. You extend the table by inserting two identical table *leaves*. The longest side length of each rectangular leaf is 3 feet. The extended table is rectangular with an area of $(18 + 6x)$ square feet.

   a. Make a diagram of the table and leaves.

   b. Write an expression that represents the length of the extended table. What does $x$ represent?

17. **STRUCTURE** The area of the trapezoid is $\left(\frac{3}{4}x - \frac{1}{4}\right)$ square centimeters. Write two different pairs of expressions that represent possible lengths of the bases.

$\frac{1}{2}$ cm

You can use a **four square** to organize information about a topic. Each of the four squares can be a category, such as *definition, vocabulary, example, non-example, words, algebra, table, numbers, visual, graph,* or *equation.* Here is an example of a four square for like terms.

| Definition<br>Terms that have the same variables raised to the same exponents | Examples<br>2 and $-3$, $3x$ and $-7x$,<br>$x^2$ and $6x^2$ |
|---|---|
| **Like Terms** | |
| Words<br>To combine like terms that have variables, use the Distributive Property to add or subtract the coefficients. | Non-Examples<br>$y$ and 4, $3x$ and $-4y$,<br>$6x^2$ and $2x$ |

## On Your Own

**Make four squares to help you study these topics.**

1. simplest form

2. linear expression

3. factoring expressions

**After you complete this chapter, make four squares for the following topics.**

4. equivalent equations

5. solving equations using addition or subtraction

6. solving equations using multiplication or division

7. solving two-step equations

"My four square shows that my new red skateboard is faster than my old blue skateboard."

**Check It Out**
Progress Check
BigIdeasMath ✓com

**Identify the terms and like terms in the expression.** *(Section 3.1)*

**1.** $11x + 2x$

**2.** $9x - 5x$

**3.** $21x + 6 - x - 5$

**4.** $8x + 14 - 3x + 1$

**Simplify the expression.** *(Section 3.1)*

**5.** $2(3x + x)$

**6.** $-7 + 3x + 4x$

**7.** $2x + 4 - 3x + 2 + 3x$

**8.** $7x + 6 + 3x - 2 - 5x$

**Find the sum or difference.** *(Section 3.2)*

**9.** $(s + 12) + (3s - 8)$

**10.** $(9t + 5) + (3t - 6)$

**11.** $(2 - k) + 3(-4k + 2)$

**12.** $\frac{1}{4}(q - 12) + \frac{1}{3}(q + 9)$

**13.** $(n - 8) - (-2n + 2)$

**14.** $-3(h - 4) - 2(-6h + 5)$

**Factor out the coefficient of the variable.** *(Section 3.2)*

**15.** $5c - 15$

**16.** $\frac{2}{9}j + \frac{2}{3}$

**17.** $2.4n + 9.6$

**18.** $-6z + 12$

Paint $21.79
Brush $3.99
Paint roller $6.89

**19. PAINTING** You buy the same number of brushes, rollers, and paint cans. Write an expression in simplest form that represents the total amount of money you spend for painting supplies. *(Section 3.1)*

**20. APPLES** A basket holds $n$ apples. You pick $2n - 3$ apples, and your friend picks $n + 4$ apples. Write an expression that represents the number of apples you and your friend picked. Interpret the expression. *(Section 3.2)*

**21. EXERCISE** Write an expression in simplest form for the perimeter of the exercise mat. *(Section 3.1)*

$w$
$3w$

**Essential Question** How can you use algebra tiles to solve addition or subtraction equations?

**1** **ACTIVITY: Solving Equations**

**Work with a partner. Use algebra tiles to model and solve the equation.**

**a.** $x - 3 = -4$

Model the equation $x - 3 = -4$.

To get the variable tile by itself, remove the ▢ tiles on the left side by adding ▢ ▢ tiles to each side.

How many *zero pairs* can you remove from each side? ▢
Circle them.

The remaining tile shows the value of *x*.

∴ So, $x = $ ▢ .

**b.** $z - 6 = 2$     **c.** $p - 7 = -3$     **d.** $-15 = t - 5$

**2** **ACTIVITY: Solving Equations**

**Work with a partner. Use algebra tiles to model and solve the equation.**

**a.** $-5 = n + 2$

Model the equation $-5 = n + 2$.

Remove the ▢ tiles on the right side by adding ▢ ▢ tiles to each side.

How many *zero pairs* can you remove from the right side? ▢
Circle them.

The remaining tiles show the value of *n*.

∴ So, $n = $ ▢ .

**b.** $y + 10 = -5$     **c.** $7 + b = -1$     **d.** $8 = 12 + z$

COMMON CORE

**Solving Equations**
In this lesson, you will
- write simple equations.
- solve equations using addition or subtraction.
- solve real-life problems.
Learning Standard
7.EE.4a

## Math Practice 4

**Interpret Results**

How can you add tiles to make zero pairs? Explain how this helps you solve the equation.

### 3 ACTIVITY: Writing and Solving Equations

**Work with a partner. Write an equation shown by the algebra tiles. Then solve.**

a.

b.

c.

d.

### 4 ACTIVITY: Using a Different Method to Find a Solution

**Work with a partner.** The *melting point* of a solid is the temperature at which the solid melts to become a liquid. The melting point of the element bromine is about 19°F. This is about 57°F more than the melting point of mercury.

a. Which of the following equations can you use to find the melting point of mercury? What is the melting point of mercury?

$$x + 57 = 19 \qquad x - 57 = 19 \qquad x + 19 = 57 \qquad x + 19 = -57$$

b. **CHOOSE TOOLS** How can you solve this problem without using an equation? Explain. How are these two methods related?

## What Is Your Answer?

5. **IN YOUR OWN WORDS** How can you use algebra tiles to solve addition or subtraction equations? Give an example of each.

6. **STRUCTURE** Explain how you could use inverse operations to solve addition or subtraction equations without using algebra tiles.

7. What makes the cartoon funny?

8. The word *variable* comes from the word *vary*. For example, the temperature in Maine varies a lot from winter to summer.

Write two other English sentences that use the word *vary*.

"To vary or not to vary." That is the question.

"Dear Sir: Yesterday you said x = 2. Today you are saying x = 3. Please make up your mind."

**Practice**

Use what you learned about solving addition or subtraction equations to complete Exercises 5–8 on page 100.

**Key Vocabulary** ◀))
equivalent equations,
p. 98

Two equations are **equivalent equations** if they have the same solutions. The Addition and Subtraction Properties of Equality can be used to write equivalent equations.

 **Key Ideas**

### Addition Property of Equality

**Words**   Adding the same number to each side of an equation produces an equivalent equation.

**Algebra**   If $a = b$, then $a + c = b + c$.

### Subtraction Property of Equality

**Words**   Subtracting the same number from each side of an equation produces an equivalent equation.

**Algebra**   If $a = b$, then $a - c = b - c$.

**Remember**

Addition and subtraction are inverse operations.

---

**EXAMPLE** ① **Solving Equations**

**a.  Solve $x - 5 = -1$.**

| | | |
|---|---|---|
| | $x - 5 = -1$ | Write the equation. |
| Undo the subtraction. → | $\underline{+5 \quad +5}$ | Addition Property of Equality |
| | $x = 4$ | Simplify. |

∴ The solution is $x = 4$.

**Check**

$$x - 5 = -1$$
$$4 - 5 \overset{?}{=} -1$$
$$-1 = -1 \checkmark$$

**b.  Solve $z + \dfrac{3}{2} = \dfrac{1}{2}$.**

| | | |
|---|---|---|
| | $z + \dfrac{3}{2} = \dfrac{1}{2}$ | Write the equation. |
| Undo the addition. → | $\underline{-\dfrac{3}{2} \quad -\dfrac{3}{2}}$ | Subtraction Property of Equality |
| | $z = -1$ | Simplify. |

∴ The solution is $z = -1$.

**Check**

$$z + \frac{3}{2} = \frac{1}{2}$$
$$-1 + \frac{3}{2} \overset{?}{=} \frac{1}{2}$$
$$\frac{1}{2} = \frac{1}{2} \checkmark$$

● **On Your Own**

**Now You're Ready**
Exercises 5–20

**Solve the equation. Check your solution.**

**1.**  $p - 5 = -2$          **2.**  $w + 13.2 = 10.4$          **3.**  $x - \dfrac{5}{6} = -\dfrac{1}{6}$

---

## EXAMPLE  2 Writing an Equation

**A company has a profit of $750 this week. This profit is $900 more than the profit _P_ last week. Which equation can be used to find _P_?**

  Ⓐ  $750 = 900 - P$           Ⓑ  $750 = P + 900$

  Ⓒ  $900 = P - 750$           Ⓓ  $900 = P + 750$

| **Words** | The profit this week is $900 more than the profit last week. |
|---|---|
| **Equation** | 750    =   _P_    +       900 |

:•: The equation is $750 = P + 900$. The correct answer is Ⓑ.

### On Your Own

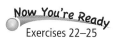
*Now You're Ready*
Exercises 22–25

4. A company has a profit of $120.50 today. This profit is $145.25 less than the profit _P_ yesterday. Write an equation that can be used to find _P_.

## EXAMPLE  3 Real-Life Application

**Video Game Scoring**

**The line graph shows the scoring while you and your friend played a video game. Write and solve an equation to find your score after Level 4.**

You can determine the following from the graph.

| **Words** | Your friend's score is 33 points less than your score. |
|---|---|
| **Variable** | Let _s_ be your score after Level 4. |
| **Equation** | −8    =   _s_    −    33 |

$$-8 = s - 33 \qquad \text{Write equation.}$$
$$\underline{+\,33} \quad \underline{+\,33} \qquad \text{Addition Property of Equality}$$
$$25 = s \qquad \text{Simplify.}$$

:•: Your score after Level 4 is 25 points.

**Reasonable?** From the graph, your score after Level 4 is between 20 points and 30 points. So, 25 points is a reasonable answer.

### On Your Own

5. **WHAT IF?** You have −12 points after Level 1. Your score is 27 points less than your friend's score. What is your friend's score?

## Vocabulary and Concept Check

1. **VOCABULARY** What property would you use to solve $m + 6 = -4$?

2. **VOCABULARY** Name two inverse operations.

3. **WRITING** Are the equations $m + 3 = -5$ and $m = -2$ equivalent? Explain.

4. **WHICH ONE DOESN'T BELONG?** Which equation does *not* belong with the other three? Explain your reasoning.

$$x + 3 = -1 \qquad x + 1 = -5 \qquad x - 2 = -6 \qquad x - 9 = -13$$

## Practice and Problem Solving

**Solve the equation. Check your solution.**

① 5. $a - 6 = 13$  6. $-3 = z - 8$  7. $-14 = k + 6$  8. $x + 4 = -14$

9. $c - 7.6 = -4$  10. $-10.1 = w + 5.3$  11. $\dfrac{1}{2} = q + \dfrac{2}{3}$  12. $p - 3\dfrac{1}{6} = -2\dfrac{1}{2}$

13. $g - 9 = -19$  14. $-9.3 = d - 3.4$  15. $4.58 + y = 2.5$  16. $x - 5.2 = -18.73$

17. $q + \dfrac{5}{9} = \dfrac{1}{6}$  18. $-2\dfrac{1}{4} = r - \dfrac{4}{5}$  19. $w + 3\dfrac{3}{8} = 1\dfrac{5}{6}$  20. $4\dfrac{2}{5} + k = -3\dfrac{2}{11}$

21. **ERROR ANALYSIS** Describe and correct the error in finding the solution.

$$\begin{aligned} x + 8 &= 10 \\ \underline{+8} \quad & \quad \underline{+8} \\ x &= 18 \end{aligned}$$

**Write the word sentence as an equation. Then solve.**

② 22. 4 less than a number $n$ is $-15$.  23. 10 more than a number $c$ is 3.

24. The sum of a number $y$ and $-3$ is $-8$.

25. The difference between a number $p$ and 6 is $-14$.

**In Exercises 26–28, write an equation. Then solve.**

26. **DRY ICE** The temperature of dry ice is $-109.3°F$. This is $184.9°F$ less than the outside temperature. What is the outside temperature?

27. **PROFIT** A company makes a profit of $1.38 million. This is $2.54 million more than last year. What was the profit last year?

28. **HELICOPTER** The difference in elevation of a helicopter and a submarine is $18\dfrac{1}{2}$ meters. The elevation of the submarine is $-7\dfrac{3}{4}$ meters. What is the elevation of the helicopter?

**GEOMETRY** Write and solve an equation to find the unknown side length.

**29.** Perimeter = 12 cm

? 3 cm
5 cm

**30.** Perimeter = 24.2 in.

8.3 in.
? 3.8 in.
8.3 in.

**31.** Perimeter = 34.6 ft

?
5.2 ft 6.4 ft
11.1 ft

**In Exercises 32–36, write an equation. Then solve.**

305 ft

**32. STATUE OF LIBERTY** The total height of the Statue of Liberty and its pedestal is 153 feet more than the height of the statue. What is the height of the statue?

**33. BUNGEE JUMPING** Your first jump is $50\frac{1}{6}$ feet higher than your second jump. Your first jump reaches $-200\frac{2}{5}$ feet. What is the height of your second jump?

**34. TRAVEL** Boatesville is $65\frac{3}{5}$ kilometers from Stanton. A bus traveling from Stanton is $24\frac{1}{3}$ kilometers from Boatesville. How far has the bus traveled?

**35. GEOMETRY** The sum of the measures of the angles of a triangle equals 180°. What is the measure of the missing angle?

$m°$
30.3° 40.8°

**36. SKATEBOARDING** The table shows your scores in a skateboarding competition. The leader has 311.62 points. What score do you need in the fourth round to win?

| Round | 1 | 2 | 3 | 4 |
|-------|-----|-----|-----|-----|
| Points | 63.43 | 87.15 | 81.96 | ? |

**37. CRITICAL THINKING** Find the value of $2x - 1$ when $x + 6 = 2$.

 **Find the values of x.**

**38.** $|x| = 2$

**39.** $|x| - 2 = 4$

**40.** $|x| + 5 = 18$

 **Fair Game Review** *What you learned in previous grades & lessons*

**Multiply or divide.** *(Section 1.4 and Section 1.5)*

**41.** $-7 \times 8$

**42.** $6 \times (-12)$

**43.** $18 \div (-2)$

**44.** $-26 \div 4$

**45. MULTIPLE CHOICE** A class of 144 students voted for a class president. Three-fourths of the students voted for you. Of the students who voted for you, $\frac{5}{9}$ are female. How many female students voted for you? *(Section 2.4)*

(A) 50      (B) 60      (C) 80      (D) 108

## 3.4 Solving Equations Using Multiplication or Division

**Essential Question** How can you use multiplication or division to solve equations?

### 1 ACTIVITY: Using Division to Solve Equations

**Work with a partner. Use algebra tiles to model and solve the equation.**

**a.** $3x = -12$

Model the equation $3x = -12$.

Your goal is to get one variable tile by itself. Because there are ___ variable tiles, divide the ___ tiles into ___ equal groups. Circle the groups.

Keep one of the groups. This shows the value of $x$.

∴ So, $x =$ ___.

**b.** $2k = -8$

**c.** $-15 = 3t$

**d.** $-20 = 5m$

**e.** $4h = -16$

### 2 ACTIVITY: Writing and Solving Equations

**Work with a partner. Write an equation shown by the algebra tiles. Then solve.**

**a.**

**b.**

**c.**

**d.**

**COMMON CORE**

**Solving Equations**

In this lesson, you will

• solve equations using multiplication or division.
• solve real-life problems.

Learning Standard
7.EE.4a

**Work with a partner. Choose the equation you can use to solve each problem. Solve the equation. Then explain how to solve the problem without using an equation. How are the two methods related?**

**Math Practice**

**Analyze Givens**
How can you use the given information to decide which equation represents the situation?

**a.** For the final part of a race, a handcyclist travels 32 feet each second across a distance of 400 feet. How many seconds does it take for the handcyclist to travel the last 400 feet of the race?

$$32x = 400 \qquad 400x = 32$$

$$\frac{x}{32} = 400 \qquad \frac{x}{400} = 32$$

**b.** The melting point of the element radon is about $-96°F$. The melting point of nitrogen is about 3.6 times the melting point of radon. What is the melting point of nitrogen?

$$3.6x = -96 \qquad x + 96 = 3.6$$

$$\frac{x}{3.6} = -96 \qquad -96x = 3.6$$

**c.** This year, a hardware store has a profit of $-\$6.0$ million. This profit is $\frac{3}{4}$ of last year's profit. What is last year's profit?

$$\frac{x}{-6} = \frac{3}{4} \qquad -6x = \frac{3}{4}$$

$$\frac{3}{4} + x = -6 \qquad \frac{3}{4}x = -6$$

## What Is Your Answer?

**4. IN YOUR OWN WORDS** How can you use multiplication or division to solve equations? Give an example of each.

**Practice**

Use what you learned about solving equations to complete Exercises 7–10 on page 106.

Key Ideas

**Multiplication Property of Equality**

**Words** Multiplying each side of an equation by the same number produces an equivalent equation.

**Algebra** If $a = b$, then $a \cdot c = b \cdot c$.

**Remember**

Multiplication and division are inverse operations.

**Division Property of Equality**

**Words** Dividing each side of an equation by the same number produces an equivalent equation.

**Algebra** If $a = b$, then $a \div c = b \div c$, $c \neq 0$.

**EXAMPLE** 1 **Solving Equations**

a. **Solve** $\dfrac{x}{3} = -6$.

$$\frac{x}{3} = -6 \qquad \text{Write the equation.}$$

Undo the division. → $3 \cdot \dfrac{x}{3} = 3 \cdot (-6)$    Multiplication Property of Equality

$$x = -18 \qquad \text{Simplify.}$$

∴ The solution is $x = -18$.

**Check**

$$\frac{x}{3} = -6$$

$$\frac{-18}{3} \overset{?}{=} -6$$

$$-6 = -6 \checkmark$$

b. **Solve** $18 = -4y$.

$$18 = -4y \qquad \text{Write the equation.}$$

Undo the multiplication. → $\dfrac{18}{-4} = \dfrac{-4y}{-4}$    Division Property of Equality

$$-4.5 = y \qquad \text{Simplify.}$$

∴ The solution is $y = -4.5$.

**Check**

$$18 = -4y$$

$$18 \overset{?}{=} -4(-4.5)$$

$$18 = 18 \checkmark$$

● **On Your Own**

**Now You're Ready**
Exercises 7–18

**Solve the equation. Check your solution.**

1. $\dfrac{x}{5} = -2$      2. $-a = -24$      3. $3 = -1.5n$

**EXAMPLE** ② **Solving an Equation Using a Reciprocal**

Solve $-\dfrac{4}{5}x = -8$.

$$-\frac{4}{5}x = -8 \qquad \text{Write the equation.}$$

Multiply each side by $-\dfrac{5}{4}$, the reciprocal of $-\dfrac{4}{5}$.

$$-\frac{5}{4} \cdot \left(-\frac{4}{5}x\right) = -\frac{5}{4} \cdot (-8) \qquad \text{Multiplicative Inverse Property}$$

$$x = 10 \qquad \text{Simplify.}$$

⸫ The solution is $x = 10$.

**On Your Own**

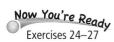

Exercises 19–22

Solve the equation. Check your solution.

4. $-14 = \dfrac{2}{3}x$     5. $-\dfrac{8}{5}b = 5$     6. $\dfrac{3}{8}h = -9$

---

**EXAMPLE** ③ **Real-Life Application**

Record low temperature in Arizona

The record low temperature in Arizona is 1.6 times the record low temperature in Rhode Island. What is the record low temperature in Rhode Island?

**Words**   The record low in Arizona is 1.6 times the record low in Rhode Island.

**Variable**   Let $t$ be the record low in Rhode Island.

**Equation**                $-40$      $= 1.6$   $\times$        $t$

$$-40 = 1.6t \qquad \text{Write equation.}$$

$$-\frac{40}{1.6} = \frac{1.6t}{1.6} \qquad \text{Division Property of Equality}$$

$$-25 = t \qquad \text{Simplify.}$$

⸫ The record low temperature in Rhode Island is $-25°$F.

**On Your Own**

Now You're Ready
Exercises 24–27

7. The record low temperature in Hawaii is –0.15 times the record low temperature in Alaska. The record low temperature in Hawaii is 12°F. What is the record low temperature in Alaska?

**Vocabulary and Concept Check**

1. **WRITING** Explain why you can use multiplication to solve equations involving division.

2. **OPEN-ENDED** Turning a light on and then turning the light off are considered to be inverse operations. Describe two other real-life situations that can be thought of as inverse operations.

**Describe the inverse operation that will undo the given operation.**

3. multiplying by 5     4. subtracting 12     5. dividing by $-8$     6. adding $-6$

 **Practice and Problem Solving**

**Solve the equation. Check your solution.**

① 7. $3h = 15$     8. $-5t = -45$     9. $\dfrac{n}{2} = -7$     10. $\dfrac{k}{-3} = 9$

11. $5m = -10$     12. $8t = -32$     13. $-0.2x = 1.6$     14. $-10 = -\dfrac{b}{4}$

15. $-6p = 48$     16. $-72 = 8d$     17. $\dfrac{n}{1.6} = 5$     18. $-14.4 = -0.6p$

② 19. $\dfrac{3}{4}g = -12$     20. $8 = -\dfrac{2}{5}c$     21. $-\dfrac{4}{9}f = -3$     22. $26 = -\dfrac{8}{5}y$

23. **ERROR ANALYSIS** Describe and correct the error in finding the solution.

$$-4.2x = 21$$
$$\frac{-4.2x}{4.2} = \frac{21}{4.2}$$
$$x = 5$$

**Write the word sentence as an equation. Then solve.**

③ 24. A number divided by $-9$ is $-16$.     25. A number multiplied by $\dfrac{2}{5}$ is $\dfrac{3}{20}$.

26. The product of 15 and a number is $-75$.

27. The quotient of a number and $-1.5$ is 21.

**In Exercises 28 and 29, write an equation. Then solve.**

28. **NEWSPAPERS** You make $0.75 for every newspaper you sell. How many newspapers do you have to sell to buy the soccer cleats?

29. **ROCK CLIMBING** A rock climber averages $12\dfrac{3}{5}$ feet per minute.

   How many feet does the rock climber climb in 30 minutes?

Soccer Cleats $36.00

**OPEN-ENDED** (a) Write a multiplication equation that has the given solution. (b) Write a division equation that has the same solution.

**30.** $-3$

**31.** $-2.2$

**32.** $-\dfrac{1}{2}$

**33.** $-1\dfrac{1}{4}$

**34. REASONING** Which of the methods can you use to solve $-\dfrac{2}{3}c = 16$?

| Multiply each side by $-\dfrac{2}{3}$. | Multiply each side by $-\dfrac{3}{2}$. |
|---|---|
| Divide each side by $-\dfrac{2}{3}$. | Multiply each side by 3, then divide each side by $-2$. |

**35. STOCK** A stock has a return of $-\$1.26$ per day. Write and solve an equation to find the number of days until the total return is $-\$10.08$.

**36. ELECTION** In a school election, $\dfrac{3}{4}$ of the students vote. There are 1464 ballots. Write and solve an equation to find the number of students.

**37. OCEANOGRAPHY** Aquarius is an underwater ocean laboratory located in the Florida Keys National Marine Sanctuary. Solve the equation $\dfrac{31}{25}x = -62$ to find the value of $x$.

**38. SHOPPING** The price of a bike at Store A is $\dfrac{5}{6}$ the price at Store B. The price at Store A is $\$150.60$. Write and solve an equation to find how much you save by buying the bike at Store A.

**39. CRITICAL THINKING** Solve $-2|m| = -10$.

**40.**  **Number Sense** In four days, your family drives $\dfrac{5}{7}$ of a trip. Your rate of travel is the same throughout the trip. The total trip is 1250 miles. In how many more days will you reach your destination?

## Fair Game Review  What you learned in previous grades & lessons

**Subtract.** *(Section 1.3)*

**41.** $5 - 12$

**42.** $-7 - 2$

**43.** $4 - (-8)$

**44.** $-14 - (-5)$

**45. MULTIPLE CHOICE** Of the 120 apartments in a building, 75 have been scheduled to receive new carpet. What fraction of the apartments have not been scheduled to receive new carpet? *(Skills Review Handbook)*

　Ⓐ $\dfrac{1}{4}$ 　　Ⓑ $\dfrac{3}{8}$ 　　Ⓒ $\dfrac{5}{8}$ 　　Ⓓ $\dfrac{3}{4}$

## 3.5 Solving Two-Step Equations

**Essential Question** How can you use algebra tiles to solve a two-step equation?

### 1 ACTIVITY: Solving a Two-Step Equation

**Work with a partner. Use algebra tiles to model and solve $2x - 3 = -5$.**

Model the equation $2x - 3 = -5$.

Remove the ⬜ red tiles on the left side by adding ⬜ yellow tiles to each side.

How many *zero pairs* can you remove from each side? ⬜
Circle them.

Because there are ⬜ green tiles, divide the red tiles into ⬜ equal groups. Circle the groups.

$$\boxed{\phantom{xx}} = \boxed{\phantom{xx}}$$

Keep one of the groups. This shows the value of $x$.

$$\boxed{+} = \boxed{\phantom{xx}}$$

∴ So, $x = $ ⬜ .

### 2 ACTIVITY: The Math behind the Tiles

**Work with a partner. Solve $2x - 3 = -5$ without using algebra tiles. Complete each step. Then answer the questions.**

Use the steps in Activity 1 as a guide.

| | |
|---|---|
| $2x - 3 = -5$ | Write the equation. |
| $2x - 3 + \boxed{\phantom{x}} = -5 + \boxed{\phantom{x}}$ | Add ⬜ to each side. |
| $2x = \boxed{\phantom{x}}$ | Simplify. |
| $\dfrac{2x}{\boxed{\phantom{x}}} = \dfrac{\boxed{\phantom{x}}}{\boxed{\phantom{x}}}$ | Divide each side by ⬜ . |
| $x = \boxed{\phantom{x}}$ | Simplify. |

∴ So, $x = $ ⬜ .

**a.** Which step is first, adding 3 to each side or dividing each side by 2?

**b.** How are the above steps related to the steps in Activity 1?

**COMMON CORE**

**Solving Equations**
In this lesson, you will
• solve two-step equations.
• solve real-life problems.
Learning Standard
7.EE.4a

**Work with a partner.**

- **Write an equation shown by the algebra tiles.**
- **Use algebra tiles to model and solve the equation.**
- **Check your answer by solving the equation without using algebra tiles.**

a.

b.

**4 ACTIVITY: Working Backwards**

**Work with a partner.**

**Math Practice** 8

**Maintain Oversight**

How does working backwards help you decide which operation to do first? Explain.

a. **Sample:** Your friend pauses a video game to get a drink. You continue the game. You double the score by saving a princess. Then you lose 75 points because you do not collect the treasure. You finish the game with $-25$ points. How many points did you have when you started?

One way to solve the problem is to work backwards. To do this, start with the end result and retrace the events.

| | |
|---|---|
| You have $-25$ points at the end of the game. | $-25$ |
| You lost 75 points for not collecting the treasure, so add 75 to $-25$. | $-25 + 75 = 50$ |
| You doubled your score for saving the princess, so find half of 50. | $50 \div 2 = 25$ |

So, you started the game with 25 points.

b. You triple your account balance by making a deposit. Then you withdraw $127.32 to buy groceries. Your account is now overdrawn by $10.56. By working backwards, find your account balance before you made the deposit.

## What Is Your Answer?

5. **IN YOUR OWN WORDS** How can you use algebra tiles to solve a two-step equation?

6. When solving the equation $4x + 1 = -11$, what is the first step?

7. **REPEATED REASONING** Solve the equation $2x - 75 = -25$. How do your steps compare with the strategy of working backwards in Activity 4?

**Practice**

Use what you learned about solving two-step equations to complete Exercises 6–11 on page 112.

**EXAMPLE 1** Solving a Two-Step Equation

Solve $-3x + 5 = 2$. Check your solution.

| | | |
|---|---|---|
| $-3x + 5 = 2$ | | Write the equation. |

Undo the addition. → $\underline{-5 \quad -5}$  Subtraction Property of Equality

$-3x = -3$  Simplify.

Undo the multiplication. → $\dfrac{-3x}{-3} = \dfrac{-3}{-3}$  Division Property of Equality

$x = 1$  Simplify.

**Check**

$-3x + 5 = 2$

$-3(1) + 5 \stackrel{?}{=} 2$

$-3 + 5 \stackrel{?}{=} 2$

$2 = 2$ ✓

∴ The solution is $x = 1$.

### On Your Own

**Now You're Ready**
Exercises 6–17

Solve the equation. Check your solution.

**1.** $2x + 12 = 4$  **2.** $-5c + 9 = -16$  **3.** $3(x - 4) = 9$

**EXAMPLE 2** Solving a Two-Step Equation

Solve $\dfrac{x}{8} - \dfrac{1}{2} = -\dfrac{7}{2}$. Check your solution.

**Study Tip**

You can simplify the equation in Example 2 before solving. Multiply each side by the LCD of the fractions, 8.

$\dfrac{x}{8} - \dfrac{1}{2} = -\dfrac{7}{2}$

$x - 4 = -28$

$x = -24$

$\dfrac{x}{8} - \dfrac{1}{2} = -\dfrac{7}{2}$  Write the equation.

$\underline{+\dfrac{1}{2} \qquad +\dfrac{1}{2}}$  Addition Property of Equality

$\dfrac{x}{8} = -3$  Simplify.

$8 \cdot \dfrac{x}{8} = 8 \cdot (-3)$  Multiplication Property of Equality

$x = -24$  Simplify.

**Check**

$\dfrac{x}{8} - \dfrac{1}{2} = -\dfrac{7}{2}$

$\dfrac{-24}{8} - \dfrac{1}{2} \stackrel{?}{=} -\dfrac{7}{2}$

$-3 - \dfrac{1}{2} \stackrel{?}{=} -\dfrac{7}{2}$

$-\dfrac{7}{2} = -\dfrac{7}{2}$ ✓

∴ The solution is $x = -24$.

### On Your Own

**Now You're Ready**
Exercises 20–25

Solve the equation. Check your solution.

**4.** $\dfrac{m}{2} + 6 = 10$  **5.** $-\dfrac{z}{3} + 5 = 9$  **6.** $\dfrac{2}{5} + 4a = -\dfrac{6}{5}$

EXAMPLE 3 **Combining Like Terms Before Solving**

**Solve 3y − 8y = 25.**

$$3y - 8y = 25 \qquad \text{Write the equation.}$$
$$-5y = 25 \qquad \text{Combine like terms.}$$
$$y = -5 \qquad \text{Divide each side by } -5.$$

∴ The solution is $y = -5$.

EXAMPLE 4 **Real-Life Application**

**The height at the top of a roller coaster hill is 10 times the height h of the starting point. The height decreases 100 feet from the top to the bottom of the hill. The height at the bottom of the hill is −10 feet. Find h.**

| Location | Verbal Description | Expression |
|---|---|---|
| Start | The height at the start is $h$. | $h$ |
| Top of hill | The height at the top of the hill is 10 times the starting height $h$. | $10h$ |
| Bottom of hill | The height decreases by 100 feet. So, subtract 100. | $10h - 100$ |

The height at the bottom of the hill is −10 feet. Solve $10h - 100 = -10$ to find $h$.

$$10h - 100 = -10 \qquad \text{Write equation.}$$
$$10h = 90 \qquad \text{Add 100 to each side.}$$
$$h = 9 \qquad \text{Divide each side by 10.}$$

∴ So, the height at the start is 9 feet.

**On Your Own**

*Now You're Ready*
Exercises 29–34

**Solve the equation. Check your solution.**

**7.** $4 - 2y + 3 = -9$     **8.** $7x - 10x = 15$     **9.** $-8 = 1.3m - 2.1m$

**10. WHAT IF?** In Example 4, the height at the bottom of the hill is −5 feet. Find the height $h$.

**Vocabulary and Concept Check**

1. **WRITING** How do you solve two-step equations?

**Match the equation with the first step to solve it.**

2. $4 + 4n = -12$     3. $4n = -12$     4. $\dfrac{n}{4} = -12$     5. $\dfrac{n}{4} - 4 = -12$

   A. Add 4.       B. Subtract 4.       C. Multiply by 4.       D. Divide by 4.

 **Practice and Problem Solving**

**Solve the equation. Check your solution.**

6. $2v + 7 = 3$     7. $4b + 3 = -9$     8. $17 = 5k - 2$

9. $-6t - 7 = 17$     10. $8n + 16.2 = 1.6$     11. $-5g + 2.3 = -18.8$

12. $2t - 5 = -10$     13. $-4p + 9 = -5$     14. $11 = -5x - 2$

15. $4 + 2.2h = -3.7$     16. $-4.8f + 6.4 = -8.48$     17. $7.3y - 5.18 = -51.9$

**ERROR ANALYSIS** Describe and correct the error in finding the solution.

18.
$$-6 + 2x = -10$$
$$-6 + \frac{2x}{2} = -\frac{10}{2}$$
$$-6 + x = -5$$
$$x = 1$$

19.
$$-3x + 2 = -7$$
$$-3x = -9$$
$$-\frac{3x}{3} = \frac{-9}{3}$$
$$x = -3$$

**Solve the equation. Check your solution.**

20. $\dfrac{3}{5}g - \dfrac{1}{3} = -\dfrac{10}{3}$     21. $\dfrac{a}{4} - \dfrac{5}{6} = -\dfrac{1}{2}$     22. $-\dfrac{1}{3} + 2z = -\dfrac{5}{6}$

23. $2 - \dfrac{b}{3} = -\dfrac{5}{2}$     24. $-\dfrac{2}{3}x + \dfrac{3}{7} = \dfrac{1}{2}$     25. $-\dfrac{9}{4}v + \dfrac{4}{5} = \dfrac{7}{8}$

**In Exercises 26–28, write an equation. Then solve.**

26. **WEATHER** Starting at 1:00 P.M., the temperature changes $-4$ degrees per hour. How long will it take to reach $-1°$?

27. **BOWLING** It costs $2.50 to rent bowling shoes. Each game costs $2.25. You have $9.25. How many games can you bowl?

28. **CELL PHONES** A cell phone company charges a monthly fee plus $0.25 for each text message. The monthly fee is $30.00 and you owe $59.50. How many text messages did you have?

Temperature
at 1:00 P.M.

35°F

**Solve the equation. Check your solution.**

③ **29.** $3v - 9v = 30$

**30.** $12t - 8t = -52$

**31.** $-8d - 5d + 7d = 72$

**32.** $6(x - 2) = -18$

**33.** $-4(m + 3) = 24$

**34.** $-8(y + 9) = -40$

**35. WRITING** Write a real-world problem that can be modeled by $\frac{1}{2}x - 2 = 8$. Then solve the equation.

**36. GEOMETRY** The perimeter of the parallelogram is 102 feet. Find $m$.

**REASONING** Exercises 37 and 38 are missing information. Tell what information you need to solve the problem.

**37. TAXI** A taxi service charges an initial fee plus $1.80 per mile. How far can you travel for $12?

**38. EARTH** The coldest surface temperature on the Moon is 57 degrees colder than twice the coldest surface temperature on Earth. What is the coldest surface temperature on Earth?

**39. PROBLEM SOLVING** On Saturday, you catch insects for your science class. Five of the insects escape. The remaining insects are divided into three groups to share in class. Each group has nine insects. How many insects did you catch on Saturday?

    **a.** Solve the problem by working backwards.

    **b.** Solve the equation $\frac{x - 5}{3} = 9$. How does the answer compare with the answer to part (a)?

**40. UNDERWATER HOTEL** You must scuba dive to the entrance of your room at Jules' Undersea Lodge in Key Largo, Florida. The diver is 1 foot deeper than $\frac{2}{3}$ of the elevation of the entrance. What is the elevation of the entrance?

**41.** **Geometry** How much should you change the length of the rectangle so that the perimeter is 54 centimeters? Write an equation that shows how you found your answer.

**Fair Game Review** What you learned in previous grades & lessons

**Multiply or divide.** *(Section 2.4)*

**42.** $-6.2 \times 5.6$

**43.** $\frac{8}{3} \times \left(-2\frac{1}{2}\right)$

**44.** $\frac{5}{2} \div \left(-\frac{4}{5}\right)$

**45.** $-18.6 \div (-3)$

**46. MULTIPLE CHOICE** Which fraction is *not* equivalent to 0.75? *(Skills Review Handbook)*

   Ⓐ $\frac{15}{20}$
   Ⓑ $\frac{9}{12}$
   Ⓒ $\frac{6}{9}$
   Ⓓ $\frac{3}{4}$

**Solve the equation. Check your solution.**   *(Section 3.3, Section 3.4, and Section 3.5)*

1. $-6.5 + x = -4.12$

2. $4\frac{1}{2} + p = -5\frac{3}{4}$

3. $-\dfrac{b}{7} = 4$

4. $-2w + 3.7 = -0.5$

**Write the word sentence as an equation. Then solve.**   *(Section 3.3 and Section 3.4)*

5. The difference between a number $b$ and 7.4 is $-6.8$.

6. $5\frac{2}{5}$ more than a number $a$ is $7\frac{1}{2}$.

7. A number $x$ multiplied by $\dfrac{3}{8}$ is $-\dfrac{15}{32}$.

8. The quotient of two times a number $k$ and $-2.6$ is 12.

**Write and solve an equation to find the value of $x$.**   *(Section 3.3 and Section 3.5)*

9. Perimeter = 26

10. Perimeter = 23.59

11. Perimeter = 33

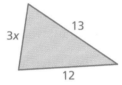

12. **BANKING** You withdraw $29.79 from your bank account. Now your balance is $-\$20.51$. Write and solve an equation to find the amount of money in your bank account before you withdrew the money.   *(Section 3.3)*

13. **WATER LEVEL** During a drought, the water level of a lake changes $-3\frac{1}{5}$ feet per day. Write and solve an equation to find how long it takes for the water level to change $-16$ feet.   *(Section 3.4)*

14. **BASKETBALL** A basketball game has four quarters. The length of a game is 32 minutes. You play the entire game except for $4\frac{1}{2}$ minutes. Write and solve an equation to find the mean time you play per quarter.   *(Section 3.5)*

15. **SCRAPBOOKING** The mat needs to be cut to have a 0.5-inch border on all four sides.   *(Section 3.5)*

    a. How much should you cut from the left and right sides?

    b. How much should you cut from the top and bottom?

Check It Out
Vocabulary Help
BigIdeasMath ✓ com

## Review Key Vocabulary

like terms, *p. 82*  
simplest form, *p. 82*  
linear expression, *p. 88*

factoring an expression, *p. 92*  
equivalent equations, *p. 98*

## Review Examples and Exercises

**3.1** **Algebraic Expressions** *(pp. 80–85)*

**a.** **Identify the terms and like terms in the expression $6y + 9 + 3y - 7$.**

Rewrite as a sum of terms.

$$6y + 9 + 3y + (-7)$$

**Terms:** $6y$, $\quad 9$, $\quad 3y$, $\quad -7$

**Like terms:** $6y$ and $3y$, $9$ and $-7$

**b.** **Simplify $\dfrac{2}{3}y + 14 - \dfrac{1}{6}y - 8$.**

$$\dfrac{2}{3}y + 14 - \dfrac{1}{6}y - 8 = \dfrac{2}{3}y + 14 + \left(-\dfrac{1}{6}y\right) + (-8) \qquad \text{Rewrite as a sum.}$$

$$= \dfrac{2}{3}y + \left(-\dfrac{1}{6}y\right) + 14 + (-8) \qquad \text{Commutative Property of Addition}$$

$$= \left[\dfrac{2}{3} + \left(-\dfrac{1}{6}\right)\right]y + 14 + (-8) \qquad \text{Distributive Property}$$

$$= \dfrac{1}{2}y + 6 \qquad \text{Combine like terms.}$$

### Exercises

**Identify the terms and like terms in the expression.**

**1.** $z + 8 - 4z$

**2.** $3n + 7 - n - 3$

**3.** $10x^2 - y + 12 - 3x^2$

**Simplify the expression.**

**4.** $4h - 8h$

**5.** $6.4r - 7 - 2.9r$

**6.** $\dfrac{3}{5}x + 19 - \dfrac{3}{20}x - 7$

**7.** $3(2 + q) + 15$

**8.** $\dfrac{1}{8}(16m - 8) - 17$

**9.** $-1.5(4 - n) + 2.8$

**Adding and Subtracting Linear Expressions** *(pp. 86–93)*

a. Find $(5z + 4) + (3z - 6)$.

$$5z + 4$$
$$\underline{+\ 3z - 6} \qquad \text{Align like terms vertically and add.}$$
$$8z - 2$$

b. Factor $\dfrac{1}{4}$ out of $\dfrac{1}{4}x - \dfrac{3}{4}$.

Write each term as a product of $\dfrac{1}{4}$ and another factor.

$$\frac{1}{4}x = \frac{1}{4} \cdot x \qquad\qquad -\frac{3}{4} = \frac{1}{4} \cdot (-3)$$

Use the Distributive Property to factor out $\dfrac{1}{4}$.

$$\frac{1}{4}x - \frac{3}{4} = \frac{1}{4} \cdot x + \frac{1}{4} \cdot (-3) = \frac{1}{4}(x - 3)$$

∴ So, $\dfrac{1}{4}x - \dfrac{3}{4} = \dfrac{1}{4}(x - 3)$.

### Exercises

**Find the sum or difference.**

**10.** $(c - 4) + (3c + 9)$

**11.** $\dfrac{2}{5}(d - 10) - \dfrac{2}{3}(d + 6)$

**Factor out the coefficient of the variable.**

**12.** $2b + 8$

**13.** $\dfrac{1}{4}y + \dfrac{3}{8}$

**14.** $1.7j - 3.4$

**15.** $-5p + 20$

**Solving Equations Using Addition or Subtraction** *(pp. 96–101)*

Solve $x - 9 = -6$.

| | | **Check** |
|---|---|---|
| $x - 9 = -6$ | Write the equation. | $x - 9 = -6$ |
| $\underline{+9 \quad +9}$ | Addition Property of Equality | $3 - 9 \overset{?}{=} -6$ |
| $x = 3$ | Simplify. | $-6 = -6$ ✓ |

Undo the subtraction. →

### Exercises

**Solve the equation. Check your solution.**

**16.** $p - 3 = -4$

**17.** $6 + q = 1$

**18.** $-2 + j = -22$

**19.** $b - 19 = -11$

**20.** $n + \dfrac{3}{4} = \dfrac{1}{4}$

**21.** $v - \dfrac{5}{6} = -\dfrac{7}{8}$

**22.** $t - 3.7 = 1.2$

**23.** $\ell + 15.2 = -4.5$

## 3.4 Solving Equations Using Multiplication or Division *(pp. 102–107)*

Solve $\dfrac{x}{5} = -7$.

| | | |
|---|---|---|
| $\dfrac{x}{5} = -7$ | Write the equation. | |

**Undo the division.** → $5 \cdot \dfrac{x}{5} = 5 \cdot (-7)$     Multiplication Property of Equality

$x = -35$     Simplify.

**Check**

$\dfrac{x}{5} = -7$

$\dfrac{-35}{5} \overset{?}{=} -7$

$-7 = -7$ ✓

### Exercises

**Solve the equation. Check your solution.**

**24.** $\dfrac{x}{3} = -8$      **25.** $-7 = \dfrac{y}{7}$      **26.** $-\dfrac{z}{4} = -\dfrac{3}{4}$      **27.** $-\dfrac{w}{20} = -2.5$

**28.** $4x = -8$      **29.** $-10 = 2y$      **30.** $-5.4z = -32.4$      **31.** $-6.8w = 3.4$

**32.** **TEMPERATURE** The mean temperature change is $-3.2°$F per day for 5 days. What is the total change over the 5-day period?

## 3.5 Solving Two-Step Equations *(pp. 108–113)*

Solve $-6y + 7 = -5$. Check your solution.

$-6y + 7 = -5$     Write the equation.

$\underline{\phantom{-6y} -7 \phantom{=} -7}$     Subtraction Property of Equality

$-6y = -12$     Simplify.

$\dfrac{-6y}{-6} = \dfrac{-12}{-6}$     Division Property of Equality

$y = 2$     Simplify.

**Check**

$-6y + 7 = -5$

$-6(2) + 7 \overset{?}{=} -5$

$-12 + 7 \overset{?}{=} -5$

$-5 = -5$ ✓

∴ The solution is $y = 2$.

### Exercises

**Solve the equation. Check your solution.**

**33.** $-2c + 6 = -8$         **34.** $3(3w - 4) = -20$

**35.** $\dfrac{w}{6} + \dfrac{5}{8} = -1\dfrac{3}{8}$         **36.** $-3x - 4.6 = 5.9$

**37.** **EROSION** The floor of a canyon has an elevation of $-14.5$ feet. Erosion causes the elevation to change by $-1.5$ feet per year. How many years will it take for the canyon floor to have an elevation of $-31$ feet?

**Check It Out**
Test Practice
BigIdeasMath.com

**Simplify the expression.**

**1.** $8x - 5 + 2x$

**2.** $2.5w - 3y + 4w$

**3.** $3(5 - 2n) + 9n$

**4.** $\dfrac{5}{7}x + 15 - \dfrac{9}{14}x - 9$

**Find the sum or difference.**

**5.** $(3j + 11) + (8j - 7)$

**6.** $\dfrac{3}{4}(8p + 12) + \dfrac{3}{8}(16p - 8)$

**7.** $(2r - 13) - (-6r + 4)$

**8.** $-2.5(2s - 5) - 3(4.5s - 5.2)$

**Factor out the coefficient of the variable.**

**9.** $3n - 24$

**10.** $\dfrac{1}{2}q + \dfrac{5}{2}$

**Solve the equation. Check your solution.**

**11.** $7x = -3$

**12.** $2(x + 1) = -2$

**13.** $\dfrac{2}{9}g = -8$

**14.** $z + 14.5 = 5.4$

**15.** $-14 = 6c$

**16.** $\dfrac{2}{7}k - \dfrac{3}{8} = -\dfrac{19}{8}$

**17. HAIR SALON** Write an expression in simplest form that represents the income from $w$ women and $m$ men getting a haircut and a shampoo.

|          | Women | Men  |
|----------|-------|------|
| Haircut  | $45   | $15  |
| Shampoo  | $12   | $7   |

**18. RECORD** A runner is compared with the world record holder during a race. A negative number means the runner is ahead of the time of the world record holder. A positive number means that the runner is behind the time of the world record holder. The table shows the time difference between the runner and the world record holder for each lap. What time difference does the runner need for the fourth lap to match the world record?

| Lap | Time Difference |
|-----|-----------------|
| 1   | −1.23           |
| 2   | 0.45            |
| 3   | 0.18            |
| 4   | ?               |

**19. GYMNASTICS** You lose 0.3 point for stepping out of bounds during a floor routine. Your final score is 9.124. Write and solve an equation to find your score before the penalty.

**20. PERIMETER** The perimeter of the triangle is 45. Find the value of $x$.

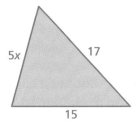

5x    17

15

1. Which equation represents the word sentence shown below? *(7.EE.4a)*

> The quotient of a number $b$ and 0.3 equals negative 10.

**A.** $0.3b = 10$

**C.** $\dfrac{0.3}{b} = -10$

**B.** $\dfrac{b}{0.3} = -10$

**D.** $\dfrac{b}{0.3} = 10$

2. What is the value of the expression below when $c = 0$ and $d = -6$? *(7.NS.2c)*

$$\dfrac{cd - d^2}{4}$$

3. What is the value of the expression below? *(7.NS.1c)*

$$-38 - (-14)$$

**F.** $-52$

**H.** $24$

**G.** $-24$

**I.** $52$

4. The daily low temperatures last week are shown below.

What is the mean low temperature of last week? *(7.NS.3)*

**A.** $-2°F$

**C.** $8°F$

**B.** $6°F$

**D.** $10°F$

**5.** Which equation is equivalent to the equation shown below?  *(7.EE.4a)*

$$-\frac{3}{4}x + \frac{1}{8} = -\frac{3}{8}$$

**F.** $-\frac{3}{4}x = -\frac{3}{8} - \frac{1}{8}$

**G.** $-\frac{3}{4}x = -\frac{3}{8} + \frac{1}{8}$

**H.** $x + \frac{1}{8} = -\frac{3}{8} \cdot \left(-\frac{4}{3}\right)$

**I.** $x + \frac{1}{8} = -\frac{3}{8} \cdot \left(-\frac{3}{4}\right)$

**6.** What is the value of the expression below?  *(7.NS.2c)*

$$-0.28 \div (-0.07)$$

**7.** Karina was solving the equation in the box below.

$$-96 = -6(x - 15)$$
$$-96 = -6x - 90$$
$$-96 + 90 = -6x - 90 + 90$$
$$-6 = -6x$$
$$\frac{-6}{-6} = \frac{-6x}{-6}$$
$$1 = x$$

What should Karina do to correct the error that she made?  *(7.EE.4a)*

**A.** First add 6 to both sides of the equation.

**B.** First subtract $x$ from both sides of the equation.

**C.** Distribute the $-6$ to get $6x - 90$.

**D.** Distribute the $-6$ to get $-6x + 90$.

**8.** The perimeter of the rectangle is 400 inches. What is the value of $j$?
(All measurements are in inches.)  *(7.EE.4a)*

**F.** 35

**G.** 85

**H.** 140

**I.** 200

**9.** Jacob was evaluating the expression below when $x = -2$ and $y = 4$.

$$3 + x^2 \div y$$

His work is in the box below.

$$3 + x^2 \div y = 3 + (-2^2) \div 4$$
$$= 3 - 4 \div 4$$
$$= 3 - 1$$
$$= 3$$

What should Jacob do to correct the error that he made?   *(7.NS.3)*

**A.** Divide 3 by 4 before subtracting.

**B.** Square $-2$, then divide.

**C.** Square then divide.

**D.** Subtract 4 from 3 before dividing.

**10.** Which number is equivalent to the expression shown below?   *(7.NS.3)*

$$-2\frac{1}{4} - \left(-8\frac{3}{8}\right)$$

**F.** $-10\frac{5}{8}$

**G.** $-10\frac{1}{3}$

**H.** $6\frac{1}{8}$

**I.** $6\frac{1}{2}$

**11.** You want to buy the bicycle. You already have \$43.50 saved and plan to save an additional \$7.25 every week.   *(7.EE.4a)*

*Part A*   Write and solve an equation to find the number of weeks you need to save before you can purchase the bicycle.

*Part B*   How much sooner could you purchase the bicycle if you had a coupon for \$20 off and saved \$8.75 every week? Explain your reasoning.

# 4 Inequalities

"If you reached into your water bowl and found more than $20..."

"And then reached into your cat food bowl and found more than $40..."

Someone else's bowls!

"What would you have?"

"Dear Precious Pet World: Your ad says 'Up to 75% off on selected items.'"

Hey, it didn't say who's doing the selecting.

"I select Yummy Tummy Bacon-Flavored Dog Biscuits."

# What You Learned Before

Why do I always have to be farther from the fulcrum?

"Move farther back. We still have an inequality."

## ● Graphing Inequalities (6.EE.8)

**Example 1** Graph $x \geq 2$.

Use a closed circle because 2 is a solution. | Shade the number line on the side where you found the solution.

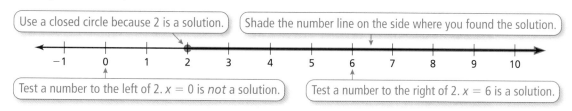

Test a number to the left of 2. $x = 0$ is *not* a solution. | Test a number to the right of 2. $x = 6$ is a solution.

### Try It Yourself
**Graph the inequality.**

**1.** $x \geq 1$       **2.** $x < 5$       **3.** $x \leq 20$       **4.** $x > 13$

## ● Comparing Numbers (6.NS.7a)

**Example 2** Compare $-\dfrac{1}{3}$ and $-\dfrac{5}{6}$.

Graph $-\dfrac{5}{6}$.      Graph $-\dfrac{1}{3}$.

$-\dfrac{5}{6}$ is to the left of $-\dfrac{1}{3}$.

∴ So, $-\dfrac{5}{6} < -\dfrac{1}{3}$.

### Try It Yourself
**Copy and complete the statement using < or >.**

**5.** $-\dfrac{2}{3} \quad\rule{1cm}{0.4pt}\quad \dfrac{3}{8}$       **6.** $-\dfrac{1}{2} \quad\rule{1cm}{0.4pt}\quad -\dfrac{7}{8}$

**7.** $-\dfrac{1}{5} \quad\rule{1cm}{0.4pt}\quad \dfrac{1}{10}$       **8.** $-1.4 \quad\rule{1cm}{0.4pt}\quad 1.2$

**9.** $-2.2 \quad\rule{1cm}{0.4pt}\quad -4.6$       **10.** $-1.9 \quad\rule{1cm}{0.4pt}\quad -1.1$

**Essential Question** How can you use a number line to represent solutions of an inequality?

> ### 1   ACTIVITY: Understanding Inequality Statements

**Work with a partner. Read the statement. Circle each number that makes the statement true, and then answer the questions.**

**a.** "You are in at least 5 of the photos."

$$-3 \quad -2 \quad -1 \quad 0 \quad 1 \quad 2 \quad 3 \quad 4 \quad 5 \quad 6$$

- What do you notice about the numbers that you circled?
- Is the number 5 included? Why or why not?
- Write four other numbers that make the statement true.

**b.** "The temperature is less than −4 degrees Fahrenheit."

$$-7 \quad -6 \quad -5 \quad -4 \quad -3 \quad -2 \quad -1 \quad 0 \quad 1 \quad 2$$

- What do you notice about the numbers that you circled?
- Can the temperature be exactly −4 degrees Fahrenheit? Explain.
- Write four other numbers that make the statement true.

**c.** "More than 3 students from our school are in the chess tournament."

$$-3 \quad -2 \quad -1 \quad 0 \quad 1 \quad 2 \quad 3 \quad 4 \quad 5 \quad 6$$

- What do you notice about the numbers that you circled?
- Is the number 3 included? Why or why not?
- Write four other numbers that make the statement true.

**d.** "The balance in a yearbook fund is no more than −$5."

$$-7 \quad -6 \quad -5 \quad -4 \quad -3 \quad -2 \quad -1 \quad 0 \quad 1 \quad 2$$

- What do you notice about the numbers that you circled?
- Is the number −5 included? Why or why not?
- Write four other numbers that make the statement true.

**COMMON CORE**

**Inequalities**
In this lesson, you will
- write and graph inequalities.
- use substitution to check whether a number is a solution of an inequality.

Preparing for Standard 7.EE.4b

**Work with a partner.**

**a.** Consider the statement "$x$ is a number such that $x > -1.5$."

- Can the number be exactly $-1.5$? Explain.

- Make a number line. Shade the part of the number line that shows the numbers that make the statement true.

- Write four other numbers that are not integers that make the statement true.

**b.** Consider the statement "$x$ is a number such that $x \le \dfrac{5}{2}$."

- Can the number be exactly $\dfrac{5}{2}$? Explain.

- Make a number line. Shade the part of the number line that shows the numbers that make the statement true.

- Write four other numbers that are not integers that make the statement true.

## 3 ACTIVITY: Writing and Graphing Inequalities

**Math Practice 1**

**Check Progress**

All the graphs are similar. So, what can you do to make sure that you have correctly written each inequality?

**Work with a partner. Write an inequality for each graph. Then, in words, describe all the values of $x$ that make the inequality true.**

**a.**

**b.**

**c.**

**d.**

## What Is Your Answer?

**4. IN YOUR OWN WORDS** How can you use a number line to represent solutions of an inequality?

**5. STRUCTURE** Is $x \ge -1.4$ the same as $-1.4 \le x$? Explain.

**Practice**

Use what you learned about writing and graphing inequalities to complete Exercises 4 and 5 on page 128.

**Check It Out**
Lesson Tutorials
BigIdeasMath.com

An **inequality** is a mathematical sentence that compares expressions. It contains the symbols <, >, ≤, or ≥. To write an inequality, look for the following phrases to determine where to place the inequality symbol.

| | | Inequality Symbols | | |
|---|---|---|---|---|
| **Symbol** | < | > | ≤ | ≥ |
| **Key Phrases** | • is less than<br><br>• is fewer than | • is greater than<br><br>• is more than | • is less than or equal to<br><br>• is at most<br><br>• is no more than | • is greater than or equal to<br><br>• is at least<br><br>• is no less than |

**EXAMPLE 1 Writing an Inequality**

**A number *q* plus 5 is greater than or equal to −7.9. Write this word sentence as an inequality.**

A <u>number *q* plus 5</u> <u>is greater than or equal to</u> −7.9.

$$q + 5 \qquad\qquad \geq \qquad\qquad -7.9$$

∴ An inequality is $q + 5 \geq -7.9$.

**On Your Own**

Now You're Ready
Exercises 6−9

**Write the word sentence as an inequality.**

1. A number *x* is at most −10.
2. Twice a number *y* is more than $-\dfrac{5}{2}$.

A **solution of an inequality** is a value that makes the inequality true. An inequality can have more than one solution. The set of all solutions of an inequality is called the **solution set**.

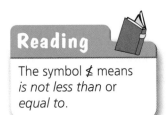
**Reading**

The symbol ≰ means *is not less than or equal to*.

| Value of *x* | $x + 2 \leq -1$ | Is the inequality true? |
|---|---|---|
| −2 | $-2 + 2 \overset{?}{\leq} -1$<br>$0 \not\leq -1$  ✗ | no |
| −3 | $-3 + 2 \overset{?}{\leq} -1$<br>$-1 \leq -1$  ✓ | yes |
| −4 | $-4 + 2 \overset{?}{\leq} -1$<br>$-2 \leq -1$  ✓ | yes |

🔊 Multi-Language Glossary at BigIdeasMath.com

**EXAMPLE** **2** **Checking Solutions**

**Tell whether −2 is a solution of each inequality.**

**a.** $y - 5 \geq -6$

$y - 5 \geq -6$    Write the inequality.

$-2 - 5 \overset{?}{\geq} -6$    Substitute −2 for y.

$-7 \ngeq -6$ ✗    Simplify.

−7 is *not* greater than or equal to −6.

   So, −2 is *not* a solution of the inequality.

**b.** $-5.5y < 14$

$-5.5y < 14$

$-5.5(-2) \overset{?}{<} 14$

$11 < 14$ ✓

11 is less than 14.

   So, −2 is a solution of the inequality.

● **On Your Own**

Now You're Ready
Exercises 11–16

**Tell whether −5 is a solution of the inequality.**

**3.** $x + 12 > 7$      **4.** $1 - 2p \leq -9$      **5.** $n \div 2.5 \geq -3$

---

The **graph of an inequality** shows all the solutions of the inequality on a number line. An open circle ○ is used when a number is *not* a solution. A closed circle ● is used when a number is a solution. An arrow to the left or right shows that the graph continues in that direction.

**EXAMPLE** **3** **Graphing an Inequality**

**Graph $y > -8$.**

Use an open circle because −8 is *not* a solution.

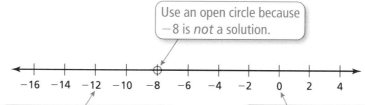

Test a number to the left of −8.
$y = -12$ is *not* a solution.

Test a number to the right of −8.
$y = 0$ is a solution.

**Study Tip**

The graph in Example 3 shows that the inequality has *infinitely many* solutions.

Shade the number line on the side where you found the solution.

● **On Your Own**

Now You're Ready
Exercises 17–20

**Graph the inequality on a number line.**

**6.** $x < -1$      **7.** $z \geq 4$      **8.** $s \leq 1.4$      **9.** $-\dfrac{1}{2} < t$

**Check It Out**
Help with Homework
BigIdeasMath.com

 ## Vocabulary and Concept Check

1. **PRECISION** Should you use an open circle or a closed circle in the graph of the inequality $b \geq -42$? Explain.

2. **DIFFERENT WORDS, SAME QUESTION** Which is different? Write "both" inequalities.

   $k$ is less than or equal to $-3$.    $k$ is no more than $-3$.

   $k$ is at most $-3$.    $k$ is at least $-3$.

3. **REASONING** Do $x < 5$ and $5 < x$ represent the same inequality? Explain.

 ## Practice and Problem Solving

Write an inequality for the graph. Then, in words, describe all the values of $x$ that make the inequality true.

4.

5. 

Write the word sentence as an inequality.

6. A number $y$ is no more than $-8$.

7. A number $w$ added to 2.3 is more than 18.

8. A number $t$ multiplied by $-4$ is at least $-\frac{2}{5}$.

9. A number $b$ minus 4.2 is less than $-7.5$.

10. **ERROR ANALYSIS** Describe and correct the error in writing the word sentence as an inequality.

Twice a number x is at most −24.

$2x \geq -24$

Tell whether the given value is a solution of the inequality.

11. $n + 8 \leq 13$; $n = 4$    12. $5h > -15$; $h = -5$    13. $p + 1.4 \leq 0.5$; $p = 0.1$

14. $\dfrac{a}{6} > -4$; $a = -18$    15. $-\dfrac{2}{3}s \geq 6$; $s = -9$    16. $\dfrac{7}{8} - 3k < -\dfrac{1}{2}$; $k = \dfrac{1}{4}$

Graph the inequality on a number line.

17. $r \leq -9$    18. $g > 2.75$    19. $x \geq -3\dfrac{1}{2}$    20. $z < 1\dfrac{1}{4}$

21. **FOOD TRUCK** Each day at lunchtime, at least 53 people buy food from a food truck. Write an inequality that represents this situation.

**Tell whether the given value is a solution of the inequality.**

**22.** $4k < k + 8; k = 3$

**23.** $\dfrac{w}{3} \geq w - 12; w = 15$

**24.** $7 - 2y > 3y + 13; y = -1$

**25.** $\dfrac{3}{4}b - 2 \leq 2b + 8; b = -4$

**26. MODELING** A subway ride for a student costs $1.25. A monthly pass costs $35.

    **a.** Write an inequality that represents the number of times you must ride the subway for the monthly pass to be a better deal.

    **b.** You ride the subway about 45 times per month. Should you buy the monthly pass? Explain.

**27. LOGIC** Consider the inequality $b > -2$.

    **a.** Describe the values of $b$ that are solutions of the inequality.

    **b.** Describe the values of $b$ that are *not* solutions of the inequality. Write an inequality for these values.

    **c.** What do all the values in parts (a) and (b) represent? Is this true for any inequality?

**28.** **Critical Thinking** A postal service says that a rectangular package can have a maximum combined length and *girth* of 108 inches. The girth of a package is the distance around the perimeter of a face that does not include the length.

    **a.** Write an inequality that represents the allowable dimensions for the package.

    **b.** Find three different sets of allowable dimensions that are reasonable for the package. Find the volume of each package.

 **Fair Game Review** What you learned in previous grades & lessons

**Solve the equation. Check your solution.** *(Section 3.3)*

**29.** $p - 8 = 3$        **30.** $8.7 + w = 5.1$        **31.** $x - 2 = -9$

**32. MULTIPLE CHOICE** Which expression has a value less than $-5$? *(Section 1.2)*

    **Ⓐ** $5 + 8$        **Ⓑ** $-9 + 5$        **Ⓒ** $1 + (-8)$        **Ⓓ** $7 + (-2)$

## Essential Question How can you use addition or subtraction to solve an inequality?

### 1 ACTIVITY: Writing an Inequality

**Work with a partner. Members of the Boy Scouts must be less than 18 years old. In 4 years, your friend will still be eligible to be a scout.**

**a.** Which of the following represents your friend's situation? What does $x$ represent? Explain your reasoning.

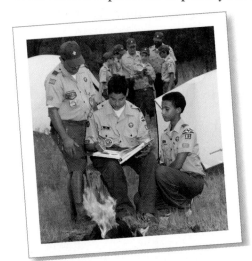

$$x + 4 > 18 \qquad x + 4 < 18$$

$$x + 4 \geq 18 \qquad x + 4 \leq 18$$

**b.** Graph the possible ages of your friend on a number line. Explain how you decided what to graph.

### 2 ACTIVITY: Writing an Inequality

**Work with a partner. Supercooling is the process of lowering the temperature of a liquid or a gas below its freezing point without it becoming a solid. Water can be supercooled to 86°F below its normal freezing point (32°F) and still not freeze.**

**COMMON CORE**

**Inequalities**

In this lesson, you will
- solve inequalities using addition or subtraction.
- solve real-life problems.

Learning Standard
7.EE.4b

**a.** Let $x$ represent the temperature of water. Which inequality represents the temperature at which water can be a liquid or a gas? Explain your reasoning.

$$x - 32 > -86 \qquad x - 32 < -86$$

$$x - 32 \geq -86 \qquad x - 32 \leq -86$$

**b.** On a number line, graph the possible temperatures at which water can be a liquid or a gas. Explain how you decided what to graph.

**ACTIVITY: Solving Inequalities**

**Math Practice 4**

**Interpret Results**
What does the solution of the inequality represent?

**Work with a partner. Complete the following steps for Activity 1. Then repeat the steps for Activity 2.**

- Use your inequality from part (a). Replace the inequality symbol with an equal sign.
- Solve the equation.
- Replace the equal sign with the original inequality symbol.
- Graph this new inequality.
- Compare the graph with your graph in part (b). What do you notice?

**4** **ACTIVITY: Temperatures of Continents**

**Work with a partner. The table shows the lowest recorded temperature on each continent. Write an inequality that represents each statement. Then solve and graph the inequality.**

a. The temperature at a weather station in Asia is more than 150°F greater than the record low in Asia.

b. The temperature at a research station in Antarctica is at least 80°F greater than the record low in Antarctica.

| Continent | Lowest Temperature |
|---|---|
| Africa | −11°F |
| Antarctica | −129°F |
| Asia | −90°F |
| Australia | −9.4°F |
| Europe | −67°F |
| North America | −81.4°F |
| South America | −27°F |

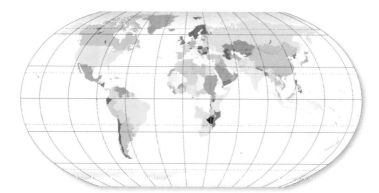

# What Is Your Answer?

5. **IN YOUR OWN WORDS** How can you use addition or subtraction to solve an inequality?

6. Describe a real-life situation that you can represent with an inequality. Write the inequality. Graph the solution on a number line.

**Practice**

Use what you learned about solving inequalities to complete Exercises 3–5 on page 134.

Key Ideas

**Study Tip**

You can solve inequalities in the same way you solve equations. Use inverse operations to get the variable by itself.

**Addition Property of Inequality**

**Words**  When you add the same number to each side of an inequality, the inequality remains true.

**Numbers**

$$\begin{array}{r} -4 < 3 \\ +2 \quad +2 \\ \hline -2 < 5 \end{array}$$

**Algebra**  If $a < b$, then $a + c < b + c$.

If $a > b$, then $a + c > b + c$.

**Subtraction Property of Inequality**

**Words**  When you subtract the same number from each side of an inequality, the inequality remains true.

**Numbers**

$$\begin{array}{r} -2 < 2 \\ -3 \quad -3 \\ \hline -5 < -1 \end{array}$$

**Algebra**  If $a < b$, then $a - c < b - c$.

If $a > b$, then $a - c > b - c$.

These properties are also true for ≤ and ≥.

**EXAMPLE** ① **Solving an Inequality Using Addition**

Solve $x - 5 < -3$. Graph the solution.

$$x - 5 < -3$$   Write the inequality.

Undo the subtraction. ⟶ $$+5 \quad +5$$   Addition Property of Inequality

$$x < 2$$   Simplify.

⋰ The solution is $x < 2$.

**Check:**

$x = 0$:  $0 - 5 \overset{?}{<} -3$

$-5 < -3$ ✓

$x = 5$:  $5 - 5 \overset{?}{<} -3$

$0 \not< -3$ ✗

$x < 2$

$x = 0$ is a solution.

$x = 5$ is *not* a solution.

⬤ **On Your Own**

Solve the inequality. Graph the solution.

**1.** $y - 6 > -7$    **2.** $b - 3.8 \le 1.7$    **3.** $-\dfrac{1}{2} > z - \dfrac{1}{4}$

EXAMPLE 2  Solving an Inequality Using Subtraction

**Solve 13 ≤ x + 14. Graph the solution.**

$$13 \le x + 14 \qquad \text{Write the inequality.}$$

 Undo the addition. $\qquad \underline{-14 \qquad -14} \qquad$ Subtraction Property of Inequality

$$-1 \le x \qquad \text{Simplify.}$$

⋮ The solution is $x \ge -1$.

 **Reading**

The inequality $-1 \le x$ is the same as $x \ge -1$.

$$x \ge -1$$

number line from $-4$ to $6$ with a closed circle at $-1$ and arrow to the right

**On Your Own**

*Now You're Ready*
Exercises 3–17

**Solve the inequality. Graph the solution.**

4. $w - 7 \le -10$  5. $-7.5 \ge d - 10$  6. $x + \dfrac{3}{4} > 1\dfrac{1}{2}$

---

EXAMPLE 3 **Real-Life Application**

**A person can be no taller than 6.25 feet to become an astronaut pilot for NASA. Your friend is 5 feet 9 inches tall. Write and solve an inequality that represents how much your friend can grow and still meet the requirement.**

| **Words** | Current height | plus | amount your friend can grow | is no more than | the height limit. |
|---|---|---|---|---|---|

**Variable** Let $h$ be the possible amounts your friend can grow.

| **Inequality** | 5.75 | + | $h$ | ≤ | 6.25 |
|---|---|---|---|---|---|

5 ft 9 in. = 60 + 9 = 69 in.

$69 \text{ in.} \times \dfrac{1 \text{ ft}}{12 \text{ in.}} = 5.75 \text{ ft}$

$$5.75 + h \le 6.25 \qquad \text{Write the inequality.}$$
$$\underline{-5.75 \qquad\quad -5.75} \qquad \text{Subtraction Property of Inequality}$$
$$h \le 0.5 \qquad \text{Simplify.}$$

 So, your friend can grow no more than 0.5 foot, or 6 inches.

**On Your Own**

7. Your cousin is 5 feet 3 inches tall. Write and solve an inequality that represents how much your cousin can grow and still meet the requirement.

**Vocabulary and Concept Check**

1. **REASONING** Is the inequality $c + 3 > 5$ the same as $c > 5 - 3$? Explain.

2. **WHICH ONE DOESN'T BELONG?** Which inequality does *not* belong with the other three? Explain your reasoning.

$$w + \frac{7}{4} > \frac{3}{4} \qquad w - \frac{3}{4} > -\frac{7}{4} \qquad w + \frac{7}{4} < \frac{3}{4} \qquad \frac{3}{4} < w + \frac{7}{4}$$

  **Practice and Problem Solving**

**Solve the inequality. Graph the solution.**

**① ②**   **3.** $x + 7 \geq 18$     **4.** $a - 2 > 4$     **5.** $3 \leq 7 + g$

    **6.** $8 + k \leq -3$     **7.** $-12 < y - 6$     **8.** $n - 4 < 5$

    **9.** $t - 5 \leq -7$     **10.** $p + \frac{1}{4} \geq 2$     **11.** $\frac{2}{7} > b + \frac{5}{7}$

    **12.** $z - 4.7 \geq -1.6$     **13.** $-9.1 < d - 6.3$     **14.** $\frac{8}{5} > s + \frac{12}{5}$

    **15.** $-\frac{7}{8} \geq m - \frac{13}{8}$     **16.** $r + 0.2 < -0.7$     **17.** $h - 6 \leq -8.4$

**ERROR ANALYSIS** Describe and correct the error in solving the inequality or graphing the solution of the inequality.

**18.**

**19.**

**20. AIRPLANE** A small airplane can hold 44 passengers. Fifteen passengers board the plane.

    **a.** Write and solve an inequality that represents the additional number of passengers that can board the plane.

    **b.** Can 30 more passengers board the plane? Explain.

**Write and solve an inequality that represents $x$.**

**21.** The perimeter is less than 28 feet.

7 ft  $x$
7 ft

**22.** The base is greater than the height.

8 in.
$x + 3$

**23.** The perimeter is less than or equal to 51 meters.

8 m   8 m
10 m   10 m
$x$

**24. REASONING** The solution of $d + s > -3$ is $d > -7$. What is the value of $s$?

**25. BIRDFEEDER** The hole for a birdfeeder post is 3 feet deep. The top of the post needs to be at least 5 feet above the ground. Write and solve an inequality that represents the required length of the post.

**26. SHOPPING** You can spend up to $35 on a shopping trip.

**a.** You want to buy a shirt that costs $14. Write and solve an inequality that represents the amount of money you will have left if you buy the shirt.

**b.** You notice that the shirt is on sale for 30% off. How does this change the inequality?

**c.** Do you have enough money to buy the shirt that is on sale and a pair of pants that costs $23? Explain.

**27. POWER** A circuit overloads at 2400 watts of electricity. A portable heater that uses 1050 watts of electricity is plugged into the circuit.

**a.** Write and solve an inequality that represents the additional number of watts you can plug in without overloading the circuit.

**b.** In addition to the portable heater, what two other items in the table can you plug in at the same time without overloading the circuit? Is there more than one possibility? Explain.

| Item | Watts |
|---|---|
| Aquarium | 200 |
| Hair dryer | 1200 |
| Television | 150 |
| Vacuum cleaner | 1100 |

**28.** **Number Sense** The possible values of $x$ are given by $x + 8 \le 6$. What is the greatest possible value of $7x$?

**Fair Game Review** What you learned in previous grades & lessons

**Solve the equation. Check your solution.** *(Section 3.4)*

**29.** $4x = 36$

**30.** $\dfrac{w}{3} = -9$

**31.** $-2b = 44$

**32.** $60 = \dfrac{3}{4}h$

**33. MULTIPLE CHOICE** Which fraction is equivalent to $-2.4$? *(Section 2.1)*

(A) $-\dfrac{12}{5}$   (B) $-\dfrac{51}{25}$   (C) $-\dfrac{8}{5}$   (D) $-\dfrac{6}{25}$

You can use a **Y chart** to compare two topics. List differences in the branches and similarities in the base of the Y. Here is an example of a Y chart that compares solving equations and solving inequalities.

Solving
Equations

Solving
Inequalities

- The sign between two expressions is an equal sign, =.

- One number is the solution.

- The sign between two expressions is an inequality symbol: <, >, ≤, or ≥.

- More than one number can be a solution.

- Use inverse operations to group numbers on one side.
- Use inverse operations to group variables on one side.
- Solve for the variable.

## On Your Own

**Make Y charts to help you study and compare these topics.**

1. writing equations and writing inequalities

2. graphing the solution of an equation and graphing the solution of an inequality

3. graphing inequalities that use > and graphing inequalities that use <

4. graphing inequalities that use > or < and graphing inequalities that use ≥ or ≤

5. solving inequalities using addition and solving inequalities using subtraction

**"Hey Descartes, do you have any suggestions for the Y chart I am making?"**

**After you complete this chapter, make Y charts for the following topics.**

6. solving inequalities using multiplication and solving inequalities using division

7. solving two-step equations and solving two-step inequalities

8. Pick two other topics that you studied earlier in this course and make a Y chart to compare them.

**Write the word sentence as an inequality.** *(Section 4.1)*

**1.** A number $y$ plus 2 is greater than $-5$.

**2.** A number $s$ minus 2.4 is at least 8.

**Tell whether the given value is a solution of the inequality.** *(Section 4.1)*

**3.** $8p < -3;\ p = -2$

**4.** $z + 2 > -4;\ z = -8$

**Graph the inequality on a number line.** *(Section 4.1)*

**5.** $x < -12$

**6.** $v > \dfrac{5}{4}$

**7.** $b \geq -\dfrac{1}{3}$

**8.** $q \leq 4.2$

**Solve the inequality. Graph the solution.** *(Section 4.2)*

**9.** $n + 2 \leq -6$

**10.** $t - \dfrac{3}{7} > \dfrac{6}{7}$

**11.** $-\dfrac{3}{4} \geq w + 1$

**12.** $y - 2.6 < -3.4$

**13.** **STUDYING** You plan to study at least 1.5 hours for a geography test. Write an inequality that represents this situation. *(Section 4.1)*

Fitness Test
- Jog at least 2 kilometers
- Perform 25 or more push-ups
- Perform at least 10 pull-ups

**14.** **FITNESS TEST** The three requirements to pass a fitness test are shown. *(Section 4.1)*

    **a.** Write and graph three inequalities that represent the requirements.

    **b.** You can jog 2500 meters, perform 30 push-ups, and perform 20 pull-ups. Do you satisfy the requirements of the test? Explain.

**15.** **NUMBER LINE** Use tape on the floor to make the number line shown. All units are in feet. You are standing at $-\dfrac{7}{2}$. You want to move to a number greater than $-\dfrac{3}{2}$. Write and solve an inequality that represents the distance you must move. *(Section 4.2)*

# 4.3 Solving Inequalities Using Multiplication or Division

**Essential Question** How can you use multiplication or division to solve an inequality?

## 1 ACTIVITY: Using a Table to Solve an Inequality

**Work with a partner.**

- Copy and complete the table.
- Decide which graph represents the solution of the inequality.
- Write the solution of the inequality.

**a.** $4x > 12$

| x | −1 | 0 | 1 | 2 | 3 | 4 | 5 |
|---|---|---|---|---|---|---|---|
| 4x | | | | | | | |
| 4x $\overset{?}{>}$ 12 | | | | | | | |

**b.** $-3x \leq 9$

| x | −5 | −4 | −3 | −2 | −1 | 0 | 1 |
|---|---|---|---|---|---|---|---|
| −3x | | | | | | | |
| −3x $\overset{?}{\leq}$ 9 | | | | | | | |

**COMMON CORE**

**Inequalities**

In this lesson, you will

- solve inequalities using multiplication or division.
- solve real-life problems.

Learning Standard
7.EE.4b

## 2 ACTIVITY: Solving an Inequality

**Work with a partner.**

**a.** Solve $-3x \leq 9$ by adding $3x$ to each side of the inequality first. Then solve the resulting inequality.

**b.** Compare the solution in part (a) with the solution in Activity 1(b).

**138　Chapter 4　Inequalities**

## 3  ACTIVITY: Using a Table to Solve an Inequality

**Work with a partner.**

- **Copy and complete the table.**
- **Decide which graph represents the solution of the inequality.**
- **Write the solution of the inequality.**

**a.**  $\dfrac{x}{3} < 1$

| x | −1 | 0 | 1 | 2 | 3 | 4 | 5 |
|---|---|---|---|---|---|---|---|
| $\dfrac{x}{3}$ | | | | | | | |
| $\dfrac{x}{3} \overset{?}{<} 1$ | | | | | | | |

**b.**  $\dfrac{x}{-4} \geq \dfrac{3}{4}$

| x | −5 | −4 | −3 | −2 | −1 | 0 | 1 |
|---|---|---|---|---|---|---|---|
| $\dfrac{x}{-4}$ | | | | | | | |
| $\dfrac{x}{-4} \overset{?}{\geq} \dfrac{3}{4}$ | | | | | | | |

## 4  ACTIVITY: Writing Rules

**Work with a partner. Use a table to solve each inequality.**

**a.**  $-2x \leq 10$    **b.**  $-6x > 0$    **c.**  $\dfrac{x}{-4} < 1$    **d.**  $\dfrac{x}{-8} \geq \dfrac{1}{8}$

Write a set of rules that describes how to solve inequalities like those in
Activities 1 and 3. Then use your rules to solve each of the four inequalities above.

**Math Practice 3**

**Analyze Conjectures**

When you apply your rules to parts (a)–(d), do you get the same solutions? Explain.

## What Is Your Answer?

**5. IN YOUR OWN WORDS**  How can you use multiplication or division to solve an inequality?

**Practice**  Use what you learned about solving inequalities using multiplication or division to complete Exercises 4–9 on page 143.

## Key Idea

### Multiplication and Division Properties of Inequality (Case 1)

**Words**    When you multiply or divide each side of an inequality by the same *positive* number, the inequality remains true.

**Remember**

Multiplication and division are inverse operations.

**Numbers**     $-4 < 6$                     $4 > -6$

$2 \cdot (-4) < 2 \cdot 6$              $\dfrac{4}{2} > \dfrac{-6}{2}$

$-8 < 12$                     $2 > -3$

**Algebra**    If $a < b$ and $c$ is positive, then

$$a \cdot c < b \cdot c \qquad \text{and} \qquad \frac{a}{c} < \frac{b}{c}.$$

If $a > b$ and $c$ is positive, then

$$a \cdot c > b \cdot c \qquad \text{and} \qquad \frac{a}{c} > \frac{b}{c}.$$

These properties are also true for $\leq$ and $\geq$.

---

**EXAMPLE** **1**    **Solving an Inequality Using Multiplication**

Solve $\dfrac{x}{5} \leq -3$. Graph the solution.

$\dfrac{x}{5} \leq -3$                     Write the inequality.

Undo the division. $\longrightarrow$ $5 \cdot \dfrac{x}{5} \leq 5 \cdot (-3)$          Multiplication Property of Inequality

$x \leq -15$                     Simplify.

The solution is $x \leq -15$.

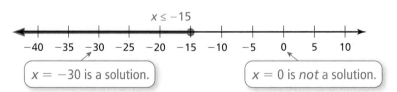

$x \leq -15$

$x = -30$ is a solution.

$x = 0$ is *not* a solution.

## On Your Own

**Solve the inequality. Graph the solution.**

**1.** $n \div 3 < 1$         **2.** $-0.5 \leq \dfrac{m}{10}$         **3.** $-3 > \dfrac{2}{3}p$

**EXAMPLE** **2** **Solving an Inequality Using Division**

Solve $6x > -18$. Graph the solution.

$$6x > -18 \qquad \text{Write the inequality.}$$

Undo the multiplication. ⟶ $\dfrac{6x}{6} > \dfrac{-18}{6} \qquad \text{Division Property of Inequality}$

$$x > -3 \qquad \text{Simplify.}$$

∴ The solution is $x > -3$.

$x > -3$

$x = -6$ is *not* a solution.          $x = 0$ is a solution.

**On Your Own**

**Now You're Ready**
Exercises 10–18

Solve the inequality. Graph the solution.

**4.** $4b \geq 2$          **5.** $12k \leq -24$          **6.** $-15 < 2.5q$

# Key Idea

**Multiplication and Division Properties of Inequality (Case 2)**

**Words**   When you multiply or divide each side of an inequality by the same *negative* number, the direction of the inequality symbol must be reversed for the inequality to remain true.

**Common Error**

A negative sign in an inequality does not necessarily mean you must reverse the inequality symbol.

Only reverse the inequality symbol when you multiply or divide both sides by a negative number.

**Numbers**          $-4 < 6$                    $4 > -6$

$-2 \cdot (-4) > -2 \cdot 6 \qquad \dfrac{4}{-2} < \dfrac{-6}{-2}$

$8 > -12 \qquad\qquad -2 < 3$

**Algebra**   If $a < b$ and $c$ is negative, then

$$a \cdot c > b \cdot c \qquad \text{and} \qquad \dfrac{a}{c} > \dfrac{b}{c}.$$

If $a > b$ and $c$ is negative, then

$$a \cdot c < b \cdot c \qquad \text{and} \qquad \dfrac{a}{c} < \dfrac{b}{c}.$$

These properties are also true for $\leq$ and $\geq$.

## EXAMPLE ③ Solving an Inequality Using Multiplication

Solve $-\dfrac{3}{2}n \le 6$. Graph the solution.

$$-\dfrac{3}{2}n \le 6$$    Write the inequality.

$$-\dfrac{2}{3} \cdot \left(-\dfrac{3}{2}n\right) \ge -\dfrac{2}{3} \cdot 6$$    Use the Multiplication Property of Inequality. Reverse the inequality symbol.

$$n \ge -4$$    Simplify.

∴ The solution is $n \ge -4$.

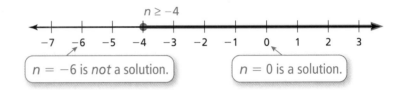

### On Your Own

Solve the inequality. Graph the solution.

**7.** $\dfrac{x}{-3} > -4$

**8.** $0.5 \le -\dfrac{y}{2}$

**9.** $-12 \ge \dfrac{6}{5}m$

**10.** $-\dfrac{2}{5}h \le -8$

## EXAMPLE ④ Solving an Inequality Using Division

Solve $-3z > -4.5$. Graph the solution.

$$-3z > -4.5$$    Write the inequality.

Undo the multiplication. →  $$\dfrac{-3z}{-3} < \dfrac{-4.5}{-3}$$    Use the Division Property of Inequality. Reverse the inequality symbol.

$$z < 1.5$$    Simplify.

∴ The solution is $z < 1.5$.

### On Your Own

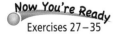

Solve the inequality. Graph the solution.

**11.** $-5z < 35$

**12.** $-2a > -9$

**13.** $-1.5 < 3n$

**14.** $-4.2 \ge -0.7w$

## Vocabulary and Concept Check

1. **WRITING** Explain how to solve $\frac{x}{3} < -2$.

2. **PRECISION** Explain how solving $4x < -16$ is different from solving $-4x < 16$.

3. **OPEN-ENDED** Write an inequality that you can solve using the Division Property of Inequality where the direction of the inequality symbol must be reversed.

 ## Practice and Problem Solving

**Use a table to solve the inequality.**

4. $2x < 2$

5. $-3x \leq 3$

6. $-6x > 18$

7. $\frac{x}{-5} \geq 7$

8. $\frac{x}{-1} > \frac{2}{5}$

9. $\frac{x}{3} \leq \frac{1}{2}$

**Solve the inequality. Graph the solution.**

① ② 10. $2n > 20$

11. $\frac{c}{9} \leq -4$

12. $2.2m < 11$

13. $-16 > x \div 2$

14. $\frac{1}{6}w \geq 2.5$

15. $7 < 3.5k$

16. $3x \leq -\frac{5}{4}$

17. $4.2y \leq -12.6$

18. $11.3 > \frac{b}{4.3}$

19. **ERROR ANALYSIS** Describe and correct the error in solving the inequality.

$$\frac{x}{3} < -9$$

$$3 \cdot \frac{x}{3} > 3 \cdot (-9)$$

$$x > -27$$

**Write the word sentence as an inequality. Then solve the inequality.**

20. The quotient of a number and 4 is at most 5.

21. A number divided by 7 is less than $-3$.

22. Six times a number is at least $-24$.

23. The product of $-2$ and a number is greater than 30.

24. **SMART PHONE** You earn $9.20 per hour at your summer job. Write and solve an inequality that represents the number of hours you need to work in order to buy a smart phone that costs $299.

25. **AVOCADOS** You have $9.60 to buy avocados for a guacamole recipe. Avocados cost $2.40 each.

    **a.** Write and solve an inequality that represents the number of avocados you can buy.

    **b.** Are there infinitely many solutions in this context? Explain.

26. **SCIENCE PROJECT** Students in a science class are divided into 6 equal groups with at least 4 students in each group for a project. Write and solve an inequality that represents the number of students in the class.

**Solve the inequality. Graph the solution.**

③ ④ **27.** $-5n \leq 15$

    **28.** $-7w > 49$

    **29.** $-\dfrac{1}{3}h \geq 8$

**30.** $-9 < -\dfrac{1}{5}x$

    **31.** $-3y < -14$

    **32.** $-2d \geq 26$

**33.** $4.5 > -\dfrac{m}{6}$

    **34.** $\dfrac{k}{-0.25} \leq 36$

    **35.** $-2.4 > \dfrac{b}{-2.5}$

36. **ERROR ANALYSIS** Describe and correct the error in solving the inequality.

$$-3m \geq 9$$
$$\dfrac{-3m}{-3} \geq \dfrac{9}{-3}$$
$$m \geq -3$$

37. **TEMPERATURE** It is currently 0°C outside. The temperature is dropping 2.5°C every hour. Write and solve an inequality that represents the number of hours that must pass for the temperature to drop below −20°C.

12.5 ft

27 in.

38. **STORAGE** You are moving some of your belongings into a storage facility.

    **a.** Write and solve an inequality that represents the number of boxes that you can stack vertically in the storage unit.

    **b.** Can you stack 6 boxes vertically in the storage unit? Explain.

**Write and solve an inequality that represents x.**

39. Area $\geq 120$ cm$^2$

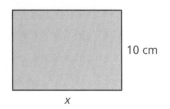

10 cm

$x$

40. Area $< 20$ ft$^2$

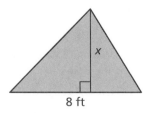

$x$

8 ft

41. **AMUSEMENT PARK** You and four friends are planning a visit to an amusement park. You want to keep the cost below $100 per person. Write and solve an inequality that represents the total cost of visiting the amusement park.

42. **LOGIC** When you multiply or divide each side of an inequality by the same negative number, you must reverse the direction of the inequality symbol. Explain why.

43. **PROJECT** Choose two novels to research.

   **a.** Use the Internet or a magazine to complete the table.

   **b.** Use the table to find and compare the average number of copies sold per month for each novel. Which novel do you consider to be the most successful? Explain.

   **c.** Assume each novel continues to sell at the average rate. Write and solve an inequality that represents the number of months it will take for the total number of copies sold to exceed twice the current number sold.

| | Author | Name of Novel | Release Date | Current Number of Copies Sold |
|---|---|---|---|---|
| 1. | | | | |
| 2. | | | | |

 **Number Sense** Describe all numbers that satisfy *both* inequalities. Include a graph with your description.

44. $4m > -4$ and $3m < 15$

45. $\dfrac{n}{3} \geq -4$ and $\dfrac{n}{-5} \geq 1$

46. $2x \geq -6$ and $2x \geq 6$

47. $-\dfrac{1}{2}s > -7$ and $\dfrac{1}{3}s < 12$

**Fair Game Review** What you learned in previous grades & lessons

Solve the equation. Check your solution. *(Section 3.5)*

48. $-2w + 4 = -12$

49. $\dfrac{v}{5} - 6 = 3$

50. $3(x - 1) = 18$

51. $\dfrac{m + 200}{4} = 51$

52. **MULTIPLE CHOICE** What is the value of $\dfrac{2}{3} + \left(-\dfrac{5}{7}\right)$? *(Section 2.2)*

   Ⓐ $-\dfrac{3}{4}$       Ⓑ $-\dfrac{1}{21}$       Ⓒ $\dfrac{7}{10}$       Ⓓ $1\dfrac{8}{21}$

## Essential Question

How can you use an inequality to describe the dimensions of a figure?

### 1 ACTIVITY: Areas and Perimeters of Figures

**Work with a partner.**

- Use the given condition to choose the inequality that you can use to find the possible values of the variable. Justify your answer.

- Write four values of the variable that satisfy the inequality you chose.

**a.** You want to find the values of $x$ so that the area of the rectangle is more than 22 square units.

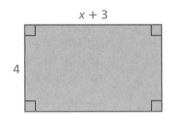

$x + 3$

$4$

| $4x + 12 > 22$ | $4x + 3 > 22$ |
| $4x + 12 \geq 22$ | $2x + 14 > 22$ |

**b.** You want to find the values of $x$ so that the perimeter of the rectangle is greater than or equal to 28 units.

| $x + 7 \geq 28$ | $4x + 12 \geq 28$ | $2x + 14 \geq 28$ | $2x + 14 \leq 28$ |

**c.** You want to find the values of $y$ so that the area of the parallelogram is fewer than 41 square units.

| $5y + 7 < 41$ | $5y + 35 < 41$ |
| $5y + 7 \leq 41$ | $5y + 35 \leq 41$ |

$y + 7$

$5$

**COMMON CORE**

**Inequalities**

In this lesson, you will
- solve multi-step inequalities.
- solve real-life problems.

Learning Standard
7.EE.4b

**d.** You want to find the values of $z$ so that the area of the trapezoid is at most 100 square units.

$10$

$z$    $6$

| $5z + 30 \leq 100$ | $10z + 30 \leq 100$ |
| $5z + 30 < 100$ | $10z + 30 < 100$ |

## 2

**ACTIVITY: Volumes of Rectangular Prisms**

**Work with a partner.**

- **Use the given condition to choose the inequality that you can use to find the possible values of the variable. Justify your answer.**

- **Write four values of the variable that satisfy the inequality you chose.**

**Math Practice 6**

**State the Meaning of Symbols**

What inequality symbols do the phrases *at least* and *no more than* represent? Explain.

**a.** You want to find the values of $x$ so that the volume of the rectangular prism is at least 50 cubic units.

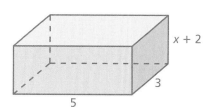

| $15x + 30 > 50$ | $x + 10 \geq 50$ | $15x + 30 \geq 50$ | $15x + 2 \geq 50$ |

**b.** You want to find the values of $x$ so that the volume of the rectangular prism is no more than 36 cubic units.

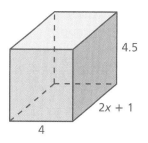

| $8x + 4 < 36$ | $36x + 18 < 36$ | $2x + 9.5 \leq 36$ | $36x + 18 \leq 36$ |

## What Is Your Answer?

**3. IN YOUR OWN WORDS** How can you use an inequality to describe the dimensions of a figure?

**4.** Use what you know about solving equations and inequalities to describe how you can solve a two-step inequality. Give an example to support your explanation.

**Practice**

Use what you learned about solving two-step inequalities to complete Exercises 3 and 4 on page 150.

You can solve two-step inequalities in the same way you solve two-step equations.

**EXAMPLE 1** **Solving Two-Step Inequalities**

a. **Solve $5x - 4 \geq 11$. Graph the solution.**

| | | |
|---|---|---|
| | $5x - 4 \geq 11$ | Write the inequality. |
| Step 1: Undo the subtraction. → | $\underline{+\,4 \quad +\,4}$ | Addition Property of Inequality |
| | $5x \geq 15$ | Simplify. |
| Step 2: Undo the multiplication. → | $\dfrac{5x}{5} \geq \dfrac{15}{5}$ | Division Property of Inequality |
| | $x \geq 3$ | Simplify. |

∴ The solution is $x \geq 3$.

$x \geq 3$

$x = 0$ is *not* a solution.

$x = 4$ is a solution.

b. **Solve $\dfrac{b}{-3} + 4 < 13$. Graph the solution.**

| | | |
|---|---|---|
| | $\dfrac{b}{-3} + 4 < 13$ | Write the inequality. |
| Step 1: Undo the addition. → | $\underline{\quad -\,4 \qquad -\,4\quad}$ | Subtraction Property of Inequality |
| | $\dfrac{b}{-3} < 9$ | Simplify. |
| Step 2: Undo the division. → | $-3 \cdot \dfrac{b}{-3} > -3 \cdot 9$ | Use the Multiplication Property of Inequality. Reverse the inequality symbol. |
| | $b > -27$ | Simplify. |

∴ The solution is $b > -27$.

$b > -27$

● **On Your Own**

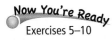
**Now You're Ready**
Exercises 5–10

**Solve the inequality. Graph the solution.**

1. $6y - 7 > 5$    2. $4 - 3d \geq 19$    3. $\dfrac{w}{-4} + 8 > 9$

**Which graph represents the solution of $-7(x + 3) \le 28$?**

**(A)**

-10  -9  -8  -7  -6  -5  -4

**(B)**
-10  -9  -8  -7  -6  -5  -4

**(C)**
4  5  6  7  8  9  10

**(D)**
4  5  6  7  8  9  10

| | | |
|---|---|---|
| $-7(x + 3) \le 28$ | | Write the inequality. |
| $-7x - 21 \le 28$ | | Distributive Property |

**Step 1: Undo the subtraction.** → $\underline{+\,21 \quad +\,21}$     Addition Property of Inequality

$-7x \le 49$     Simplify.

**Step 2: Undo the multiplication.** → $\dfrac{-7x}{-7} \ge \dfrac{49}{-7}$     Use the Division Property of Inequality. Reverse the inequality symbol.

$x \ge -7$     Simplify.

⋮• The correct answer is **(B)**.

---

**Progress Report**

| Month | Pounds Lost |
|-------|-------------|
| 1 | 12 |
| 2 | 9 |
| 3 | 5 |
| 4 | 8 |

**A contestant in a weight-loss competition wants to lose an average of at least 8 pounds per month during a 5-month period. How many pounds must the contestant lose in the fifth month to meet the goal?**

Write and solve an inequality. Let $x$ be the number of pounds lost in the fifth month.

$\dfrac{12 + 9 + 5 + 8 + x}{5} \ge 8$

> The phrase *at least* means *greater than or equal to*.

$\dfrac{34 + x}{5} \ge 8$     Simplify.

$5 \cdot \dfrac{34 + x}{5} \ge 5 \cdot 8$     Multiplication Property of Inequality

$34 + x \ge 40$     Simplify.

$x \ge 6$     Subtract 34 from each side.

**Remember**

In Example 3, the average is equal to the sum of the pounds lost divided by the number of months.

⋮• So, the contestant must lose at least 6 pounds to meet the goal.

**On Your Own**

*Now You're Ready*
Exercises 12–17

**Solve the inequality. Graph the solution.**

**4.** $2(k - 5) < 6$     **5.** $-4(n - 10) < 32$     **6.** $-3 \le 0.5(8 + y)$

**7. WHAT IF?** In Example 3, the contestant wants to lose an average of at least 9 pounds per month. How many pounds must the contestant lose in the fifth month to meet the goal?

## Vocabulary and Concept Check

1. **WRITING** Compare and contrast solving two-step inequalities and solving two-step equations.

2. **OPEN-ENDED** Describe how to solve the inequality $3(a + 5) < 9$.

## Practice and Problem Solving

**Match the inequality with its graph.**

3. $\dfrac{t}{3} - 1 \geq -3$

A. 
![number line with open circle at -6, shaded left, marks -9 to -2]

B. 
![number line with closed circle at -6, shaded left, marks -9 to -2]

C. 
![number line with closed circle at -6, shaded right, marks -9 to -2]

4. $5x + 7 \leq 32$

A. 
![number line with closed circle at 5, shaded left, marks 2 to 9]

B. 
![number line with open circle at 5, shaded right, marks 2 to 9]

C. 
![number line with closed circle at 5, shaded left, marks 2 to 9]

**Solve the inequality. Graph the solution.**

① 5. $8y - 5 < 3$

6. $3p + 2 \geq -10$

7. $2 > 8 - \dfrac{4}{3}h$

8. $-2 > \dfrac{m}{6} - 7$

9. $-1.2b - 5.3 \geq 1.9$

10. $-1.3 \geq 2.9 - 0.6r$

11. **ERROR ANALYSIS** Describe and correct the error in solving the inequality.

$\dfrac{x}{3} + 4 < 6$

$x + 4 < 18$

$x < 14$

**Solve the inequality. Graph the solution.**

② 12. $5(g + 4) > 15$

13. $4(w - 6) \leq -12$

14. $-8 \leq \dfrac{2}{5}(k - 2)$

15. $-\dfrac{1}{4}(d + 1) < 2$

16. $7.2 > 0.9(n + 8.6)$

17. $20 \geq -3.2(c - 4.3)$

10 cm

18. **UNICYCLE** The first jump in a unicycle high-jump contest is shown. The bar is raised 2 centimeters after each jump. Solve the inequality $2n + 10 \geq 26$ to find the number of additional jumps needed to meet or exceed the goal of clearing a height of 26 centimeters.

**Solve the inequality. Graph the solution.**

**19.** $9x - 4x + 4 \geq 36 - 12$

**20.** $3d - 7d + 2.8 < 5.8 - 27$

**21.** **SCUBA DIVER** A scuba diver is at an elevation of $-38$ feet. The diver starts moving at a rate of $-12$ feet per minute. Write and solve an inequality that represents how long it will take the diver to reach an elevation deeper than $-200$ feet.

**22.** **KILLER WHALES** A killer whale has eaten 75 pounds of fish today. It needs to eat at least 140 pounds of fish each day.

    **a.** A bucket holds 15 pounds of fish. Write and solve an inequality that represents how many more buckets of fish the whale needs to eat.

    **b.** Should the whale eat *four* or *five* more buckets of fish? Explain.

**23.** **REASONING** A student theater charges $9.50 per ticket.

    **a.** The theater has already sold 70 tickets. Write and solve an inequality that represents how many more tickets the theater needs to sell to earn at least $1000.

    **b.** The theater increases the ticket price by $1. Without solving an inequality, describe how this affects the total number of tickets needed to earn at least $1000.

**24.** **Problem Solving** For what values of $r$ will the area of the shaded region be greater than or equal to 12 square units?

 **Fair Game Review** *What you learned in previous grades & lessons*

**Find the missing values in the ratio table. Then write the equivalent ratios.**
*(Skills Review Handbook)*

**25.**

| Flutes | 7 | | 28 |
|---|---|---|---|
| Clarinets | 4 | 12 | |

**26.**

| Boys | 6 | 3 | |
|---|---|---|---|
| Girls | 10 | | 50 |

**27.** **MULTIPLE CHOICE** What is the volume of the cube?
*(Skills Review Handbook)*

    **Ⓐ** 8 ft$^3$         **Ⓑ** 16 ft$^3$

    **Ⓒ** 24 ft$^3$       **Ⓓ** 32 ft$^3$

**Check It Out**
Progress Check
BigIdeasMath.com

**Solve the inequality. Graph the solution.**    *(Section 4.3 and Section 4.4)*

**1.** $3p \leq 18$

**2.** $2x > -\dfrac{3}{5}$

**3.** $\dfrac{r}{3} \geq -5$

**4.** $-\dfrac{z}{8} < 1.5$

**5.** $3n + 2 \leq 11$

**6.** $-2 < 5 - \dfrac{k}{2}$

**7.** $1.3m - 3.8 < -1.2$

**8.** $4.8 \geq 0.3(12 - y)$

**Write the word sentence as an inequality. Then solve the inequality.**    *(Section 4.3)*

**9.** The quotient of a number and 5 is less than 4.

**10.** Six times a number is at least $-14$.

**11.** **PEPPERS** You have $18 to buy peppers. Peppers cost $1.50 each. Write and solve an inequality that represents the number of peppers you can buy.    *(Section 4.3)*

**12.** **MOVIES** You have a gift card worth $90. You want to buy several movies that cost $12 each. Write and solve an inequality that represents the number of movies you can buy and still have at least $30 on the gift card.    *(Section 4.4)*

**13.** **ORANGES** Your class sells boxes of oranges to raise $500 for a field trip. You earn $6.25 for each box of oranges sold. Write and solve an inequality that represents the number of boxes your class must sell to meet or exceed the fundraising goal.    *(Section 4.3)*

**14.** **FENCE** You want to put up a fence that encloses a triangular region with an area greater than or equal to 60 square feet. What is the least possible value of $c$? Explain.    *(Section 4.3)*

c ft

12 ft

## Review Key Vocabulary

inequality, *p. 126*

solution set, *p. 126*

solution of an inequality, *p. 126*

graph of an inequality, *p. 127*

## Review Examples and Exercises

**4.1** **Writing and Graphing Inequalities** *(pp. 124–129)*

a. **Six plus a number *x* is at most** $-\dfrac{1}{4}$**. Write this word sentence as an inequality.**

Six plus a number *x*   is at most   $-\dfrac{1}{4}$.

$$6 + x \qquad \leq \qquad -\dfrac{1}{4}$$

∴ An inequality is $6 + x \leq -\dfrac{1}{4}$.

b. **Graph *m* > 3.**

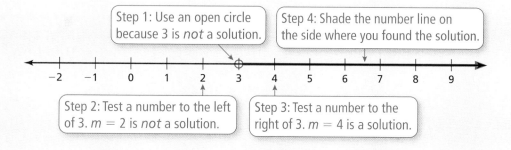

Step 1: Use an open circle because 3 is *not* a solution.

Step 4: Shade the number line on the side where you found the solution.

Step 2: Test a number to the left of 3. *m* = 2 is *not* a solution.

Step 3: Test a number to the right of 3. *m* = 4 is a solution.

### Exercises

**Write the word sentence as an inequality.**

1. A number *w* is greater than −3.

2. A number *y* minus $\dfrac{1}{2}$ is no more than $-\dfrac{3}{2}$.

**Tell whether the given value is a solution of the inequality.**

3. $5 + j > 8; j = 7$

4. $6 \div n \leq -5; n = -3$

**Graph the inequality on a number line.**

5. $q > -1.3$

6. $s < 1\dfrac{3}{4}$

7. **BUMPER CARS** You must be at least 42 inches tall to ride the bumper cars at an amusement park. Write an inequality that represents this situation.

## 4.2 Solving Inequalities Using Addition or Subtraction (pp. 130–135)

**Solve $-5 < m - 3$. Graph the solution.**

| | |
|---|---|
| $-5 < m - 3$ | Write the inequality. |
| $\underline{+\ 3 \qquad\quad +\ 3}$ | Addition Property of Inequality |
| $-2 < m$ | Simplify. |

*Undo the subtraction.*

The solution is $m > -2$.

$m > -2$

$m = -3$ is *not* a solution.

$m = 3$ is a solution.

### Exercises

**Solve the inequality. Graph the solution.**

**8.** $d + 12 < 19$    **9.** $t - 4 \leq -14$    **10.** $-8 \leq z + 6.4$

## 4.3 Solving Inequalities Using Multiplication or Division (pp. 138–145)

**Solve $\dfrac{c}{-3} \geq -2$. Graph the solution.**

| | |
|---|---|
| $\dfrac{c}{-3} \geq -2$ | Write the inequality. |
| $-3 \cdot \dfrac{c}{-3} \leq -3 \cdot (-2)$ | Use the Multiplication Property of Inequality. Reverse the inequality symbol. |
| $c \leq 6$ | Simplify. |

*Undo the division.*

The solution is $c \leq 6$.

$c \leq 6$

$c = 3$ is a solution.

$c = 9$ is *not* a solution.

### Exercises

**Solve the inequality. Graph the solution.**

**11.** $6q < -18$    **12.** $-\dfrac{r}{3} \leq 6$    **13.** $-4 > -\dfrac{4}{3}s$

## 4.4 Solving Two-Step Inequalities (pp. 146–151)

**a.** **Solve $6x - 8 \leq 10$. Graph the solution.**

$$6x - 8 \leq 10$$  Write the inequality.

Step 1: Undo the subtraction. $\longrightarrow$ $\underline{+\ 8 \quad +\ 8}$  Addition Property of Inequality

$$6x \leq 18$$  Simplify.

Step 2: Undo the multiplication. $\longrightarrow$ $\dfrac{6x}{6} \leq \dfrac{18}{6}$  Division Property of Inequality

$$x \leq 3$$  Simplify.

∴ The solution is $x \leq 3$.

$x = 0$ is a solution.  $x = 5$ is *not* a solution.

**b.** **Solve $\dfrac{q}{-4} + 7 < 11$. Graph the solution.**

$$\dfrac{q}{-4} + 7 < 11$$  Write the inequality.

Step 1: Undo the addition. $\longrightarrow$ $\underline{-\ 7 \quad -\ 7}$  Subtraction Property of Inequality

$$\dfrac{q}{-4} < 4$$  Simplify.

Step 2: Undo the division. $\longrightarrow$ $-4 \cdot \dfrac{q}{-4} > -4 \cdot 4$  Use the Multiplication Property of Inequality. Reverse the inequality symbol.

$$q > -16$$  Simplify.

∴ The solution is $q > -16$.

$q = -20$ is *not* a solution.  $q = -12$ is a solution.

### Exercises

**Solve the inequality. Graph the solution.**

**14.** $3x + 4 > 16$

**15.** $\dfrac{z}{-2} - 6 \leq -2$

**16.** $-2t - 5 < 9$

**17.** $7(q + 2) < -77$

**18.** $-\dfrac{1}{3}(p + 9) \leq 4$

**19.** $1.2(j + 3.5) \geq 4.8$

**Write the word sentence as an inequality.**

1. A number $k$ plus 19.5 is less than or equal to 40.

2. A number $q$ multiplied by $\frac{1}{4}$ is greater than $-16$.

**Tell whether the given value is a solution of the inequality.**

3. $n - 3 \le 4$; $n = 7$

4. $-\frac{3}{7}m < 1$; $m = -7$

5. $-4c \ge 7$; $c = -2$

6. $-2.4m > -6.8$; $m = -3$

**Solve the inequality. Graph the solution.**

7. $w + 4 \le 3$

8. $x - 4 > -6$

9. $-\frac{2}{9} + y \le \frac{5}{9}$

10. $-6z \ge 36$

11. $-5.2 \ge \frac{p}{4}$

12. $4k - 8 \ge 20$

13. $\frac{4}{7} - b \ge -\frac{1}{7}$

14. $-0.6 > -0.3(d + 6)$

15. **SUGAR-FREE GUMBALLS** You have \$2.50. Each sugar-free gumball in a gumball machine costs \$0.25. Write and solve an inequality that represents the number of gumballs you can buy.

16. **PARTY** You can spend no more than \$100 on a party you are hosting. The cost per guest is \$8.

   a. Write and solve an inequality that represents the number of guests you can invite to the party.

   b. What is the greatest number of guests that you can invite to the party? Explain your reasoning.

17. **BASEBALL CARDS** You have \$30 to buy baseball cards. Each pack of cards costs \$5. Write and solve an inequality that represents the number of packs of baseball cards you can buy and still have at least \$10 left.

**1.** What is the value of the expression below when $x = -5$, $y = 3$, and $z = -1$? *(7.NS.3)*

$$\frac{x^2 - 3y}{z}$$

**A.** $-34$

**C.** $16$

**B.** $-16$

**D.** $34$

**2.** What is the value of the expression below? *(7.NS.2a)*

$$-\frac{3}{8} \cdot \frac{2}{5}$$

**F.** $-\dfrac{20}{3}$

**H.** $-\dfrac{15}{16}$

**G.** $-\dfrac{16}{15}$

**I.** $-\dfrac{3}{20}$

**3.** Which graph represents the inequality below? *(7.EE.4b)*

$$\frac{x}{-4} - 8 \geq -9$$

**A.**

**C.**

**B.**

**D.**

**4.** Which value of $p$ makes the equation below true? *(7.EE.4a)*

$$5(p + 6) = 25$$

**F.** $-1$

**H.** $11$

**G.** $3\dfrac{4}{5}$

**I.** $14$

**5.** You set up the lemonade stand. Your profit is equal to your revenue from lemonade sales minus your cost to operate the stand. Your cost is $8. How many cups of lemonade must you sell to earn a profit of $30? *(7.EE.4a)*

LEMONADE

Lemonade
50¢
per cup

**A.** 4

**C.** 60

**B.** 44

**D.** 76

**6.** Which value is a solution of the inequality below? *(7.EE.4b)*

$$3 - 2y < 7$$

**F.** −6

**H.** −2

**G.** −3

**I.** −1

**7.** What value of *y* makes the equation below true? *(7.EE.4a)*

$$12 - 3y = -6$$

**8.** What is the mean distance of the four points from −3? *(7.NS.3)*

**A.** $-\dfrac{1}{2}$

**C.** 3

**B.** $2\dfrac{1}{2}$

**D.** $7\dfrac{1}{8}$

**9.** Martin graphed the solution of the inequality $-4x + 18 > 6$ in the box below.

What should Martin do to correct the error that he made? *(7.EE.4b)*

**F.** Use an open circle at 3 and shade to the left of 3.

**G.** Use an open circle at 3 and shade to the right of 3.

**H.** Use a closed circle and shade to the right of 3.

**I.** Use an open circle and shade to the left of $-3$.

**10.**  What is the value of the expression below? *(7.NS.1c)*

$$\frac{5}{12} - \frac{7}{8}$$

**11.** You are selling T-shirts to raise money for a charity. You sell the T-shirts for $10 each. *(7.EE.4b)*

*Part A*  You have already sold 2 T-shirts. How many more T-shirts must you sell to raise at least $500? Explain.

*Part B*  Your friend is raising money for the same charity. He sells the T-shirts for $8 each. What is the total number of T-shirts he must sell to raise at least $500? Explain.

*Part C*  Who has to sell more T-shirts in total? How many more? Explain.

**12.** Which expression is equivalent to the expression below? *(7.NS.3)*

$$-\frac{2}{3} - \left(-\frac{4}{9}\right)$$

**A.** $-\frac{1}{3} + \frac{1}{9}$

**C.** $-\frac{1}{3} - \frac{7}{9}$

**B.** $-\frac{2}{3} \times \left(-\frac{1}{3}\right)$

**D.** $\frac{3}{2} \div \left(-\frac{1}{3}\right)$

# 5 Ratios and Proportions

"I am doing an experiment with slope. I want you to run up and down the board 10 times."

"Now with 2 more dog biscuits, do it again and we'll compare your rates."

"Dear Sir: I counted the number of bacon, cheese, and chicken dog biscuits in the box I bought."

"There were 16 bacon, 12 cheese, and only 8 chicken. That's a ratio of 4:3:2. Please go back to the original ratio of 1:1:1."

# What You Learned Before

...or possibly proportional to our stupidity!

"I wonder if our rate is proportional to the slope of the hill."

## Simplifying Fractions (4.NF.1)

**Example 1** Simplify $\dfrac{4}{8}$.

$$\dfrac{4 \div 4}{8 \div 4} = \dfrac{1}{2}$$

**Example 2** Simplify $\dfrac{10}{15}$.

$$\dfrac{10 \div 5}{15 \div 5} = \dfrac{2}{3}$$

## Identifying Equivalent Fractions (4.NF.1)

**Example 3** Is $\dfrac{1}{4}$ equivalent to $\dfrac{13}{52}$?

$$\dfrac{13 \div 13}{52 \div 13} = \dfrac{1}{4}$$

∴ $\dfrac{1}{4}$ is equivalent to $\dfrac{13}{52}$.

**Example 4** Is $\dfrac{30}{54}$ equivalent to $\dfrac{5}{8}$?

$$\dfrac{30 \div 6}{54 \div 6} = \dfrac{5}{9}$$

∴ $\dfrac{30}{54}$ is *not* equivalent to $\dfrac{5}{8}$.

## Solving Equations (6.EE.7)

**Example 5** Solve $12x = 168$.

| | |
|---|---|
| $12x = 168$ | Write the equation. |
| $\dfrac{12x}{12} = \dfrac{168}{12}$ | Division Property of Equality |
| $x = 14$ | Simplify. |

**Check**

$$12x = 168$$
$$12(14) \stackrel{?}{=} 168$$
$$168 = 168 \ \checkmark$$

### Try It Yourself

**Simplify.**

1. $\dfrac{12}{144}$
2. $\dfrac{15}{45}$
3. $\dfrac{75}{100}$
4. $\dfrac{16}{24}$

**Are the fractions equivalent? Explain.**

5. $\dfrac{15}{60} \stackrel{?}{=} \dfrac{3}{4}$
6. $\dfrac{2}{5} \stackrel{?}{=} \dfrac{24}{144}$
7. $\dfrac{15}{20} \stackrel{?}{=} \dfrac{3}{5}$
8. $\dfrac{2}{8} \stackrel{?}{=} \dfrac{16}{64}$

**Solve the equation. Check your solution.**

9. $\dfrac{y}{-5} = 3$
10. $0.6 = 0.2a$
11. $-2w = -9$
12. $\dfrac{1}{7}n = -4$

## Essential Question How do rates help you describe real-life problems?

### The Meaning of a Word ● Rate

When you rent snorkel gear at the beach, you should pay attention to the rental **rate**. The rental rate is in dollars per hour.

Snorkel Rentals
$8.75 per hour

Snorkel Rentals
$7.25 per hour

### 1 ACTIVITY: Finding Reasonable Rates

**Work with a partner.**

a. Match each description with a verbal rate.

b. Match each verbal rate with a numerical rate.

c. Give a reasonable numerical rate for each description. Then give an unreasonable rate.

| Description | Verbal Rate | Numerical Rate |
|---|---|---|
| Your running rate in a 100-meter dash | Dollars per year | $\dfrac{\text{in.}}{\text{yr}}$ |
| The fertilization rate for an apple orchard | Inches per year | $\dfrac{\text{lb}}{\text{acre}}$ |
| The average pay rate for a professional athlete | Meters per second | $\dfrac{\$}{\text{yr}}$ |
| The average rainfall rate in a rain forest | Pounds per acre | $\dfrac{\text{m}}{\text{sec}}$ |

**COMMON CORE**

**Ratios and Rates**

In this lesson, you will
- find ratios, rates, and unit rates.
- find ratios and rates involving ratios of fractions.

Learning Standards
7.RP.1
7.RP.3

### 2 ACTIVITY: Simplifying Expressions That Contain Fractions

**Work with a partner. Describe a situation where the given expression may apply. Show how you can rewrite each expression as a division problem. Then simplify and interpret your result.**

a. $\dfrac{\frac{1}{2}\,c}{4\,\text{fl oz}}$

b. $\dfrac{2\,\text{in.}}{\frac{3}{4}\,\text{sec}}$

c. $\dfrac{\frac{3}{8}\,c\text{ sugar}}{\frac{3}{5}\,c\text{ flour}}$

d. $\dfrac{\frac{5}{6}\,\text{gal}}{\frac{2}{3}\,\text{sec}}$

**ACTIVITY: Using Ratio Tables to Find Equivalent Rates**

Work with a partner. A communications satellite in orbit travels about 18 miles every 4 seconds.

a. Identify the rate in this problem.

b. Recall that you can use *ratio tables* to find and organize equivalent ratios and rates. Complete the ratio table below.

| Time (seconds) | 4 | 8 | 12 | 16 | 20 |
|---|---|---|---|---|---|
| Distance (miles) | | | | | |

c. How can you use a ratio table to find the speed of the satellite in miles per minute? miles per hour?

d. How far does the satellite travel in 1 second? Solve this problem (1) by using a ratio table and (2) by evaluating a quotient.

e. How far does the satellite travel in $\frac{1}{2}$ second? Explain your steps.

4 **ACTIVITY: Unit Analysis**

**Math Practice 7**

**View as Components**

What is the product of the numbers? What is the product of the units? Explain.

Work with a partner. Describe a situation where the product may apply. Then find each product and list the units.

a. $10 \text{ gal} \times \dfrac{22 \text{ mi}}{\text{gal}}$

b. $\dfrac{7}{2} \text{ lb} \times \dfrac{\$3}{\frac{1}{2} \text{ lb}}$

c. $\dfrac{1}{2} \text{ sec} \times \dfrac{30 \text{ ft}^2}{\text{sec}}$

## What Is Your Answer?

5. **IN YOUR OWN WORDS** How do rates help you describe real-life problems? Give two examples.

6. To estimate the annual salary for a given hourly pay rate, multiply by 2 and insert "000" at the end.

   **Sample:** $10 per hour is about $20,000 per year.

   a. Explain why this works. Assume the person is working 40 hours a week.

   b. Estimate the annual salary for an hourly pay rate of $8 per hour.

   c. You earn $1 million per month. What is your annual salary?

   d. Why is the cartoon funny?

"We had someone apply for the job. He says he would like $1 million a month, but will settle for $8 an hour."

**Practice**

Use what you discovered about ratios and rates to complete Exercises 7–10 on page 167.

## 5.1 Lesson

**Check It Out**
Lesson Tutorials
BigIdeasMath  com

**Key Vocabulary** 🔊
ratio, *p. 164*
rate, *p. 164*
unit rate, *p. 164*
complex fraction,
   *p. 165*

A **ratio** is a comparison of two quantities using division.

$\frac{3}{4}$, 3 to 4, 3 : 4

A **rate** is a ratio of two quantities with different units.

$\frac{60 \text{ miles}}{2 \text{ hours}}$

A rate with a denominator of 1 is called a **unit rate**.

$\frac{30 \text{ miles}}{1 \text{ hour}}$

---

**EXAMPLE** ❶ **Finding Ratios and Rates**

There are 45 males and 60 females in a subway car. The subway car travels 2.5 miles in 5 minutes.

**a. Find the ratio of males to females.**

$$\frac{\text{males}}{\text{females}} = \frac{45}{60} = \frac{3}{4}$$

⋮• The ratio of males to females is $\frac{3}{4}$.

**b. Find the speed of the subway car.**

$$2.5 \text{ miles in 5 minutes} = \frac{2.5 \text{ mi}}{5 \text{ min}} = \frac{2.5 \text{ mi} \div 5}{5 \text{ min} \div 5} = \frac{0.5 \text{ mi}}{1 \text{ min}}$$

⋮• The speed is 0.5 mile per minute.

---

**EXAMPLE** ❷ **Finding a Rate from a Ratio Table**

The ratio table shows the costs for different amounts of artificial turf. Find the unit rate in dollars per square foot.

|                          | × 4 | × 4 | × 4 |     |
|--------------------------|-----|-----|-----|-----|
| **Amount (square feet)** | 25  | 100 | 400 | 1600 |
| **Cost (dollars)**       | 100 | 400 | 1600 | 6400 |
|                          | × 4 | × 4 | × 4 |     |

Use a ratio from the table to find the unit rate.

$$\frac{\text{cost}}{\text{amount}} = \frac{\$100}{25 \text{ ft}^2} \qquad \text{Use the first ratio in the table.}$$

$$= \frac{\$4}{1 \text{ ft}^2} \qquad \text{Simplify.}$$

**Remember**

The abbreviation ft² means *square feet*.

⋮• So, the unit rate is $4 per square foot.

## On Your Own

Now You're Ready
Exercises 11–24

**1.** In Example 1, find the ratio of females to males.

**2.** In Example 1, find the ratio of females to total passengers.

**3.** The ratio table shows the distance that the *International Space Station* travels while orbiting Earth. Find the speed in miles per second.

| Time (seconds) | 3 | 6 | 9 | 12 |
|---|---|---|---|---|
| Distance (miles) | 14.4 | 28.8 | 43.2 | 57.6 |

A **complex fraction** has at least one fraction in the numerator, denominator, or both. You may need to simplify complex fractions when finding ratios and rates.

**EXAMPLE** ③ **Finding a Rate from a Graph**

**The graph shows the speed of a subway car. Find the speed in miles per minute. Compare the speed to the speed of the subway car in Example 1.**

**Subway Car Speed**

**Step 1:** Choose and interpret a point on the line.

The point $\left(\frac{1}{2}, \frac{1}{4}\right)$ indicates that the subway car travels

$\frac{1}{4}$ mile in $\frac{1}{2}$ minute.

**Step 2:** Find the speed.

$$\frac{\text{distance traveled}}{\text{elapsed time}} = \frac{\frac{1}{4} \leftarrow \text{miles}}{\frac{1}{2} \leftarrow \text{minutes}}$$

$$= \frac{1}{4} \div \frac{1}{2} \qquad \text{Rewrite the quotient.}$$

$$= \frac{1}{4} \cdot 2 = \frac{1}{2} \qquad \text{Simplify.}$$

∴ The speed of the subway car is $\frac{1}{2}$ mile per minute.

Because $\frac{1}{2}$ mile per minute = 0.5 mile per minute, the speeds of the two subway cars are the same.

## On Your Own

Now You're Ready
Exercise 28

**4.** You use the point $\left(3, 1\frac{1}{2}\right)$ to find the speed of the subway car. Does your answer change? Explain your reasoning.

EXAMPLE 4 Solving a Ratio Problem

You mix $\frac{1}{2}$ cup of yellow paint for every $\frac{3}{4}$ cup of blue paint to make 15 cups of green paint. How much yellow paint and blue paint do you use?

**Math Practice**

**Analyze Givens**
What information is given in the problem? How does this help you know that the ratio table needs a "total" column? Explain.

**Method 1:** The ratio of yellow paint to blue paint is $\frac{1}{2}$ to $\frac{3}{4}$. Use a ratio table to find an equivalent ratio in which the total amount of yellow paint and blue paint is 15 cups.

| Yellow (cups) | Blue (cups) | Total (cups) |
|---|---|---|
| $\frac{1}{2}$ | $\frac{3}{4}$ | $\frac{1}{2}+\frac{3}{4}=\frac{5}{4}$ |
| 2 | 3 | 5 |
| 6 | 9 | 15 |

$\times 4$    $\times 3$    $\times 4$    $\times 3$

So, you use 6 cups of yellow paint and 9 cups of blue paint.

**Method 2:** Use the fraction of the green paint that is made from yellow paint and the fraction of the green paint that is made from blue paint. You use $\frac{1}{2}$ cup of yellow paint for every $\frac{3}{4}$ cup of blue paint, so the fraction of the green paint that is made from yellow paint is

yellow → green

$$\frac{\frac{1}{2}}{\frac{1}{2}+\frac{3}{4}}=\frac{\frac{1}{2}}{\frac{5}{4}}=\frac{1}{2}\cdot\frac{4}{5}=\frac{2}{5}.$$

Similarly, the fraction of the green paint that is made from blue paint is

blue → green

$$\frac{\frac{3}{4}}{\frac{1}{2}+\frac{3}{4}}=\frac{\frac{3}{4}}{\frac{5}{4}}=\frac{3}{4}\cdot\frac{4}{5}=\frac{3}{5}.$$

So, you use $\frac{2}{5}\cdot15=6$ cups of yellow paint and $\frac{3}{5}\cdot15=9$ cups of blue paint.

## On Your Own

Now You're Ready
Exercises 33 and 34

5. How much yellow paint and blue paint do you use to make 20 cups of green paint?

## Vocabulary and Concept Check

1. **VOCABULARY** How can you tell when a rate is a unit rate?

2. **WRITING** Why do you think rates are usually written as unit rates?

3. **OPEN-ENDED** Write a real-life rate that applies to you.

**Estimate the unit rate.**

4. $74.75

Gloss White PAINT 5 gal

5. $1.19

GRAPE JUICE 12 fl oz

6. $2.35

12 Grade AA Eggs

 ## Practice and Problem Solving

**Find the product. List the units.**

7. $8 \text{ h} \times \dfrac{\$9}{\text{h}}$

8. $8 \text{ lb} \times \dfrac{\$3.50}{\text{lb}}$

9. $\dfrac{29}{2} \text{ sec} \times \dfrac{60 \text{ MB}}{\text{sec}}$

10. $\dfrac{3}{4} \text{ h} \times \dfrac{19 \text{ mi}}{\frac{1}{4} \text{ h}}$

**Write the ratio as a fraction in simplest form.**

① 11. 25 to 45

12. $63 : 28$

13. 35 girls : 15 boys

14. 51 correct : 9 incorrect

15. 16 dogs to 12 cats

16. $2\dfrac{1}{3}$ feet : $4\dfrac{1}{2}$ feet

**Find the unit rate.**

17. 180 miles in 3 hours

18. 256 miles per 8 gallons

19. $9.60 for 4 pounds

20. $4.80 for 6 cans

21. 297 words in 5.5 minutes

22. $21\dfrac{3}{4}$ meters in $2\dfrac{1}{2}$ hours

**Use the ratio table to find the unit rate with the specified units.**

② 23. servings per package

| Packages | 3 | 6 | 9 | 12 |
|---|---|---|---|---|
| Servings | 13.5 | 27 | 40.5 | 54 |

24. feet per year

| Years | 2 | 6 | 10 | 14 |
|---|---|---|---|---|
| Feet | 7.2 | 21.6 | 36 | 50.4 |

25. **DOWNLOAD** At 1:00 P.M., you have 24 megabytes of a movie. At 1:15 P.M., you have 96 megabytes. What is the download rate in megabytes per minute?

26. **POPULATION** In 2007, the U.S. population was 302 million people. In 2012, it was 314 million. What was the rate of population change per year?

27. **PAINTING** A painter can paint 350 square feet in 1.25 hours. What is the painting rate in square feet per hour?

③ 28. **TICKETS** The graph shows the cost of buying tickets to a concert.

    **a.** What does the point (4, 122) represent?

    **b.** What is the unit rate?

    **c.** What is the cost of buying 10 tickets?

29. **CRITICAL THINKING** Are the two statements equivalent? Explain your reasoning.

    ● The ratio of boys to girls is 2 to 3.

    ● The ratio of girls to boys is 3 to 2.

30. **TENNIS** A sports store sells three different packs of tennis balls. Which pack is the best buy? Explain.

31. **FLOORING** It costs $68 for 16 square feet of flooring. How much does it cost for 12 square feet of flooring?

32. **OIL SPILL** An oil spill spreads 25 square meters every $\frac{1}{6}$ hour. How much area does the oil spill cover after 2 hours?

④ 33. **JUICE** You mix $\frac{1}{4}$ cup of juice concentrate for every 2 cups of water to make 18 cups of juice. How much juice concentrate and water do you use?

34. **LANDSCAPING** A supplier sells $2\frac{1}{4}$ pounds of mulch for every $1\frac{1}{3}$ pounds of gravel. The supplier sells 172 pounds of mulch and gravel combined. How many pounds of each item does the supplier sell?

35. **HEART RATE** Your friend's heart beats 18 times in 15 seconds when at rest. While running, your friend's heart beats 25 times in 10 seconds.

    **a.** Find the heart rate in beats per minute at rest and while running.

    **b.** How many more times does your friend's heart beat in 3 minutes while running than while at rest?

**36. PRECISION** The table shows nutritional information for three beverages.

| Beverage | Serving Size | Calories | Sodium |
|---|---|---|---|
| Whole milk | 1 c | 146 | 98 mg |
| Orange juice | 1 pt | 210 | 10 mg |
| Apple juice | 24 fl oz | 351 | 21 mg |

    **a.** Which has the most calories per fluid ounce?

    **b.** Which has the least sodium per fluid ounce?

**37. RESEARCH** Fire hydrants are painted one of four different colors to indicate the rate at which water comes from the hydrant.

    **a.** Use the Internet to find the ranges of the rates for each color.

    **b.** Research why a firefighter needs to know the rate at which water comes out of a hydrant.

**38. PAINT** You mix $\frac{2}{5}$ cup of red paint for every $\frac{1}{4}$ cup of blue paint to make $1\frac{5}{8}$ gallons of purple paint.

    **a.** How much red paint and blue paint do you use?

    **b.** You decide that you want to make a lighter purple paint. You make the new mixture by adding $\frac{1}{10}$ cup of white paint for every $\frac{2}{5}$ cup of red paint and $\frac{1}{4}$ cup of blue paint. How much red paint, blue paint, and white paint do you use to make $\frac{3}{8}$ gallon of lighter purple paint?

**39.** **Critical Thinking** You and a friend start hiking toward each other from opposite ends of a 17.5-mile hiking trail. You hike $\frac{2}{3}$ mile every $\frac{1}{4}$ hour. Your friend hikes $2\frac{1}{3}$ miles per hour.

    **a.** Who hikes faster? How much faster?

    **b.** After how many hours do you meet?

    **c.** When you meet, who hiked farther? How much farther?

 **Fair Game Review** What you learned in previous grades & lessons

**Copy and complete the statement using <, >, or =.** *(Section 2.1)*

**40.** $\dfrac{9}{2}$ ⬜ $\dfrac{8}{3}$

**41.** $-\dfrac{8}{15}$ ⬜ $\dfrac{10}{18}$

**42.** $\dfrac{-6}{24}$ ⬜ $\dfrac{-2}{8}$

**43. MULTIPLE CHOICE** Which fraction is greater than $-\dfrac{2}{3}$ and less than $-\dfrac{1}{2}$? *(Section 2.1)*

    Ⓐ $-\dfrac{3}{4}$      Ⓑ $-\dfrac{7}{12}$      Ⓒ $-\dfrac{5}{12}$      Ⓓ $-\dfrac{3}{8}$

**Essential Question** How can proportions help you decide when things are "fair"?

### The Meaning of a Word ● Proportional

When you work toward a goal, your success is usually **proportional** to the amount of work you put in.

An equation stating that two ratios are equal is a **proportion**.

---

**1 ACTIVITY: Determining Proportions**

**Work with a partner. Tell whether the two ratios are equivalent. If they are not equivalent, change the next day to make the ratios equivalent. Explain your reasoning.**

**a.** On the first day, you pay $5 for 2 boxes of popcorn. The next day, you pay $7.50 for 3 boxes.

First Day   Next Day

$$\frac{\$5.00}{2 \text{ boxes}} \stackrel{?}{=} \frac{\$7.50}{3 \text{ boxes}}$$

**b.** On the first day, it takes you $3\frac{1}{2}$ hours to drive 175 miles. The next day, it takes you 5 hours to drive 200 miles.

First Day   Next Day

$$\frac{3\frac{1}{2} \text{ h}}{175 \text{ mi}} \stackrel{?}{=} \frac{5 \text{ h}}{200 \text{ mi}}$$

**COMMON CORE**

**Proportions**

In this lesson, you will
- use equivalent ratios to determine whether two ratios form a proportion.
- use the Cross Products Property to determine whether two ratios form a proportion.

Learning Standard
7.RP.2a

**c.** On the first day, you walk 4 miles and burn 300 calories. The next day, you walk $3\frac{1}{3}$ miles and burn 250 calories.

First Day   Next Day

$$\frac{4 \text{ mi}}{300 \text{ cal}} \stackrel{?}{=} \frac{3\frac{1}{3} \text{ mi}}{250 \text{ cal}}$$

**d.** On the first day, you paint 150 square feet in $2\frac{1}{2}$ hours. The next day, you paint 200 square feet in 4 hours.

First Day   Next Day

$$\frac{150 \text{ ft}^2}{2\frac{1}{2} \text{ h}} \stackrel{?}{=} \frac{200 \text{ ft}^2}{4 \text{ h}}$$

## 2 · ACTIVITY: Checking a Proportion

**Work with a partner.**

**a.** It is said that "one year in a dog's life is equivalent to seven years in a human's life." Explain why Newton thinks he has a score of 105 points. Did he solve the proportion correctly?

$$\frac{1 \text{ year}}{7 \text{ years}} \stackrel{?}{=} \frac{15 \text{ points}}{105 \text{ points}}$$

**b.** If Newton thinks his score is 98 points, how many points does he actually have? Explain your reasoning.

"I got 15 on my online test. That's 105 in dog points! Isn't that an A+?"

## 3 · ACTIVITY: Determining Fairness

> **Math Practice** 3
>
> **Justify Conclusions**
> What information can you use to justify your conclusion?

**Work with a partner. Write a ratio for each sentence. Compare the ratios. If they are equal, then the answer is "It is fair." If they are not equal, then the answer is "It is not fair." Explain your reasoning.**

**a.**

| You pay $184 for 2 tickets to a concert. | & | I pay $266 for 3 tickets to the same concert. | ➤ Is this fair? |

**b.**

| You get 75 points for answering 15 questions correctly. | & | I get 70 points for answering 14 questions correctly. | ➤ Is this fair? |

**c.**

| You trade 24 football cards for 15 baseball cards. | & | I trade 20 football cards for 32 baseball cards. | ➤ Is this fair? |

## What Is Your Answer?

**4.** Find a recipe for something you like to eat. Then show how two of the ingredient amounts are proportional when you double or triple the recipe.

**5.** **IN YOUR OWN WORDS** How can proportions help you decide when things are "fair"? Give an example.

**Practice** ➤ Use what you discovered about proportions to complete Exercises 15–20 on page 174.

## 5.2 Lesson

**Check It Out**
Lesson Tutorials
BigIdeasMath ✓com

Key Vocabulary 🔊
proportion, *p. 172*
proportional, *p. 172*
cross products, *p. 173*

### 🔑 Key Idea

**Proportions**

**Words**  A **proportion** is an equation stating that two ratios are equivalent. Two quantities that form a proportion are **proportional**.

**Numbers**  $\dfrac{2}{3} = \dfrac{4}{6}$    The proportion is read "2 is to 3 as 4 is to 6."

---

**EXAMPLE 1** **Determining Whether Ratios Form a Proportion**

Tell whether $\dfrac{6}{4}$ and $\dfrac{8}{12}$ form a proportion.

Compare the ratios in simplest form.

$$\frac{6}{4} = \frac{6 \div 2}{4 \div 2} = \frac{3}{2}$$  ← The ratios are *not* equivalent.

$$\frac{8}{12} = \frac{8 \div 4}{12 \div 4} = \frac{2}{3}$$  ←

∴ So, $\dfrac{6}{4}$ and $\dfrac{8}{12}$ do *not* form a proportion.

---

**EXAMPLE 2** **Determining Whether Two Quantities Are Proportional**

Tell whether $x$ and $y$ are proportional.

| $x$ | $y$ |
|---|---|
| $\frac{1}{2}$ | 3 |
| 1 | 6 |
| $\frac{3}{2}$ | 9 |
| 2 | 12 |

Compare each ratio $x$ to $y$ in simplest form.

$$\frac{\frac{1}{2}}{3} = \frac{1}{6} \qquad \frac{1}{6} \qquad \frac{\frac{3}{2}}{9} = \frac{1}{6} \qquad \frac{2}{12} = \frac{1}{6}$$

The ratios are equivalent.

∴ So, $x$ and $y$ are proportional.

**Reading**

Two quantities that are proportional are in a *proportional relationship.*

### ⬤ On Your Own

**Now You're Ready**
Exercises 5–14

**Tell whether the ratios form a proportion.**

1. $\dfrac{1}{2}, \dfrac{5}{10}$    2. $\dfrac{4}{6}, \dfrac{18}{24}$    3. $\dfrac{10}{3}, \dfrac{5}{6}$    4. $\dfrac{25}{20}, \dfrac{15}{12}$

5. Tell whether $x$ and $y$ are proportional.

| Birdhouses Built, $x$ | 1 | 2 | 4 | 6 |
|---|---|---|---|---|
| Nails Used, $y$ | 12 | 24 | 48 | 72 |

---

**Key Ideas**

**Cross Products**

In the proportion $\dfrac{a}{b} = \dfrac{c}{d}$, the products $a \cdot d$ and $b \cdot c$ are called **cross products**.

**Cross Products Property**

**Words**   The cross products of a proportion are equal.

**Numbers**

$$\dfrac{2}{3} \bowtie \dfrac{4}{6}$$

$$2 \cdot 6 = 3 \cdot 4$$

**Algebra**

$$\dfrac{a}{b} \bowtie \dfrac{c}{d}$$

$$ad = bc,$$
where $b \neq 0$ and $d \neq 0$

**Study Tip**

You can use the Multiplication Property of Equality to show that the cross products are equal.

$$\dfrac{a}{b} = \dfrac{c}{d}$$

$$\cancel{b}d \cdot \dfrac{a}{\cancel{b}} = b\cancel{d} \cdot \dfrac{c}{\cancel{d}}$$

$$ad = bc$$

**EXAMPLE** ③ **Identifying Proportional Relationships**

1 length   1 lap

**You swim your first 4 laps in 2.4 minutes. You complete 16 laps in 12 minutes. Is the number of laps proportional to your time?**

**Method 1:**   Compare unit rates.

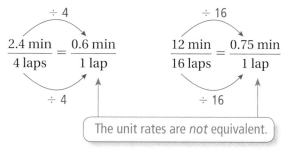

$$\dfrac{2.4 \text{ min}}{4 \text{ laps}} = \dfrac{0.6 \text{ min}}{1 \text{ lap}} \qquad \dfrac{12 \text{ min}}{16 \text{ laps}} = \dfrac{0.75 \text{ min}}{1 \text{ lap}}$$

The unit rates are *not* equivalent.

∴   So, the number of laps is *not* proportional to the time.

**Method 2:**   Use the Cross Products Property.

$$\dfrac{2.4 \text{ min}}{4 \text{ laps}} \stackrel{?}{=} \dfrac{12 \text{ min}}{16 \text{ laps}}$$   Test to see if the rates are equivalent.

$$2.4 \cdot 16 \stackrel{?}{=} 4 \cdot 12$$   Find the cross products.

$$38.4 \neq 48$$   The cross products are *not* equal.

∴   So, the number of laps is *not* proportional to the time.

**On Your Own**

**Now You're Ready**
Exercises 15–20

**6.** You read the first 20 pages of a book in 25 minutes. You read 36 pages in 45 minutes. Is the number of pages read proportional to your time?

**Vocabulary and Concept Check**

1. **VOCABULARY** What does it mean for two ratios to form a proportion?

2. **VOCABULARY** What are two ways you can tell that two ratios form a proportion?

3. **OPEN-ENDED** Write two ratios that are equivalent to $\frac{3}{5}$.

4. **WHICH ONE DOESN'T BELONG?** Which ratio does *not* belong with the other three? Explain your reasoning.

$$\frac{4}{10} \qquad \frac{2}{5} \qquad \frac{3}{5} \qquad \frac{6}{15}$$

 **Practice and Problem Solving**

**Tell whether the ratios form a proportion.**

1  **5.** $\frac{1}{3}, \frac{7}{21}$     **6.** $\frac{1}{5}, \frac{6}{30}$     **7.** $\frac{3}{4}, \frac{24}{18}$     **8.** $\frac{2}{5}, \frac{40}{16}$

   **9.** $\frac{48}{9}, \frac{16}{3}$     **10.** $\frac{18}{27}, \frac{33}{44}$     **11.** $\frac{7}{2}, \frac{16}{6}$     **12.** $\frac{12}{10}, \frac{14}{12}$

**Tell whether $x$ and $y$ are proportional.**

2  **13.**

| $x$ | 1 | 2 | 3 | 4 |
|---|---|---|---|---|
| $y$ | 7 | 8 | 9 | 10 |

**14.**

| $x$ | 2 | 4 | 6 | 8 |
|---|---|---|---|---|
| $y$ | 5 | 10 | 15 | 20 |

**Tell whether the two rates form a proportion.**

3  **15.** 7 inches in 9 hours; 42 inches in 54 hours

   **16.** 12 players from 21 teams; 15 players from 24 teams

   **17.** 440 calories in 4 servings; 300 calories in 3 servings

   **18.** 120 units made in 5 days; 88 units made in 4 days

   **19.** 66 wins in 82 games; 99 wins in 123 games

   **20.** 68 hits in 172 at bats; 43 hits in 123 at bats

   **21.** **FITNESS** You can do 90 sit-ups in 2 minutes. Your friend can do 135 sit-ups in 3 minutes. Do these rates form a proportion? Explain.

   **22.** **HEART RATES** Find the heart rates of you and your friend. Do these rates form a proportion? Explain.

|  | Heartbeats | Seconds |
|---|---|---|
| You | 22 | 20 |
| Friend | 18 | 15 |

**Tell whether the ratios form a proportion.**

**23.** $\dfrac{2.5}{4}, \dfrac{7}{11.2}$

**24.** 2 to 4, 11 to $\dfrac{11}{2}$

**25.** $2 : \dfrac{4}{5}, \dfrac{3}{4} : \dfrac{3}{10}$

**26. PAY RATE** You earn $56 walking your neighbor's dog for 8 hours. Your friend earns $36 painting your neighbor's fence for 4 hours.

   **a.** What is your pay rate?

   **b.** What is your friend's pay rate?

   **c.** Are the pay rates equivalent? Explain.

**27. GEOMETRY** Are the heights and bases of the two triangles proportional? Explain.

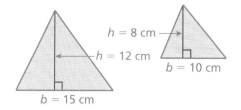

**28. BASEBALL** A pitcher coming back from an injury limits the number of pitches thrown in bull pen sessions as shown.

   **a.** Which quantities are proportional?

   **b.** How many pitches that are not curveballs do you think the pitcher will throw in Session 5?

| Session Number, $x$ | Pitches, $y$ | Curveballs, $z$ |
|---|---|---|
| 1 | 10 | 4 |
| 2 | 20 | 8 |
| 3 | 30 | 12 |
| 4 | 40 | 16 |

**29. NAIL POLISH** A specific shade of red nail polish requires 7 parts red to 2 parts yellow. A mixture contains 35 quarts of red and 8 quarts of yellow. How can you fix the mixture to make the correct shade of red?

**30. COIN COLLECTION** The ratio of quarters to dimes in a coin collection is $5 : 3$. You add the same number of new quarters as dimes to the collection.

   **a.** Is the ratio of quarters to dimes still $5 : 3$?

   **b.** If so, illustrate your answer with an example. If not, show why with a "counterexample."

**31. AGE** You are 13 years old, and your cousin is 19 years old. As you grow older, is your age proportional to your cousin's age? Explain your reasoning.

**32.** *Critical Thinking* Ratio $A$ is equivalent to Ratio $B$. Ratio $B$ is equivalent to Ratio $C$. Is Ratio $A$ equivalent to Ratio $C$? Explain.

 **Fair Game Review** What you learned in previous grades & lessons

**Add or subtract.** *(Section 1.2 and Section 1.3)*

**33.** $-28 + 15$

**34.** $-6 + (-11)$

**35.** $-10 - 8$

**36.** $-17 - (-14)$

**37. MULTIPLE CHOICE** Which fraction is not equivalent to $\dfrac{2}{6}$? *(Skills Review Handbook)*

   **Ⓐ** $\dfrac{1}{3}$

   **Ⓑ** $\dfrac{12}{36}$

   **Ⓒ** $\dfrac{4}{12}$

   **Ⓓ** $\dfrac{6}{9}$

Recall that you can graph the values from a ratio table.

The structure in the ratio table shows why the graph has a constant *rate of change*. You can use the constant rate of change to show that the graph passes through the origin. The graph of every proportional relationship is a line through the origin.

**EXAMPLE 1 Determining Whether Two Quantities Are Proportional**

**Use a graph to tell whether *x* and *y* are in a proportional relationship.**

a.

| *x* | 2 | 4 | 6 |
|---|---|---|---|
| *y* | 6 | 8 | 10 |

b.

| *x* | 1 | 2 | 3 |
|---|---|---|---|
| *y* | 2 | 4 | 6 |

Plot (2, 6), (4, 8), and (6, 10). Draw a line through the points.

Plot (1, 2), (2, 4), and (3, 6). Draw a line through the points.

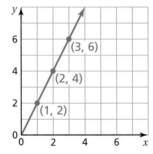

COMMON CORE

**Proportions**

In this extension, you will

• use graphs to determine whether two ratios form a proportion.

• interpret graphs of proportional relationships.

Learning Standards
7.RP.2a
7.RP.2b
7.RP.2d

The graph is a line that does not pass through the origin.

⋮ So, *x* and *y* are not in a proportional relationship.

The graph is a line that passes through the origin.

⋮ So, *x* and *y* are in a proportional relationship.

● **Practice**

**Use a graph to tell whether *x* and *y* are in a proportional relationship.**

1.

| *x* | 1 | 2 | 3 | 4 |
|---|---|---|---|---|
| *y* | 3 | 4 | 5 | 6 |

2.

| *x* | 1 | 3 | 5 | 7 |
|---|---|---|---|---|
| *y* | 0.5 | 1.5 | 2.5 | 3.5 |

EXAMPLE 2 Interpreting the Graph of a Proportional Relationship

The graph shows that the distance traveled by the Mars rover *Curiosity* is proportional to the time traveled. Interpret each plotted point in the graph.

**Curiosity Rover at Top Speed**

(0, 0): The rover travels 0 inches in 0 seconds.

**Study Tip**

In the graph of a proportional relationship, you can find the unit rate $r$ from the point $(1, r)$.

(1, 1.5): The rover travels 1.5 inches in 1 second. So, the unit rate is 1.5 inches per second.

(3, 4.5): The rover travels 4.5 inches in 3 seconds. Because the relationship is proportional, you can also use this point to find the unit rate.

$$\frac{4.5 \text{ in.}}{3 \text{ sec}} = \frac{1.5 \text{ in.}}{1 \text{ sec}}, \text{ or } 1.5 \text{ inches per second}$$

## Practice

Interpret each plotted point in the graph of the proportional relationship.

3.

Money

4.

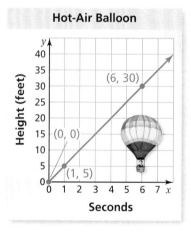

Hot-Air Balloon

**Tell whether $x$ and $y$ are in a proportional relationship. If so, find the unit rate.**

5.

| x (hours) | 1 | 4 | 7 | 10 |
|-----------|---|---|---|----|
| y (feet) | 5 | 20 | 35 | 50 |

6. Let $y$ be the temperature $x$ hours after midnight. The temperature is 60°F at midnight and decreases 2°F every $\frac{1}{2}$ hour.

7. **REASONING** The graph of a proportional relationship passes through (12, 16) and (1, $y$). Find $y$.

8. **MOVIE RENTAL** You pay $1 to rent a movie plus an additional $0.50 per day until you return the movie. Your friend pays $1.25 per day to rent a movie.

   a. Make tables showing the costs to rent a movie up to 5 days.

   b. Which person pays an amount proportional to the number of days rented?

# 5.3 Writing Proportions

**Essential Question** How can you write a proportion that solves a problem in real life?

## 1 ACTIVITY: Writing Proportions

**Work with a partner.** A rough rule for finding the correct bat length is "the bat length should be half of the batter's height." So, a 62-inch-tall batter uses a bat that is 31 inches long. Write a proportion to find the bat length for each given batter height.

a. 58 inches

b. 60 inches

c. 64 inches

$2x$

$x$

## 2 ACTIVITY: Bat Lengths

**Work with a partner.** Here is a more accurate table for determining the bat length for a batter. Find all the batter heights and corresponding weights for which the rough rule in Activity 1 is exact.

| | Height of Batter (inches) | | | | | | | |
|---|---|---|---|---|---|---|---|---|
| **Weight of Batter (pounds)** | 45–48 | 49–52 | 53–56 | 57–60 | 61–64 | 65–68 | 69–72 | Over 72 |
| Under 61 | 28 | 29 | 29 | | | | | |
| 61–70 | 28 | 29 | 30 | 30 | | | | |
| 71–80 | 28 | 29 | 30 | 30 | 31 | | | |
| 81–90 | 29 | 29 | 30 | 30 | 31 | 32 | | |
| 91–100 | 29 | 30 | 30 | 31 | 31 | 32 | | |
| 101–110 | 29 | 30 | 30 | 31 | 31 | 32 | | |
| 111–120 | 29 | 30 | 30 | 31 | 31 | 32 | | |
| 121–130 | 29 | 30 | 30 | 31 | 32 | 33 | 33 | |
| 131–140 | 30 | 30 | 31 | 31 | 32 | 33 | 33 | |
| 141–150 | 30 | 30 | 31 | 31 | 32 | 33 | 33 | |
| 151–160 | 30 | 31 | 31 | 32 | 32 | 33 | 33 | 33 |
| 161–170 | | 31 | 31 | 32 | 32 | 33 | 33 | 34 |
| 171–180 | | | | 32 | 33 | 33 | 34 | 34 |
| Over 180 | | | | | 33 | 33 | 34 | 34 |

**COMMON CORE**

**Proportions**

In this lesson, you will
- write proportions.
- solve proportions using mental math.

Learning Standards
7.RP.2c
7.RP.3

## ACTIVITY: Writing Proportions

**Math Practice 8**

**Evaluate Results**

How do you know if your results are reasonable? Explain.

**Work with a partner. The batting average of a baseball player is the number of "hits" divided by the number of "at bats."**

$$\text{batting average} = \frac{\text{hits } (H)}{\text{at bats } (A)}$$

A player whose batting average is 0.250 is said to be "batting 250."

**Write a proportion to find how many hits $H$ a player needs to achieve the given batting average. Then solve the proportion.**

a. 50 times at bat; batting average is 0.200.

b. 84 times at bat; batting average is 0.250.

c. 80 times at bat; batting average is 0.350.

d. 1 time at bat; batting average is 1.000.

## What Is Your Answer?

4. **IN YOUR OWN WORDS** How can you write a proportion that solves a problem in real life?

5. Two players have the same batting average.

| | At Bats | Hits | Batting Average |
|---|---|---|---|
| **Player 1** | 132 | 45 | |
| **Player 2** | 132 | 45 | |

Player 1 gets four hits in the next five at bats. Player 2 gets three hits in the next three at bats.

a. Who has the higher batting average?

b. Does this seem fair? Explain your reasoning.

**Practice** Use what you discovered about proportions to complete Exercises 4–7 on page 182.

One way to write a proportion is to use a table.

|  | Last Month | This Month |
|---|---|---|
| **Purchase** | 2 ringtones | 3 ringtones |
| **Total Cost** | 6 dollars | $x$ dollars |

Use the columns or the rows to write a proportion.

**Use columns:**

$$\frac{2 \text{ ringtones}}{6 \text{ dollars}} = \frac{3 \text{ ringtones}}{x \text{ dollars}}$$

> Numerators have the same units.

> Denominators have the same units.

**Use rows:**

$$\frac{2 \text{ ringtones}}{3 \text{ ringtones}} = \frac{6 \text{ dollars}}{x \text{ dollars}}$$

> The units are the same on each side of the proportion.

**EXAMPLE 1  Writing a Proportion**

**Black Bean Soup**

1.5 cups black beans
0.5 cup salsa
2 cups water
1 tomato
2 teaspoons seasoning

**A chef increases the amounts of ingredients in a recipe to make a proportional recipe. The new recipe has 6 cups of black beans. Write a proportion that gives the number $x$ of tomatoes in the new recipe.**

Organize the information in a table.

|  | Original Recipe | New Recipe |
|---|---|---|
| **Black Beans** | 1.5 cups | 6 cups |
| **Tomatoes** | 1 tomato | $x$ tomatoes |

∴ One proportion is $\dfrac{1.5 \text{ cups beans}}{1 \text{ tomato}} = \dfrac{6 \text{ cups beans}}{x \text{ tomatoes}}$.

### On Your Own

Now You're Ready
Exercises 8–11

1.  Write a different proportion that gives the number $x$ of tomatoes in the new recipe.

2.  Write a proportion that gives the amount $y$ of water in the new recipe.

**EXAMPLE** ② **Solving Proportions Using Mental Math**

Solve $\dfrac{3}{2} = \dfrac{x}{8}$.

**Step 1:** Think: The product of 2 and what number is 8?

$$\dfrac{3}{2} = \dfrac{x}{8}$$

$$2 \times ? = 8$$

**Step 2:** Because the product of 2 and 4 is 8, multiply the numerator by 4 to find $x$.

$$3 \times 4 = 12$$

$$\dfrac{3}{2} = \dfrac{x}{8}$$

$$2 \times 4 = 8$$

∴ The solution is $x = 12$.

**EXAMPLE** ③ **Solving Proportions Using Mental Math**

**In Example 1, how many tomatoes are in the new recipe?**

Solve the proportion $\dfrac{1.5}{1} = \dfrac{6}{x}$. ← cups black beans

← tomatoes

**Step 1:** Think: The product of 1.5 and what number is 6?

$$1.5 \times ? = 6$$

$$\dfrac{1.5}{1} = \dfrac{6}{x}$$

**Step 2:** Because the product of 1.5 and 4 is 6, multiply the denominator by 4 to find $x$.

$$1.5 \times 4 = 6$$

$$\dfrac{1.5}{1} = \dfrac{6}{x}$$

$$1 \times 4 = 4$$

∴ So, there are 4 tomatoes in the new recipe.

● **On Your Own**

*Now You're Ready*
Exercises 16–21

Solve the proportion.

**3.** $\dfrac{5}{8} = \dfrac{20}{d}$

**4.** $\dfrac{7}{z} = \dfrac{14}{10}$

**5.** $\dfrac{21}{24} = \dfrac{x}{8}$

**6.** A school has 950 students. The ratio of female students to all students is $\dfrac{48}{95}$. Write and solve a proportion to find the number $f$ of students who are female.

Vocabulary and Concept Check

1. **WRITING** Describe two ways you can use a table to write a proportion.

2. **WRITING** What is your first step when solving $\frac{x}{15} = \frac{3}{5}$? Explain.

3. **OPEN-ENDED** Write a proportion using an unknown value $x$ and the ratio $5:6$. Then solve it.

 Practice and Problem Solving

**Write a proportion to find how many points a student needs to score on the test to get the given score.**

4. test worth 50 points; test score of 40%

5. test worth 50 points; test score of 78%

6. test worth 80 points; test score of 80%

7. test worth 150 points; test score of 96%

**Use the table to write a proportion.**

8.

| | Game 1 | Game 2 |
|---|---|---|
| Points | 12 | 18 |
| Shots | 14 | $w$ |

9.

| | May | June |
|---|---|---|
| Winners | $n$ | 34 |
| Entries | 85 | 170 |

10.

| | Today | Yesterday |
|---|---|---|
| Miles | 15 | $m$ |
| Hours | 2.5 | 4 |

11.

| | Race 1 | Race 2 |
|---|---|---|
| Meters | 100 | 200 |
| Seconds | $x$ | 22.4 |

12. **ERROR ANALYSIS** Describe and correct the error in writing the proportion.

| ✗ | Monday | Tuesday |
|---|---|---|
| Dollars | 2.08 | $d$ |
| Ounces | 8 | 16 |

$$\frac{2.08}{16} = \frac{d}{8}$$

13. **T-SHIRTS** You can buy 3 T-shirts for $24. Write a proportion that gives the cost $c$ of buying 7 T-shirts.

14. **COMPUTERS** A school requires 2 computers for every 5 students. Write a proportion that gives the number $c$ of computers needed for 145 students.

15. **SWIM TEAM** The school team has 80 swimmers. The ratio of seventh-grade swimmers to all swimmers is $5:16$. Write a proportion that gives the number $s$ of seventh-grade swimmers.

**Solve the proportion.**

2 3 **16.** $\dfrac{1}{4} = \dfrac{z}{20}$

**17.** $\dfrac{3}{4} = \dfrac{12}{y}$

**18.** $\dfrac{35}{k} = \dfrac{7}{3}$

**19.** $\dfrac{15}{8} = \dfrac{45}{c}$

**20.** $\dfrac{b}{36} = \dfrac{5}{9}$

**21.** $\dfrac{1.4}{2.5} = \dfrac{g}{25}$

**22. ORCHESTRA** In an orchestra, the ratio of trombones to violas is 1 to 3.

    **a.** There are 9 violas. Write a proportion that gives the number $t$ of trombones in the orchestra.

    **b.** How many trombones are in the orchestra?

**23. ATLANTIS** Your science teacher has a 1 : 200 scale model of the space shuttle *Atlantis*. Which of the proportions can you use to find the actual length $x$ of *Atlantis*? Explain.

$$\dfrac{1}{200} = \dfrac{19.5}{x} \qquad \dfrac{1}{200} = \dfrac{x}{19.5} \qquad \dfrac{200}{19.5} = \dfrac{x}{1} \qquad \dfrac{x}{200} = \dfrac{1}{19.5}$$

19.5 cm

**24. YOU BE THE TEACHER** Your friend says "$48x = 6 \cdot 12$." Is your friend right? Explain.

> Solve $\dfrac{6}{x} = \dfrac{12}{48}$.

**25.** ⟨Reasoning⟩ There are 180 white lockers in the school. There are 3 white lockers for every 5 blue lockers. How many lockers are in the school?

**Fair Game Review** What you learned in previous grades & lessons

**Solve the equation.** *(Section 3.4)*

**26.** $\dfrac{x}{6} = 25$

**27.** $8x = 72$

**28.** $150 = 2x$

**29.** $35 = \dfrac{x}{4}$

**30. MULTIPLE CHOICE** What is the value of $-\dfrac{9}{4} + \left| -\dfrac{8}{5} \right| - 2\dfrac{1}{2}$? *(Section 2.3)*

    **A** $-6\dfrac{7}{20}$
    **B** $-5\dfrac{7}{20}$
    **C** $-3\dfrac{3}{20}$
    **D** $-2\dfrac{3}{20}$

You can use an **information wheel** to organize information about a concept. Here is an example of an information wheel for ratio.

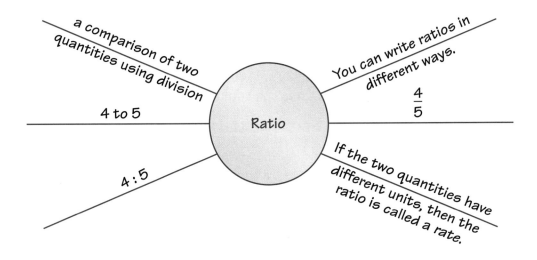

## On Your Own

**Make information wheels to help you study these topics.**

1. rate

2. unit rate

3. proportion

4. cross products

5. graphing proportional relationships

**After you complete this chapter, make information wheels for the following topics.**

6. solving proportions

7. slope

8. direct variation

"My information wheel **summarizes how cats act when they get baths.**"

**Write the ratio as a fraction in simplest form.**  *(Section 5.1)*

**1.** 18 red buttons : 12 blue buttons

**2.** $\frac{5}{4}$ inches to $\frac{2}{3}$ inch

**Use the ratio table to find the unit rate with the specified units.**  *(Section 5.1)*

**3.** cost per song

| Songs | 0 | 2 | 4 | 6 |
|---|---|---|---|---|
| Cost | $0 | $1.98 | $3.96 | $5.94 |

**4.** gallons per hour

| Hours | 3 | 6 | 9 | 12 |
|---|---|---|---|---|
| Gallons | 10.5 | 21 | 31.5 | 42 |

**Tell whether the ratios form a proportion.**  *(Section 5.2)*

**5.** $\frac{1}{8}, \frac{4}{32}$

**6.** $\frac{2}{3}, \frac{10}{30}$

**7.** $\frac{7}{4}, \frac{28}{16}$

**Tell whether the two rates form a proportion.**  *(Section 5.2)*

**8.** 75 miles in 3 hours; 140 miles in 4 hours

**9.** 12 gallons in 4 minutes; 21 gallons in 7 minutes

**10.** 150 steps in 50 feet; 72 steps in 24 feet

**11.** 3 rotations in 675 days; 2 rotations in 730 days

**Use the table to write a proportion.**  *(Section 5.3)*

**12.**

| | Monday | Tuesday |
|---|---|---|
| Dollars | 42 | 56 |
| Hours | 6 | $h$ |

**13.**

| | Series 1 | Series 2 |
|---|---|---|
| Games | $g$ | 6 |
| Wins | 4 | 3 |

**14. MUSIC DOWNLOAD** The amount of time needed to download music is shown in the table. Find the unit rate in megabytes per second.  *(Section 5.1)*

| Seconds | 6 | 12 | 18 | 24 |
|---|---|---|---|---|
| Megabytes | 2 | 4 | 6 | 8 |

**15. SOUND** The graph shows the distance that sound travels through steel. Interpret each plotted point in the graph of the proportional relationship.  *(Section 5.2)*

**16. GAMING** You advance 3 levels in 15 minutes. Your friend advances 5 levels in 20 minutes. Do these rates form a proportion? Explain.  *(Section 5.2)*

**17. CLASS TIME** You spend 150 minutes in 3 classes. Write and solve a proportion to find how many minutes you spend in 5 classes.  *(Section 5.3)*

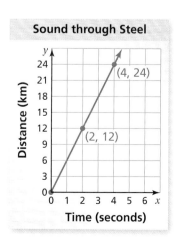

**Sound through Steel**

(4, 24)

(2, 12)

Distance (km) — Time (seconds)

# 5.4 Solving Proportions

## Essential Question How can you use ratio tables and cross products to solve proportions?

### 1 ACTIVITY: Solving a Proportion in Science

**Work with a partner. You can use ratio tables to determine the amount of a compound (like salt) that is dissolved in a solution. Determine the unknown quantity. Explain your procedure.**

**a. Salt Water**

| Salt Water | 1 L | 3 L |
|---|---|---|
| Salt | 250 g | $x$ g |

$$\frac{1\,L}{\phantom{xx}} = \frac{\phantom{xx}}{\phantom{xx}}$$  Write proportion.

$$1 \cdot \boxed{\phantom{x}} = \boxed{\phantom{x}} \cdot \boxed{\phantom{x}}$$  Set cross products equal.

$$\boxed{\phantom{x}} = \boxed{\phantom{x}}$$  Simplify.

There are $\boxed{\phantom{x}}$ grams of salt in the 3-liter solution.

1 liter      3 liters

**b. White Glue Solution**

| Water | ½ cup | 1 cup |
|---|---|---|
| White Glue | ½ cup | $x$ cups |

**c. Borax Solution**

| Borax | 1 tsp | 2 tsp |
|---|---|---|
| Water | 1 cup | $x$ cups |

**d. Slime (See recipe.)**

| Borax Solution | ½ cup | 1 cup |
|---|---|---|
| White Glue Solution | $y$ cups | $x$ cups |

### Recipe for SLIME

1. Add ½ cup of water and ½ cup white glue. Mix thoroughly. This is your white glue solution.

2. Add a couple drops of food coloring to the white glue solution. Mix thoroughly.

3. Add 1 teaspoon of borax to 1 cup of water. Mix thoroughly. This is your borax solution (about 1 cup).

4. Pour the borax solution and the glue solution into a separate bowl.

5. Place the slime that forms into a plastic bag. Squeeze the mixture repeatedly to mix it up.

**COMMON CORE**

**Proportions**
In this lesson, you will
- solve proportions using multiplication or the Cross Products Property.
- use a point on a graph to write and solve proportions.
Learning Standards
7.RP.2b
7.RP.2c

## ACTIVITY: The Game of Criss Cross

**CRISS CROSS**

**Preparation:**

- Cut index cards to make 48 playing cards.
- Write each number on a card.

  1, 1, 1, 2, 2, 2, 3, 3, 3, 4, 4, 4, 5, 5, 5, 6, 6, 6, 7, 7,

  7, 8, 8, 8, 9, 9, 9, 10, 10, 10, 12, 12, 12, 13, 13,

  13, 14, 14, 14, 15, 15, 15, 16, 16, 16, 18, 20, 25

- Make a copy of the game board.

**To Play:**

- Play with a partner.
- Deal eight cards to each player.
- Begin by drawing a card from the remaining cards. Use four of your cards to try to form a proportion.
- Lay the four cards on the game board. If you form a proportion, then say "Criss Cross." You earn 4 points. Place the four cards in a discard pile. Now it is your partner's turn.
- If you cannot form a proportion, then it is your partner's turn.
- When the original pile of cards is empty, shuffle the cards in the discard pile. Start again.
- The first player to reach 20 points wins.

## What Is Your Answer?

3. **IN YOUR OWN WORDS** How can you use ratio tables and cross products to solve proportions? Give an example.

4. **PUZZLE** Use each number once to form three proportions.

| 1 | 2 | 10 | 4 | 12 | 20 |
|---|---|----|---|----|----|

| 15 | 5 | 16 | 6 | 8 | 3 |
|----|---|----|---|---|---|

**Practice**  Use what you discovered about solving proportions to complete Exercises 10–13 on page 190.

## 5.4 Lesson

 Key Idea

**Solving Proportions**

**Method 1**   Use mental math.   *(Section 5.3)*

**Method 2**   Use the Multiplication Property of Equality.   *(Section 5.4)*

**Method 3**   Use the Cross Products Property.   *(Section 5.4)*

---

**EXAMPLE** ① **Solving Proportions Using Multiplication**

Solve $\dfrac{5}{7} = \dfrac{x}{21}$.

$$\dfrac{5}{7} = \dfrac{x}{21} \qquad \text{Write the proportion.}$$

$$21 \cdot \dfrac{5}{7} = 21 \cdot \dfrac{x}{21} \qquad \text{Multiplication Property of Equality}$$

$$15 = x \qquad \text{Simplify.}$$

⋰ The solution is 15.

### On Your Own

*Now You're Ready*
Exercises 4–9

**Use multiplication to solve the proportion.**

1. $\dfrac{w}{6} = \dfrac{6}{9}$

2. $\dfrac{12}{10} = \dfrac{a}{15}$

3. $\dfrac{y}{6} = \dfrac{2}{4}$

---

**EXAMPLE** ② **Solving Proportions Using the Cross Products Property**

**Solve each proportion.**

**a.** $\dfrac{x}{8} = \dfrac{7}{10}$

$x \cdot 10 = 8 \cdot 7$   Cross Products Property

$10x = 56$   Multiply.

$x = 5.6$   Divide.

⋰ The solution is 5.6.

**b.** $\dfrac{9}{y} = \dfrac{3}{17}$

$9 \cdot 17 = y \cdot 3$

$153 = 3y$

$51 = y$

⋰ The solution is 51.

## On Your Own

Now You're Ready
Exercises 10–21

**Use the Cross Products Property to solve the proportion.**

**4.** $\dfrac{2}{7} = \dfrac{x}{28}$

**5.** $\dfrac{12}{5} = \dfrac{6}{y}$

**6.** $\dfrac{40}{z+1} = \dfrac{15}{6}$

---

**EXAMPLE** ③ **Real-Life Application**

**TOLL PLAZA**
**½ MILE**
**REDUCE SPEED**

**The graph shows the toll $y$ due on a turnpike for driving $x$ miles. Your toll is \$7.50. How many *kilometers* did you drive?**

Turnpike

The point $(100, 7.5)$ on the graph shows that the toll is \$7.50 for driving 100 miles. Convert 100 miles to kilometers.

**Method 1:** Convert using a ratio.

$1 \text{ mi} \approx 1.61 \text{ km}$

$$100 \text{ mi} \times \dfrac{1.61 \text{ km}}{1 \text{ mi}} = 161 \text{ km}$$

∴ So, you drove about 161 kilometers.

**Method 2:** Convert using a proportion.

Let $x$ be the number of kilometers equivalent to 100 miles.

kilometers → $\dfrac{1.61}{1} = \dfrac{x}{100}$ ← kilometers

miles ↗ ↖ miles

Write a proportion. Use $1.61 \text{ km} \approx 1 \text{ mi}$.

$$1.61 \cdot 100 = 1 \cdot x \qquad \text{Cross Products Property}$$
$$161 = x \qquad \text{Simplify.}$$

∴ So, you drove about 161 kilometers.

## On Your Own

Now You're Ready
Exercises 28–30

**Write and solve a proportion to complete the statement. Round to the nearest hundredth, if necessary.**

**7.** $7.5 \text{ in.} \approx \quad \text{cm}$

**8.** $100 \text{ g} \approx \quad \text{oz}$

**9.** $2 \text{ L} \approx \quad \text{qt}$

**10.** $4 \text{ m} \approx \quad \text{ft}$

Vocabulary and Concept Check

1. **WRITING** What are three ways you can solve a proportion?

2. **OPEN-ENDED** Which way would you choose to solve $\frac{3}{x} = \frac{6}{14}$? Explain your reasoning.

3. **NUMBER SENSE** Does $\frac{x}{4} = \frac{15}{3}$ have the same solution as $\frac{x}{15} = \frac{4}{3}$? Use the Cross Products Property to explain your answer.

 Practice and Problem Solving

**Use multiplication to solve the proportion.**

(1) 4. $\frac{9}{5} = \frac{z}{20}$     5. $\frac{h}{15} = \frac{16}{3}$     6. $\frac{w}{4} = \frac{42}{24}$

7. $\frac{35}{28} = \frac{n}{12}$     8. $\frac{7}{16} = \frac{x}{4}$     9. $\frac{y}{9} = \frac{44}{54}$

**Use the Cross Products Property to solve the proportion.**

(2) 10. $\frac{a}{6} = \frac{15}{2}$     11. $\frac{10}{7} = \frac{8}{k}$     12. $\frac{3}{4} = \frac{v}{14}$     13. $\frac{5}{n} = \frac{16}{32}$

14. $\frac{36}{42} = \frac{24}{r}$     15. $\frac{9}{10} = \frac{d}{6.4}$     16. $\frac{x}{8} = \frac{3}{12}$     17. $\frac{8}{m} = \frac{6}{15}$

18. $\frac{4}{24} = \frac{c}{36}$     19. $\frac{20}{16} = \frac{d}{12}$     20. $\frac{30}{20} = \frac{w}{14}$     21. $\frac{2.4}{1.8} = \frac{7.2}{k}$

22. **ERROR ANALYSIS** Describe and correct the error in solving the proportion $\frac{m}{8} = \frac{15}{24}$.

$$\times \quad \frac{m}{8} = \frac{15}{24}$$
$$8 \cdot m = 24 \cdot 15$$
$$m = 45$$

23. **PENS** Forty-eight pens are packaged in 4 boxes. How many pens are packaged in 9 boxes?

24. **PIZZA PARTY** How much does it cost to buy 10 medium pizzas?

3 Medium Pizzas for $10.50

**Solve the proportion.**

25. $\frac{2x}{5} = \frac{9}{15}$     26. $\frac{5}{2} = \frac{d-2}{4}$     27. $\frac{4}{k+3} = \frac{8}{14}$

**Write and solve a proportion to complete the statement. Round to the nearest hundredth if necessary.**

③ **28.** $6 \text{ km} \approx$ [  ] mi

**29.** $2.5 \text{ L} \approx$ [  ] gal

**30.** $90 \text{ lb} \approx$ [  ] kg

**31. TRUE OR FALSE?** Tell whether the statement is *true* or *false*. Explain.

If $\dfrac{a}{b} = \dfrac{2}{3}$, then $\dfrac{3}{2} = \dfrac{b}{a}$.

**32. CLASS TRIP** It costs \$95 for 20 students to visit an aquarium. How much does it cost for 162 students?

**Human Hair**

**33. GRAVITY** A person who weighs 120 pounds on Earth weighs 20 pounds on the Moon. How much does a 93-pound person weigh on the Moon?

**34. HAIR** The length of human hair is proportional to the number of months it has grown.

    **a.** What is the hair length in *centimeters* after 6 months?

    **b.** How long does it take hair to grow 8 inches?

    **c.** Use a different method than the one in part (b) to find how long it takes hair to grow 20 inches.

**35. SWING SET** It takes 6 hours for 2 people to build a swing set. Can you use the proportion $\dfrac{2}{6} = \dfrac{5}{h}$ to determine the number of hours $h$ it will take 5 people to build the swing set? Explain.

**36. REASONING** There are 144 people in an audience. The ratio of adults to children is 5 to 3. How many are adults?

**37. PROBLEM SOLVING** Three pounds of lawn seed covers 1800 square feet. How many bags are needed to cover 8400 square feet?

**38.**  **Critical Thinking** Consider the proportions $\dfrac{m}{n} = \dfrac{1}{2}$ and $\dfrac{n}{k} = \dfrac{2}{5}$.

What is the ratio $\dfrac{m}{k}$? Explain your reasoning.

## Fair Game Review   What you learned in previous grades & lessons

**Plot the ordered pair in a coordinate plane.** *(Skills Review Handbook)*

**39.** $A(-5, -2)$      **40.** $B(-3, 0)$      **41.** $C(-1, 2)$      **42.** $D(1, 4)$

**43. MULTIPLE CHOICE** What is the value of $(3w - 8) - 4(2w + 3)$? *(Section 3.2)*

    Ⓐ $11w + 4$      Ⓑ $-5w - 5$      Ⓒ $-5w + 4$      Ⓓ $-5w - 20$

**Essential Question** How can you compare two rates graphically?

### 1 ACTIVITY: Comparing Unit Rates

**Work with a partner. The table shows the maximum speeds of several animals.**

**a.** Find the missing speeds. Round your answers to the nearest tenth.

**b.** Which animal is fastest? Which animal is slowest?

**c.** Explain how you convert between the two units of speed.

| Animal | Speed (miles per hour) | Speed (feet per second) |
|---|---|---|
| Antelope | 61.0 | |
| Black mamba snake | | 29.3 |
| Cheetah | | 102.6 |
| Chicken | | 13.2 |
| Coyote | 43.0 | |
| Domestic pig | | 16.0 |
| Elephant | | 36.6 |
| Elk | | 66.0 |
| Giant tortoise | 0.2 | |
| Giraffe | 32.0 | |
| Gray fox | | 61.6 |
| Greyhound | 39.4 | |
| Grizzly bear | | 44.0 |
| Human | | 41.0 |
| Hyena | 40.0 | |
| Jackal | 35.0 | |
| Lion | | 73.3 |
| Peregrine falcon | 200.0 | |
| Quarter horse | 47.5 | |
| Spider | | 1.76 |
| Squirrel | 12.0 | |
| Thomson's gazelle | 50.0 | |
| Three-toed sloth | | 0.2 |
| Tuna | 47.0 | |

**COMMON CORE**

**Slope**

In this lesson, you will
- find the slopes of lines.
- interpret the slopes of lines as rates.

Learning Standard
7.RP.2b

## ACTIVITY: Comparing Two Rates Graphically

**Math Practice** 4

**Apply Mathematics**

How can you use the graph to determine which animal has the greater speed?

**Work with a partner. A cheetah and a Thomson's gazelle run at maximum speed.**

**a.** Use the table in Activity 1 to calculate the missing distances.

| Time (seconds) | Cheetah Distance (feet) | Gazelle Distance (feet) |
|---|---|---|
| 0 | | |
| 1 | | |
| 2 | | |
| 3 | | |
| 4 | | |
| 5 | | |
| 6 | | |
| 7 | | |

**b.** Use the table to write ordered pairs. Then plot the ordered pairs and connect the points for each animal. What do you notice about the graphs?

**c.** Which graph is steeper? The speed of which animal is greater?

# What Is Your Answer?

**3. IN YOUR OWN WORDS** How can you compare two rates graphically? Explain your reasoning. Give some examples with your answer.

**4. REPEATED REASONING** Choose 10 animals from Activity 1.

    **a.** Make a table for each animal similar to the table in Activity 2.

    **b.** Sketch a graph of the distances for each animal.

    **c.** Compare the steepness of the 10 graphs. What can you conclude?

**Check It Out**
Lesson Tutorials
BigIdeasMath V com

**Key Vocabulary**
slope, *p. 194*

**Study Tip**

The slope of a line is the same between any two points on the line because lines have a *constant* rate of change.

## Key Idea

**Slope**

**Slope** is the rate of change between any two points on a line. It is a measure of the *steepness* of a line.

To find the slope of a line, find the ratio of the change in *y* (vertical change) to the change in *x* (horizontal change).

$$\text{slope} = \frac{\text{change in } y}{\text{change in } x}$$

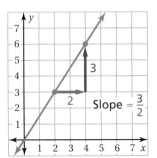

Slope $= \dfrac{3}{2}$

---

**EXAMPLE 1  Finding Slopes**

**Find the slope of each line.**

**a.**

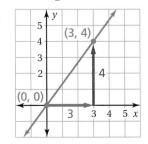

$$\text{slope} = \frac{\text{change in } y}{\text{change in } x}$$

$$= \frac{4}{3}$$

∴ The slope of the line is $\dfrac{4}{3}$.

**b.**

$$\text{slope} = \frac{\text{change in } y}{\text{change in } x}$$

$$= \frac{3}{6} = \frac{1}{2}$$

∴ The slope of the line is $\dfrac{1}{2}$.

## On Your Own

**Now You're Ready**
Exercises 4–9

**Find the slope of the line.**

**1.**

**2.**

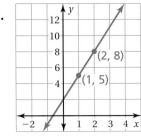

---

◀) Multi-Language Glossary at BigIdeasMath V com

EXAMPLE 2 Interpreting a Slope

The table shows your earnings for babysitting.

a. Graph the data.

b. Find and interpret the slope of the line through the points.

| Hours, x | 0 | 2 | 4 | 6 | 8 | 10 |
|---|---|---|---|---|---|---|
| Earnings, y (dollars) | 0 | 10 | 20 | 30 | 40 | 50 |

**Babysitting**

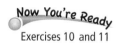

a. Graph the data. Draw a line through the points.

b. Choose any two points to find the slope of the line.

$$\text{slope} = \frac{\text{change in } y}{\text{change in } x}$$

$$= \frac{20}{4} \leftarrow \boxed{\text{dollars}}$$
$$\quad \leftarrow \boxed{\text{hours}}$$

$$= 5$$

∴ The slope of the line represents the unit rate.
The slope is 5. So, you earn $5 per hour babysitting.

**On Your Own**

Now You're Ready
Exercises 10 and 11

3. In Example 2, use two other points to find the slope. Does the slope change?

4. The graph shows the amounts you and your friend earn babysitting.

a. Compare the steepness of the lines. What does this mean in the context of the problem?

b. Find and interpret the slope of the blue line.

## Vocabulary and Concept Check

1. **VOCABULARY** Is there a connection between rate and slope? Explain.

2. **REASONING** Which line has the greatest slope?

3. **REASONING** Is it more difficult to run up a ramp with a slope of $\frac{1}{5}$ or a ramp with a slope of 5? Explain.

 ## Practice and Problem Solving

**Find the slope of the line.**

 **4.**

**5.**

**6.**

**7.**

**8.**

**9.**

**Graph the data. Then find and interpret the slope of the line through the points.**

**10.**

| Minutes, x | 3 | 5 | 7 | 9 |
|---|---|---|---|---|
| Words, y | 135 | 225 | 315 | 405 |

**11.**

| Gallons, x | 5 | 10 | 15 | 20 |
|---|---|---|---|---|
| Miles, y | 162.5 | 325 | 487.5 | 650 |

12. **ERROR ANALYSIS** Describe and correct the error in finding the slope of the line passing through (0, 0) and (4, 5).

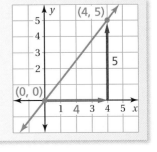

slope = $\frac{4}{5}$

**Graph the line that passes through the two points. Then find the slope of the line.**

**13.** $(0, 0)$, $\left(\dfrac{1}{3}, \dfrac{7}{3}\right)$

**14.** $\left(-\dfrac{3}{2}, -\dfrac{3}{2}\right)$, $\left(\dfrac{3}{2}, \dfrac{3}{2}\right)$

**15.** $\left(1, \dfrac{5}{2}\right)$, $\left(-\dfrac{1}{2}, -\dfrac{1}{4}\right)$

**16. CAMPING** The graph shows the amount of money you and a friend are saving for a camping trip.

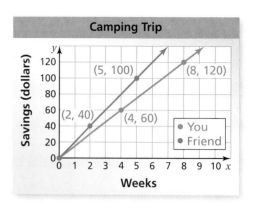

**Camping Trip**

a. Compare the steepness of the lines. What does this mean in the context of the problem?

b. Find the slope of each line.

c. How much more money does your friend save each week than you?

d. The camping trip costs $165. How long will it take you to save enough money?

**17. MAPS** An atlas contains a map of Ohio. The table shows data from the key on the map.

| Distance on Map (mm), x | 10 | 20 | 30 | 40 |
|---|---|---|---|---|
| Actual Distance (mi), y | 25 | 50 | 75 | 100 |

a. Graph the data.

b. Find the slope of the line. What does this mean in the context of the problem?

c. The map distance between Toledo and Columbus is 48 millimeters. What is the actual distance?

d. Cincinnati is about 225 miles from Cleveland. What is the distance between these cities on the map?

**18. CRITICAL THINKING** What is the slope of a line that passes through the points $(2, 0)$ and $(5, 0)$? Explain.

**19.** **Number Sense** A line has a slope of 2. It passes through the points $(1, 2)$ and $(3, y)$. What is the value of $y$?

## Fair Game Review  What you learned in previous grades & lessons

**Multiply.**  *(Section 2.4)*

**20.** $-\dfrac{3}{5} \times \dfrac{8}{6}$

**21.** $1\dfrac{1}{2} \times \left(-\dfrac{6}{15}\right)$

**22.** $-2\dfrac{1}{4} \times \left(-1\dfrac{1}{3}\right)$

**23. MULTIPLE CHOICE** You have 18 stamps from Mexico in your stamp collection. These stamps represent $\dfrac{3}{8}$ of your collection. The rest of the stamps are from the United States. How many stamps are from the United States?  *(Section 3.4)*

   Ⓐ 12       Ⓑ 24       Ⓒ 30       Ⓓ 48

**Essential Question** How can you use a graph to show the relationship between two quantities that vary directly? How can you use an equation?

### 1   ACTIVITY: Math in Literature

*Gulliver's Travels* was written by Jonathan Swift and published in 1726. Gulliver was shipwrecked on the island Lilliput, where the people were only 6 inches tall. When the Lilliputians decided to make a shirt for Gulliver, a Lilliputian tailor stated that he could determine Gulliver's measurements by simply measuring the distance around Gulliver's thumb. He said "Twice around the thumb equals once around the wrist. Twice around the wrist is once around the neck. Twice around the neck is once around the waist."

**Work with a partner. Use the tailor's statement to complete the table.**

| Thumb, $t$ | Wrist, $w$ | Neck, $n$ | Waist, $x$ |
|---|---|---|---|
| 0 in. | | | |
| 1 in. | | | |
| | 4 in. | | |
| | | 12 in. | |
| | | | 32 in. |
| | 10 in. | | |

**COMMON CORE**

**Direct Variation**

In this lesson, you will
- identify direct variation from graphs or equations.
- use direct variation models to solve problems.

Learning Standards
7.RP.2a
7.RP.2b
7.RP.2c
7.RP.2d

## 2 ACTIVITY: Drawing a Graph

**Work with a partner. Use the information from Activity 1.**

**a.** In your own words, describe the relationship between $t$ and $w$.

**b.** Use the table to write the ordered pairs $(t, w)$. Then plot the ordered pairs.

**c.** What do you notice about the graph of the ordered pairs?

**d.** Choose two points and find the slope of the line between them.

**e.** The quantities $t$ and $w$ are said to *vary directly*. An equation that describes the relationship is

$$w = \boxed{\phantom{0}}\, t.$$

## 3 ACTIVITY: Drawing a Graph and Writing an Equation

**Math Practice 6**

**Label Axes**

How do you know which labels to use for the axes? Explain.

**Work with a partner. Use the information from Activity 1 to draw a graph of the relationship. Write an equation that describes the relationship between the two quantities.**

**a.** Thumb $t$ and neck $n$    $(n = \boxed{\phantom{0}}\, t)$

**b.** Wrist $w$ and waist $x$    $(x = \boxed{\phantom{0}}\, w)$

**c.** Wrist $w$ and thumb $t$    $(t = \boxed{\phantom{0}}\, w)$

**d.** Waist $x$ and wrist $w$    $(w = \boxed{\phantom{0}}\, x)$

## What Is Your Answer?

**4. IN YOUR OWN WORDS** How can you use a graph to show the relationship between two quantities that vary directly? How can you use an equation?

**5. STRUCTURE** How are all the graphs in Activity 3 alike?

**6.** Give a real-life example of two variables that vary directly.

**7.** Work with a partner. Use string to find the distance around your thumb, wrist, and neck. Do your measurements agree with the tailor's statement in *Gulliver's Travels*? Explain your reasoning.

**Practice**

Use what you learned about quantities that vary directly to complete Exercises 4 and 5 on page 202.

**Key Vocabulary**
direct variation, p. 200
constant of proportionality, p. 200

 **Key Idea**

**Direct Variation**

**Words** Two quantities $x$ and $y$ show **direct variation** when $y = kx$, where $k$ is a number and $k \neq 0$. The number $k$ is called the **constant of proportionality**.

**Graph** The graph of $y = kx$ is a line with a slope of $k$ that passes through the origin. So, two quantities that show direct variation are in a proportional relationship.

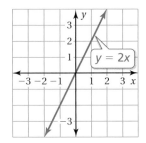

**EXAMPLE 1 Identifying Direct Variation**

**Tell whether $x$ and $y$ show direct variation. Explain your reasoning.**

a.

| x | 1 | 2 | 3 | 4 |
|---|---|---|---|---|
| y | −2 | 0 | 2 | 4 |

b.

| x | 0 | 2 | 4 | 6 |
|---|---|---|---|---|
| y | 0 | 2 | 4 | 6 |

**Study Tip**

Other ways to say that $x$ and $y$ show direct variation are "$y$ varies directly with $x$" and "$x$ and $y$ are directly proportional."

Plot the points. Draw a line through the points.

Plot the points. Draw a line through the points.

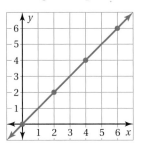

The line does *not* pass through the origin. So, $x$ and $y$ do *not* show direct variation.

The line passes through the origin. So, $x$ and $y$ show direct variation.

**EXAMPLE 2 Identifying Direct Variation**

**Tell whether $x$ and $y$ show direct variation. Explain your reasoning.**

a. $y + 1 = 2x$

$y = 2x - 1$   Solve for $y$.

The equation *cannot* be written as $y = kx$. So, $x$ and $y$ do *not* show direct variation.

b. $\frac{1}{2}y = x$

$y = 2x$   Solve for $y$.

The equation can be written as $y = kx$. So, $x$ and $y$ show direct variation.

Multi-Language Glossary at BigIdeasMath.com

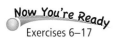
Now You're Ready
Exercises 6–17

## On Your Own

**Tell whether _x_ and _y_ show direct variation. Explain your reasoning.**

**1.**

| x | y |
|---|---|
| 0 | −2 |
| 1 | 1 |
| 2 | 4 |
| 3 | 7 |

**2.**

| x | y |
|---|---|
| 1 | 4 |
| 2 | 8 |
| 3 | 12 |
| 4 | 16 |

**3.**

| x | y |
|---|---|
| −2 | 4 |
| −1 | 2 |
| 0 | 0 |
| 1 | 2 |

**4.** $xy = 3$

**5.** $x = \dfrac{1}{3}y$

**6.** $y + 1 = x$

**EXAMPLE 3** **Real-Life Application**

| x | y |
|---|---|
| $\dfrac{1}{2}$ | 8 |
| 1 | 16 |
| $\dfrac{3}{2}$ | 24 |
| 2 | 32 |

The table shows the area _y_ (in square feet) that a robotic vacuum cleans in _x_ minutes.

**a. Graph the data. Tell whether _x_ and _y_ are directly proportional.**

Graph the data. Draw a line through the points.

⁝• The graph is a line through the origin. So, _x_ and _y_ are directly proportional.

**b. Write an equation that represents the line.**

Choose any two points to find the slope of the line.

$$\text{slope} = \frac{\text{change in } y}{\text{change in } x} = \frac{16}{1} = 16$$

⁝• The slope of the line is the constant of proportionality, _k_. So, an equation of the line is $y = 16x$.

**c. Use the equation to find the area cleaned in 10 minutes.**

$y = 16x$      Write the equation.

$\phantom{y} = 16(10)$      Substitute 10 for _x_.

$\phantom{y} = 160$      Multiply.

⁝• So, the vacuum cleans 160 square feet in 10 minutes.

## On Your Own

Now You're Ready
Exercise 19

**7. WHAT IF?** The battery weakens and the robot begins cleaning less and less area each minute. Do _x_ and _y_ show direct variation? Explain.

**Vocabulary and Concept Check**

1. **VOCABULARY** What does it mean for $x$ and $y$ to vary directly?

2. **WRITING** What point is on the graph of every direct variation equation?

3. **DIFFERENT WORDS, SAME QUESTION** Which is different? Find "both" answers.

Do $x$ and $y$ show direct variation?

Are $x$ and $y$ in a proportional relationship?

Is the graph of the relationship a line?

Does $y$ vary directly with $x$?

 **Practice and Problem Solving**

**Graph the ordered pairs in a coordinate plane. Do you think that graph shows that the quantities vary directly? Explain your reasoning.**

4. $(-1, -1)$, $(0, 0)$, $(1, 1)$, $(2, 2)$

5. $(-4, -2)$, $(-2, 0)$, $(0, 2)$, $(2, 4)$

**Tell whether $x$ and $y$ show direct variation. Explain your reasoning. If so, find $k$.**

**6.**

| x | 1 | 2 | 3 | 4 |
|---|---|---|---|---|
| y | 2 | 4 | 6 | 8 |

**7.**

| x | −2 | −1 | 0 | 1 |
|---|---|---|---|---|
| y | 0 | 2 | 4 | 6 |

**8.**

| x | −1 | 0 | 1 | 2 |
|---|---|---|---|---|
| y | −2 | −1 | 0 | 1 |

**9.**

| x | 3 | 6 | 9 | 12 |
|---|---|---|---|---|
| y | 2 | 4 | 6 | 8 |

**10.** $y - x = 4$

**11.** $x = \dfrac{2}{5}y$

**12.** $y + 3 = x + 6$

**13.** $y - 5 = 2x$

**14.** $x - y = 0$

**15.** $\dfrac{x}{y} = 2$

**16.** $8 = xy$

**17.** $x^2 = y$

18. **ERROR ANALYSIS** Describe and correct the error in telling whether $x$ and $y$ show direct variation.

The graph is a line, so it shows direct variation.

19. **RECYCLING** The table shows the profit $y$ for recycling $x$ pounds of aluminum. Graph the data. Tell whether $x$ and $y$ show direct variation. If so, write an equation that represents the line.

| Aluminum (lb), x | 10 | 20 | 30 | 40 |
|---|---|---|---|---|
| Profit, y | $4.50 | $9.00 | $13.50 | $18.00 |

**The variables *x* and *y* vary directly. Use the values to find the constant of proportionality. Then write an equation that relates *x* and *y*.**

**20.** $y = 72; x = 3$

**21.** $y = 20; x = 12$

**22.** $y = 45; x = 40$

2.54 cm

1 in.

**23. MEASUREMENT** Write a direct variation equation that relates *x* inches to *y* centimeters.

**24. MODELING** Design a waterskiing ramp. Show how you can use direct variation to plan the heights of the vertical supports.

Vertical supports

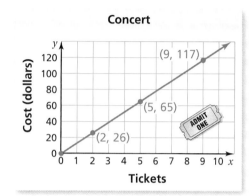

**Concert**

y

Cost (dollars)

120
100
80
60
40
20
0

(9, 117)

(5, 65)

(2, 26)

ADMIT ONE

0 1 2 3 4 5 6 7 8 9 10 x

**Tickets**

**25. REASONING** Use $y = kx$ to show why the graph of a proportional relationship always passes through the origin.

**26. TICKETS** The graph shows the cost of buying concert tickets. Tell whether *x* and *y* show direct variation. If so, find and interpret the constant of proportionality. Then write an equation and find the cost of 14 tickets.

**27. CELL PHONE PLANS** Tell whether *x* and *y* show direct variation. If so, write an equation of direct variation.

| Minutes, *x* | 500 | 700 | 900 | 1200 |
|---|---|---|---|---|
| Cost, *y* | $40 | $50 | $60 | $75 |

**28. CHLORINE** The amount of chlorine in a swimming pool varies directly with the volume of water. The pool has 2.5 milligrams of chlorine per liter of water. How much chlorine is in the pool?

8000 gallons

**29.** **Critical Thinking** Is the graph of every direct variation equation a line? Does the graph of every line represent a direct variation equation? Explain your reasoning.

## Fair Game Review What you learned in previous grades & lessons

**Write the fraction as a decimal.** *(Section 2.1)*

**30.** $\dfrac{13}{20}$

**31.** $\dfrac{9}{16}$

**32.** $\dfrac{21}{40}$

**33.** $\dfrac{24}{25}$

**34. MULTIPLE CHOICE** Which rate is *not* equivalent to 180 feet per 8 seconds? *(Section 5.1)*

Ⓐ $\dfrac{225 \text{ ft}}{10 \text{ sec}}$

Ⓑ $\dfrac{45 \text{ ft}}{2 \text{ sec}}$

Ⓒ $\dfrac{135 \text{ ft}}{6 \text{ sec}}$

Ⓓ $\dfrac{180 \text{ ft}}{1 \text{ sec}}$

## 5.4–5.6    Quiz

**Solve the proportion.**    *(Section 5.4)*

**1.** $\dfrac{7}{n} = \dfrac{42}{48}$

**2.** $\dfrac{x}{2} = \dfrac{40}{16}$

**3.** $\dfrac{3}{11} = \dfrac{27}{z}$

**Find the slope of the line.**    *(Section 5.5)*

**4.**

**5.**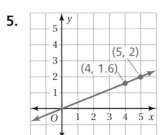

**Graph the data. Then find and interpret the slope of the line through the points.**
*(Section 5.5)*

**6.**

| Hours, *x* | 2 | 4 | 6 | 8 |
|---|---|---|---|---|
| Miles, *y* | 10 | 20 | 30 | 40 |

**7.**

| Packages, *x* | 6 | 10 | 14 | 18 |
|---|---|---|---|---|
| Servings, *y* | 9 | 15 | 21 | 27 |

**Tell whether *x* and *y* show direct variation. Explain your reasoning.**    *(Section 5.6)*

**8.** $y - 9 = 6 + x$

**9.** $x = \dfrac{5}{8}y$

**10. CONCERT** A benefit concert with three performers lasts 8 hours. At this rate, how many hours is a concert with four performers?    *(Section 5.4)*

**11. LAWN MOWING** The graph shows how much you and your friend each earn mowing lawns.    *(Section 5.5)*

**a.** Compare the steepness of the lines. What does this mean in the context of the problem?

**b.** Find and interpret the slope of each line.

**c.** How much more money do you earn per hour than your friend?

**12. PIE SALE** The table shows the profits of a pie sale. Tell whether *x* and *y* show direct variation. If so, write the equation of direct variation.    *(Section 5.6)*

| Pies Sold, *x* | 10 | 12 | 14 | 16 |
|---|---|---|---|---|
| Profit, *y* | $79.50 | $95.40 | $111.30 | $127.20 |

**Check It Out**
Vocabulary Help
BigIdeasMath ✓com

## Review Key Vocabulary

ratio, *p. 164*
rate, *p. 164*
unit rate, *p. 164*
complex fraction, *p. 165*

proportion, *p. 172*
proportional, *p. 172*
cross products, *p. 173*
slope, *p. 194*

direct variation, *p. 200*
constant of proportionality,
  *p. 200*

## Review Examples and Exercises

### 5.1 Ratios and Rates (pp. 162–169)

There are 15 orangutans and 25 gorillas in a nature preserve.
One of the orangutans swings 75 feet in 15 seconds on a rope.

a. Find the ratio of orangutans to gorillas.

b. How fast is the orangutan swinging?

a. $\dfrac{\text{orangutans}}{\text{gorillas}} = \dfrac{15}{25} = \dfrac{3}{5}$

⋮· The ratio of orangutans
to gorillas is $\dfrac{3}{5}$.

b. 75 feet in 15 seconds $= \dfrac{75 \text{ ft}}{15 \text{ sec}}$

$= \dfrac{75 \text{ ft} \div 15}{15 \text{ sec} \div 15}$

$= \dfrac{5 \text{ ft}}{1 \text{ sec}}$

⋮· The orangutan is swinging
5 feet per second.

#### Exercises

**Find the unit rate.**

1. 289 miles on 10 gallons

2. $6\dfrac{2}{5}$ revolutions in $2\dfrac{2}{3}$ seconds

3. calories per serving

| Servings | 2 | 4 | 6 | 8 |
|----------|-----|-----|-----|-----|
| Calories | 240 | 480 | 720 | 960 |

### 5.2 Proportions (pp. 170–177)

Tell whether the ratios $\dfrac{9}{12}$ and $\dfrac{6}{8}$ form a proportion.

$\dfrac{9}{12} = \dfrac{9 \div 3}{12 \div 3} = \dfrac{3}{4}$ ←

The ratios are equivalent.

$\dfrac{6}{8} = \dfrac{6 \div 2}{8 \div 2} = \dfrac{3}{4}$ ←

⋮· So, $\dfrac{9}{12}$ and $\dfrac{6}{8}$ form a proportion.

## Exercises

**Tell whether the ratios form a proportion.**

**4.** $\dfrac{4}{9}, \dfrac{2}{3}$

**5.** $\dfrac{12}{22}, \dfrac{18}{33}$

**6.** $\dfrac{8}{50}, \dfrac{4}{10}$

**7.** $\dfrac{32}{40}, \dfrac{12}{15}$

**8.** Use a graph to determine whether $x$ and $y$ are in a proportional relationship.

| x | 1 | 3 | 6 | 8 |
|---|---|---|---|---|
| y | 4 | 12 | 24 | 32 |

**5.3 Writing Proportions** *(pp. 178–183)*

**Write a proportion that gives the number $r$ of returns on Saturday.**

| | Friday | Saturday |
|---|---|---|
| **Sales** | 40 | 85 |
| **Returns** | 32 | $r$ |

⋮ One proportion is $\dfrac{40 \text{ sales}}{32 \text{ returns}} = \dfrac{85 \text{ sales}}{r \text{ returns}}$.

## Exercises

**Use the table to write a proportion.**

**9.**

| | Game 1 | Game 2 |
|---|---|---|
| **Penalties** | 6 | 8 |
| **Minutes** | 16 | $m$ |

**10.**

| | Concert 1 | Concert 2 |
|---|---|---|
| **Songs** | 15 | 18 |
| **Hours** | 2.5 | $h$ |

**5.4 Solving Proportions** *(pp. 186–191)*

**Solve $\dfrac{15}{2} = \dfrac{30}{y}$.**

| | |
|---|---|
| $15 \cdot y = 2 \cdot 30$ | Cross Products Property |
| $15y = 60$ | Multiply. |
| $y = 4$ | Divide. |

⋮ The solution is 4.

## Exercises

**Solve the proportion.**

**11.** $\dfrac{x}{4} = \dfrac{2}{5}$

**12.** $\dfrac{5}{12} = \dfrac{y}{15}$

**13.** $\dfrac{8}{20} = \dfrac{6}{w}$

**14.** $\dfrac{s+1}{4} = \dfrac{4}{8}$

## 5.5 Slope (pp. 192–197)

**The graph shows the number of visits your website received over the past 6 months. Find and interpret the slope.**

Choose any two points to find the slope of the line.

$$\text{slope} = \frac{\text{change in } y}{\text{change in } x}$$

$$= \frac{50}{1} \leftarrow \text{visits}$$
$$\leftarrow \text{months}$$

$$= 50$$

**Website Visits**

⋰ The slope of the line represents the unit rate. The slope is 50. So, the number of visits increased by 50 each month.

### Exercises

**Find the slope of the line.**

**15.**

**16.**

**17.**
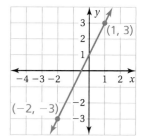

## 5.6 Direct Variation (pp. 198–203)

**Tell whether $x$ and $y$ show direct variation. Explain your reasoning.**

**a.** $x + y - 1 = 3$

$\qquad y = 4 - x \qquad$ Solve for $y$.

⋰ The equation *cannot* be written as $y = kx$. So, $x$ and $y$ do *not* show direct variation.

**b.** $x = 8y$

$\qquad \frac{1}{8}x = y \qquad$ Solve for $y$.

⋰ The equation can be written as $y = kx$. So, $x$ and $y$ show direct variation.

### Exercises

**Tell whether $x$ and $y$ show direct variation. Explain your reasoning.**

**18.** $x + y = 6$     **19.** $y - x = 0$     **20.** $\dfrac{x}{y} = 20$     **21.** $x = y + 2$

**Find the unit rate.**

**1.** 84 miles in 12 days

**2.** $2\frac{2}{5}$ kilometers in $3\frac{3}{4}$ minutes

**Tell whether the ratios form a proportion.**

**3.** $\frac{1}{9}, \frac{6}{54}$

**4.** $\frac{9}{12}, \frac{8}{72}$

**Use a graph to tell whether x and y are in a proportional relationship.**

**5.**

| x | 2 | 4 | 6 | 8 |
|---|---|---|---|---|
| y | 10 | 20 | 30 | 40 |

**6.**

| x | 1 | 3 | 5 | 7 |
|---|---|---|---|---|
| y | 3 | 7 | 11 | 15 |

**Use the table to write a proportion.**

**7.**

| | Monday | Tuesday |
|---|---|---|
| Gallons | 6 | 8 |
| Miles | 180 | $m$ |

**8.**

| | Thursday | Friday |
|---|---|---|
| Classes | 6 | $c$ |
| Hours | 8 | 4 |

**Solve the proportion.**

**9.** $\frac{x}{8} = \frac{9}{4}$

**10.** $\frac{17}{3} = \frac{y}{6}$

**Graph the line that passes through the two points. Then find the slope of the line.**

**11.** $(15, 9), (-5, -3)$

**12.** $(2, 9), (4, 18)$

**Tell whether x and y show direct variation. Explain your reasoning.**

**13.** $xy - 11 = 5$

**14.** $x = \frac{3}{y}$

**15.** $\frac{y}{x} = 8$

**16. MOVIE TICKETS** Five movie tickets cost $36.25. What is the cost of 8 movie tickets?

**17. CROSSWALK** The graph shows the number of cycles of a crosswalk signal during the day and during the night.

Don't Walk

Walk

**a.** Compare the steepness of the lines. What does this mean in the context of the problem?

**b.** Find and interpret the slope of each line.

Crosswalk Signal

**18. GLAZE** A specific shade of green glaze requires 5 parts blue to 3 parts yellow. A glaze mixture contains 25 quarts of blue and 9 quarts of yellow. How can you fix the mixture to make the specific shade of green glaze?

**1.** The school store sells 4 pencils for $0.80. What is the unit cost of a pencil? *(7.RP.1)*

   **A.** $0.20      **C.** $3.20

   **B.** $0.80      **D.** $5.00

**2.** Which expressions do *not* have a value of 3? *(7.NS.3)*

   I.   $2 + (-1)$      II.  $2 - (-1)$

   III.  $-3 \times (-1)$      IV.  $-3 \div (-1)$

   **F.** I only      **H.** II only

   **G.** III and IV      **I.** I, III, and IV

**Test-Taking Strategy**

**Read Question Before Answering**

What is **NOT** the ratio of human years to dog years?

(A) $\frac{1}{7}$   (B) 1:7   (C) 1 to 7   (D) 7

Newton the senior citizen

"Be sure to read the question before choosing your answer. You may find a word that changes the meaning."

**3.** What is the value of the expression below? *(7.NS.3)*

$$-4 \times (-6) - (-5)$$

**4.** What is the slope of the line shown? *(7.RP.2b)*

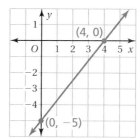

   **A.** $\frac{4}{5}$      **C.** 4

   **B.** $\frac{5}{4}$      **D.** 5

**5.** The graph below represents which inequality? *(7.EE.4b)*

   **F.** $-3 - 6x < -27$      **H.** $5 - 3x > -7$

   **G.** $2x + 6 \geq 14$      **I.** $2x + 3 \leq 11$

**6.** The quantities $x$ and $y$ are proportional. What is the missing value in the table? *(7.RP.2a)*

| $x$ | $y$ |
|-----|-----|
| $\dfrac{2}{3}$ | 6 |
| $\dfrac{4}{3}$ | 12 |
| $\dfrac{8}{3}$ | 24 |
| 5 | |

**A.** 30  **C.** 45

**B.** 36  **D.** 48

**7.** You are selling tomatoes. You have already earned $16 today. How many additional pounds of tomatoes do you need to sell to earn a total of $60? *(7.EE.4a)*

**F.** 4  **H.** 15

**G.** 11  **I.** 19

$4 per pound

**8.** The distance traveled by the a high-speed train is proportional to the number of hours traveled. Which of the following is *not* a valid interpretation of the graph below? *(7.RP.2d)*

**High-Speed Train**

(4, 800)

(1, 200)

(0, 0)

Distance (kilometers)

Hours

**A.** The train travels 0 kilometers in 0 hours.

**B.** The unit rate is 200 kilometers per hour.

**C.** After 4 hours, the train is traveling 800 kilometers per hour.

**D.** The train travels 800 kilometers in 4 hours.

**9.** Regina was evaluating the expression below. What should Regina do to correct the error she made? *(7.NS.3)*

$$-\frac{3}{2} \div \left(-\frac{8}{7}\right) = -\frac{2}{3} \times \left(-\frac{7}{8}\right)$$

$$= \frac{2 \times 7}{3 \times 8}$$

$$= \frac{14}{24}$$

$$= \frac{7}{12}$$

**F.** Rewrite $-\frac{3}{2} \div \left(-\frac{8}{7}\right)$ as $-\frac{2}{3} \times \left(-\frac{8}{7}\right)$.

**G.** Rewrite $-\frac{3}{2} \div \left(-\frac{8}{7}\right)$ as $-\frac{3}{2} \times \left(-\frac{7}{8}\right)$.

**H.** Rewrite $-\frac{3}{2} \div \left(-\frac{8}{7}\right)$ as $-\frac{3}{7} \times \left(-\frac{8}{2}\right)$.

**I.** Rewrite $-\frac{2}{3} \times \left(-\frac{7}{8}\right)$ as $-\frac{2 \times 7}{3 \times 8}$.

**10.** What is the least value of $t$ for which the inequality is true? *(7.EE.4b)*

$$3 - 6t \le -15$$

**11.** You can mow 800 square feet of lawn in 15 minutes. At this rate, how many minutes will you take to mow a lawn that measures 6000 square feet? *(7.RP.2c)*

*Part A* Write a proportion to represent the problem. Use $m$ to represent the number of minutes. Explain your reasoning.

*Part B* Solve the proportion you wrote in Part A. Then use it to answer the problem. Show your work.

**12.** What value of $p$ makes the equation below true? *(7.EE.4a)*

$$6 - 2p = -48$$

**A.** $-27$

**B.** $-21$

**C.** $21$

**D.** $27$

# 6 Percents

"Here's my sales strategy. I buy each dog bone for $0.05."

"Then I mark each one up to $1. Then, I have a 75% off sale. Cool, huh?"

I guess P.T. Barnum was right.

"At 4 a day, I have chewed 17,536 dog biscuits. At only 99.9% pure, that means that..."

"I have swallowed seventeen and a half contaminated dog biscuits during the past twelve years."

I hope the same logic doesn't apply for mice!

# What You Learned Before

"The fact that these two percents do not total 100 is a sad commentary on humans."

## ● Writing Percents as Fractions
(6.RP.3C)

**Example 1** Write 45% as a fraction in simplest form.

$$45\% = \frac{45}{100}$$

Write as a fraction with a denominator of 100.

$$= \frac{9}{20}$$

Simplify.

∴ So, $45\% = \frac{9}{20}$.

### Try It Yourself
**Write the percent as a fraction or mixed number in simplest form.**

**1.** 16%  **2.** 40%  **3.** 68%  **4.** 85%

**5.** 148%  **6.** 150%  **7.** 105%  **8.** 276%

## ● Writing Fractions as Percents (6.RP.3C)

**Example 2** Write $\frac{3}{25}$ as a percent.

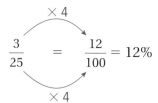

$$\frac{3}{25} = \frac{12}{100} = 12\%$$

Because $25 \times 4 = 100$, multiply the numerator and denominator by 4. Write the numerator with a percent symbol.

### Try It Yourself
**Write the fraction or mixed number as a percent.**

**9.** $\frac{9}{25}$  **10.** $\frac{43}{50}$  **11.** $\frac{11}{20}$  **12.** $\frac{3}{5}$

**13.** $1\frac{1}{4}$  **14.** $1\frac{12}{25}$  **15.** $1\frac{4}{5}$  **16.** $2\frac{3}{10}$

**Essential Question** How does the decimal point move when you rewrite a percent as a decimal and when you rewrite a decimal as a percent?

### 1 ACTIVITY: Writing Percents as Decimals

Work with a partner. Write the percent shown by the model. Write the percent as a decimal.

**a.**

$$\square \% = \frac{\square}{\square} \longleftarrow \text{per} \atop \longleftarrow \text{cent}$$

$$= \frac{\square}{\square} \qquad \text{Simplify.}$$

$$= \square \qquad \text{Write fraction as a decimal.}$$

**b.**

**c.**

**d.**

**e.**

**f.**

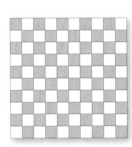

**g.**

COMMON CORE

**Percents and Decimals**

In this lesson, you will

- write percents as decimals.
- write decimals as percents.
- solve real-life problems.

Learning Standard
7.EE.3

## 2  ACTIVITY: Writing Percents as Decimals

**Math Practice 6**

**Communicate Precisely**

How can reading the fraction aloud help you write it as a decimal?

**Work with a partner. Write the percent as a decimal.**

**a.**  13.5%

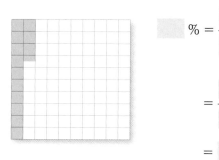

$\boxed{\phantom{x}} \% = \dfrac{\boxed{\phantom{x}}}{\boxed{\phantom{x}}}$  ← per  ← cent

$= \dfrac{\boxed{\phantom{x}}}{\boxed{\phantom{x}}}$  Multiply numerator and denominator by 10.

$= \boxed{\phantom{x}}$  Write fraction as a decimal.

**b.**  12.5%  **c.**  3.8%  **d.**  0.5%

## 3  ACTIVITY: Writing Decimals as Percents

**Work with a partner. Draw a model to represent the decimal. Write the decimal as a percent.**

**a.**  0.1

$0.1 \quad = \quad 0.10 = \dfrac{\boxed{\phantom{x}}}{\boxed{\phantom{x}}} \quad = \quad \boxed{\phantom{x}} \%$

One $\boxed{\phantom{x}}$  Ten $\boxed{\phantom{x}}$  $\boxed{\phantom{x}}$ Percent

**b.**  0.24  **c.**  0.58  **d.**  0.05

## What Is Your Answer?

**4.  IN YOUR OWN WORDS** How does the decimal point move when you rewrite a percent as a decimal and when you rewrite a decimal as a percent?

**5.** Explain why the decimal point moves when you rewrite a percent as a decimal and when you rewrite a decimal as a percent.

**Practice**

Use what you learned about percents and decimals to complete Exercises 7–12 and 19–24 on page 218.

### 🔒 Key Idea

**Writing Percents as Decimals**

**Words**    Remove the percent symbol. Then divide by 100, or just move the decimal point two places to the left.

**Numbers**    $23\% = 23.\% = 0.23$

---

**EXAMPLE**   **1**   **Writing Percents as Decimals**

**Study Tip**

When moving the decimal point, you may need to place one or more zeros in the number.

**a.**   Write 52% as a decimal.

$$52\% = 52.\% = 0.52$$

**Check**

**b.**   Write 7% as a decimal.

$$7\% = 07.\% = 0.07$$

**Check**

### ⚫ On Your Own

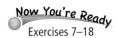
**Now You're Ready**
Exercises 7–18

Write the percent as a decimal. Use a model to check your answer.

**1.**   24%      **2.**   3%      **3.**   107%      **4.**   92.7%

---

### 🔒 Key Idea

**Writing Decimals as Percents**

**Words**    Multiply by 100, or just move the decimal point two places to the right. Then add a percent symbol.

**Numbers**    $0.36 = 0.36 = 36\%$

---

**EXAMPLE**   **2**   **Writing Decimals as Percents**

**a.**   Write 0.47 as a percent.

$$0.47 = 0.47 = 47\%$$

**b.**   Write 0.663 as a percent.

$$0.663 = 0.663 = 66.3\%$$

**c.**   Write 1.8 as a percent.

$$1.8 = 1.80 = 180\%$$

**d.**   Write 0.009 as a percent.

$$0.009 = 0.009 = 0.9\%$$

Now You're Ready
Exercises 19–30

## On Your Own

**Write the decimal as a percent. Use a model to check your answer.**

**5.** 0.94      **6.** 1.2      **7.** 0.316      **8.** 0.005

---

### EXAMPLE 3   Writing a Fraction as a Percent and a Decimal

On a math test, you get 92 out of a possible 100 points. Which of the following is *not* another way of expressing 92 out of 100?

**(A)** $\dfrac{23}{25}$      **(B)** 92%      **(C)** $\dfrac{17}{20}$      **(D)** 0.92

$$92 \text{ out of } 100 = \frac{92}{100}$$

$= 92\%$    Eliminate Choice B.

$= \dfrac{23}{25}$    Eliminate Choice A.

$= 0.92$    Eliminate Choice D.

So, the correct answer is **(C)**.

---

### EXAMPLE 4   Real-Life Application

The figure shows the portions of ultraviolet (UV) rays reflected by four different surfaces. How many times more UV rays are reflected by water than by sea foam?

Grass — 3%    Sand — 0.15    Sea foam — 25%    Water — $\dfrac{21}{25}$

Write 25% and $\dfrac{21}{25}$ as decimals.

**Sea foam:** $25\% = 25.\% = 0.25$      **Water:** $\dfrac{21}{25} = \dfrac{84}{100} - 0.84$

Divide 0.84 by 0.25:   $0.25\overline{)0.84} \longrightarrow 25\overline{)84.00}$   (quotient $3.36$)

So, water reflects about 3.4 times more UV rays than sea foam.

## On Your Own

**9.** Write "18 out of 100" as a percent, a fraction, and a decimal.

**10.** In Example 4, how many times more UV rays are reflected by water than by sand?

**Vocabulary and Concept Check**

**MATCHING** Match the decimal with its equivalent percent.

**1.** 0.42     **2.** 4.02     **3.** 0.042     **4.** 0.0402

   **A.** 4.02%     **B.** 42%     **C.** 4.2%     **D.** 402%

**5. OPEN-ENDED** Write three different decimals that are between 10% and 20%.

**6. WHICH ONE DOESN'T BELONG?** Which one does *not* belong with the other three? Explain your reasoning.

| 70% | 0.7 | $\frac{7}{10}$ | 0.07 |

 **Practice and Problem Solving**

**Write the percent as a decimal.**

**7.** 78%     **8.** 55%     **9.** 18.5%

**10.** 57.4%     **11.** 33%     **12.** 9%

**13.** 47.63%     **14.** 91.25%     **15.** 166%

**16.** 217%     **17.** 0.06%     **18.** 0.034%

**Write the decimal as a percent.**

**19.** 0.74     **20.** 0.52     **21.** 0.89

**22.** 0.768     **23.** 0.99     **24.** 0.49

**25.** 0.487     **26.** 0.128     **27.** 3.68

**28.** 5.12     **29.** 0.0371     **30.** 0.0046

**31. ERROR ANALYSIS** Describe and correct the error in writing 0.86 as a percent.

     ✗   0.86 = 00.86 = 0.0086%

**32. MUSIC** Thirty-six percent of the songs on your MP3 player are pop songs. Write this percent as a decimal.

**33. CAT** About 0.34 of the length of a cat is its tail. Write this decimal as a percent.

**34. COMPUTER** Write the percent of free space on the computer as a decimal.

| Volume | Capacity | Free Space | % Free Space |
|---|---|---|---|
| (C:) | 149 GB | 133 GB | 89 % |

**Write the percent as a fraction in simplest form and as a decimal.**

**35.** 36%      **36.** 23.5%      **37.** 16.24%

38. **SCHOOL** The percents of students who travel to school by car, bus, and bicycle are shown for a school of 825 students.

**Car: 20%**

**School bus: 48%**

**Bicycle: 8%**

a. Write the percents as decimals.

b. Write the percents as fractions.

c. What percent of students use another method to travel to school?

d. **RESEARCH** Make a bar graph that represents how the students in your class travel to school.

39. **ELECTIONS** In an election, the winning candidate receives 60% of the votes. What percent of the votes does the other candidate receive?

40. **COLORS** Students in a class were asked to tell their favorite color.

a. What percent said red, blue, or yellow?

b. How many times more students said red than yellow?

c. Use two methods to find the percent of students who said green. Which method do you prefer?

**Favorite Color**

41. **Problem Solving** In the first 42 Super Bowls, $0.1\overline{6}$ of the MVPs (most valuable players) were running backs.

a. What percent of the MVPs were running backs?

b. What fraction of the MVPs were *not* running backs?

 **Fair Game Review** What you learned in previous grades & lessons

**Write the decimal as a fraction or mixed number in simplest form.**
*(Skills Review Handbook)*

**42.** 0.46    **43.** 0.31    **44.** 2.2    **45.** 4.32

**Simplify the expression.** *(Section 3.1)*

**46.** $4x + 3 - 9x$

**47.** $5 + 3.2n - 6 - 4.8n$

**48.** $2y - 5(y - 3)$

**49.** $-\dfrac{1}{2}(8b + 3) + 3b$

50. **MULTIPLE CHOICE** Ham costs $4.48 per pound. Cheese costs $6.36 per pound. You buy 1.5 pounds of ham and 0.75 pound of cheese. How much more do you pay for the ham? *(Skills Review Handbook)*

Ⓐ $1.41    Ⓑ $1.95    Ⓒ $4.77    Ⓓ $6.18

**Essential Question** How can you order numbers that are written as fractions, decimals, and percents?

### 1 ACTIVITY: Using Fractions, Decimals, and Percents

**Work with a partner. Decide which number form (fraction, decimal, or percent) is more common. Then find which is greater.**

**a.** 7% sales tax    or    $\frac{1}{20}$ sales tax

**b.** 0.37 cup of flour    or    $\frac{1}{3}$ cup of flour

**c.** $\frac{5}{8}$-inch wrench    or    0.375-inch wrench

**d.** $12\frac{3}{5}$ dollars    or    12.56 dollars

**e.** 93% test score    or    $\frac{7}{8}$ test score

**f.** $5\frac{5}{6}$ fluid ounces    or    5.6 fluid ounces

**COMMON CORE**

Fractions, Decimals, and Percents

In this lesson, you will
- compare and order fractions, decimals, and percents.
- solve real-life problems.

Learning Standard 7.EE.3

### 2 ACTIVITY: Ordering Numbers

**Work with a partner to order the following numbers.**

$$\frac{1}{8} \qquad 11\% \qquad \frac{3}{20} \qquad 0.172 \qquad 0.32 \qquad 43\% \qquad 7\% \qquad 0.7 \qquad \frac{5}{6}$$

**a.** Decide on a strategy for ordering the numbers. Will you write them all as fractions, decimals, or percents?

**b.** Use your strategy and a number line to order the numbers from least to greatest. (Note: Label the number line appropriately.)

**Math Practice 2**

**Make Sense of Quantities**
What strategies can you use to determine which number is greater?

**Preparation:**
- Cut index cards to make 40 playing cards.
- Write each number in the table onto a card.

**To Play:**
- Play with a partner.
- Deal 20 cards facedown to each player.
- Each player turns one card faceup. The player with the greater number wins. The winner collects both cards and places them at the bottom of his or her cards.
- Suppose there is a tie. Each player lays three cards facedown, then a new card faceup. The player with the greater of these new cards wins. The winner collects all 10 cards and places them at the bottom of his or her cards.
- Continue playing until one player has all the cards. This player wins the game.

| 75% | $\frac{3}{4}$ | $\frac{1}{3}$ | $\frac{3}{10}$ | 0.3 | 25% | 0.4 | 0.25 | 100% | 0.27 |
|---|---|---|---|---|---|---|---|---|---|
| 0.75 | $66\frac{2}{3}\%$ | 12.5% | 40% | $\frac{1}{4}$ | 4% | 0.5% | 0.04 | $\frac{1}{100}$ | $\frac{2}{3}$ |
| 0 | 30% | 5% | $\frac{27}{100}$ | 0.05 | $33\frac{1}{3}\%$ | $\frac{2}{5}$ | 0.333... | 27% | 1% |
| 1 | 0.01 | $\frac{1}{20}$ | $\frac{1}{8}$ | 0.125 | $\frac{1}{25}$ | $\frac{1}{200}$ | 0.005 | 0.666... | 0% |

## What Is Your Answer?

**4. IN YOUR OWN WORDS** How can you order numbers that are written as fractions, decimals, and percents? Give an example with your answer.

**5.** All but one of the U.S. coins shown has a name that is related to its value. Which one is it? How are the names of the others related to their values?

**Practice** ▶ Use what you learned about ordering numbers to complete Exercises 4–7, 16, and 17 on page 224.

When comparing and ordering fractions, decimals, and percents, write the numbers as all fractions, all decimals, or all percents.

**EXAMPLE** 1   **Comparing Fractions, Decimals, and Percents**

a.  **Which is greater, $\frac{3}{20}$ or 16%?**

Write $\frac{3}{20}$ as a percent: $\frac{3}{20} = \frac{15}{100} = 15\%$

**Study Tip**

It is usually easier to order decimals or percents than to order fractions.

∷• 15% is less than 16%. So, 16% is the greater number.

b.  **Which is greater, 79% or 0.08?**

Write 79% as a decimal:  79% = 79.% = 0.79

∷• 0.79 is greater than 0.08. So, 79% is the greater number.

**On Your Own**

Now You're Ready
Exercises 4–15

1.  Which is greater, 25% or $\frac{7}{25}$?

2.  Which is greater, 0.49 or 94%?

**EXAMPLE** 2   **Real-Life Application**

**You, your sister, and a friend each take the same number of shots at a soccer goal. You make 72% of your shots, your sister makes $\frac{19}{25}$ of her shots, and your friend makes 0.67 of his shots. Who made the fewest shots?**

**Remember**

To order numbers from least to greatest, write them as they appear on a number line from left to right.

Write 72% and $\frac{19}{25}$ as decimals.

**You:**  72% = 72.% = 0.72

**Sister:**  $\frac{19}{25} = \frac{76}{100} = 0.76$ (× 4)

Graph the decimals on a number line.

Friend: 0.67     You: 72% = 0.72     Sister: $\frac{19}{25} = 0.76$

0.66  0.68  0.70  0.72  0.74  0.76  0.78

∷• 0.67 is the least number. So, your friend made the fewest shots.

## On Your Own

**Now You're Ready**
Exercises 16–21

**3.** You make 75% of your shots, your sister makes $\frac{13}{20}$ of her shots, and your friend makes 0.7 of his shots. Who made the most shots?

---

**EXAMPLE** ③ **Real-Life Application**

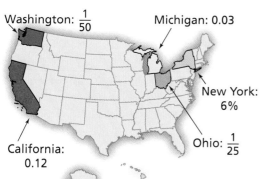

Washington: $\frac{1}{50}$    Michigan: 0.03

New York: 6%

California: 0.12    Ohio: $\frac{1}{25}$

**The map shows the portions of the U.S. population that live in five states.**

**List the five states in order by population from least to greatest.**

Begin by writing each portion as a fraction, a decimal, and a percent.

| State | Fraction | Decimal | Percent |
|-------|----------|---------|---------|
| Michigan | $\frac{3}{100}$ | 0.03 | 3% |
| New York | $\frac{6}{100}$ | 0.06 | 6% |
| Washington | $\frac{1}{50}$ | 0.02 | 2% |
| California | $\frac{12}{100}$ | 0.12 | 12% |
| Ohio | $\frac{1}{25}$ | 0.04 | 4% |

Graph the percent for each state on a number line.

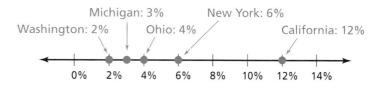

The states in order by population from least to greatest are Washington, Michigan, Ohio, New York, and California.

## On Your Own

**4.** The portion of the U.S. population that lives in Texas is $\frac{2}{25}$. The portion that lives in Illinois is 0.042. Reorder the states in Example 3 including Texas and Illinois.

## Vocabulary and Concept Check

1. **NUMBER SENSE** Copy and complete the table.

2. **NUMBER SENSE** How would you decide whether $\frac{3}{5}$ or 59% is greater? Explain.

3. **WHICH ONE DOESN'T BELONG?** Which one does *not* belong with the other three? Explain your reasoning.

| 40% | $\frac{2}{5}$ |
|-----|-----|
| 0.4 | 0.04 |

| Fraction | Decimal | Percent |
|----------|---------|---------|
| $\frac{18}{25}$ | 0.72 | |
| $\frac{17}{20}$ | | 85% |
| $\frac{13}{50}$ | | |
| | 0.62 | |
| | | 45% |

## Practice and Problem Solving

**Tell which number is greater.**

① 4. 0.9, 95%          5. 20%, 0.02          6. $\frac{37}{50}$, 37%          7. 50%, $\frac{13}{25}$

8. 0.086, 86%          9. 76%, 0.67          10. 60%, $\frac{5}{8}$          11. 0.12, 1.2%

12. 17%, $\frac{4}{25}$          13. 140%, 0.14          14. $\frac{1}{3}$, 30%          15. 80%, $\frac{7}{9}$

**Use a number line to order the numbers from least to greatest.**

② 16. 38%, $\frac{8}{25}$, 0.41          17. 68%, 0.63, $\frac{13}{20}$

18. $\frac{43}{50}$, 0.91, $\frac{7}{8}$, 84%          19. 0.15%, $\frac{3}{20}$, 0.015

20. 2.62, $2\frac{2}{5}$, 26.8%, 2.26, 271%          21. $\frac{87}{200}$, 0.44, 43.7%, $\frac{21}{50}$

22. **TEST** You answered 21 out of 25 questions correctly on a test. Did you reach your goal of getting at least 80%?

23. **POPULATION** The table shows the portions of the world population that live in four countries. Order the countries by population from least to greatest.

| Country | Brazil | India | Russia | United States |
|---------|--------|-------|--------|---------------|
| **Portion of World Population** | 2.8% | $\frac{7}{40}$ | $\frac{1}{50}$ | 0.044 |

**PRECISION** Order the numbers from least to greatest.

24. 66.1%, 0.66, $\frac{2}{3}$, 0.667

25. $\frac{2}{9}$, 21%, 0.2$\bar{1}$, $\frac{11}{50}$

**Tell which letter shows the graph of the number.**

26. $\frac{2}{5}$ 　　　　27. 45.2% 　　　　28. 0.435 　　　　29. $\frac{4}{9}$

30. **TOUR DE FRANCE** The Tour de France is a bicycle road race. The whole race is made up of 21 small races called *stages*. The table shows how several stages compare to the whole Tour de France in a recent year. Order the stages from shortest to longest.

| Stage | 1 | 7 | 8 | 17 | 21 |
|---|---|---|---|---|---|
| Portion of Total Distance | $\frac{11}{200}$ | 0.044 | $\frac{6}{125}$ | 0.06 | 4% |

31. **SLEEP** The table shows the portions of the day that several animals sleep.

　a. Order the animals by sleep time from least to greatest.

　b. Estimate the portion of the day that you sleep.

　c. Where do you fit on the ordered list?

| Animal | Portion of Day Sleeping |
|---|---|
| Dolphin | 0.433 |
| Lion | 56.3% |
| Rabbit | $\frac{19}{40}$ |
| Squirrel | $\frac{31}{50}$ |
| Tiger | 65.8% |

32.  Tell what whole number you can substitute for $a$ in each list so the numbers are ordered from least to greatest. If there is none, explain why.

　a. $\frac{2}{a}$, $\frac{a}{22}$, 33%　　　　b. $\frac{1}{a}$, $\frac{a}{8}$, 33%

**Fair Game Review** What you learned in previous grades & lessons

**Tell whether the ratios form a proportion.** *(Section 5.2)*

33. $\frac{6}{10}$, $\frac{9}{15}$ 　　　　34. $\frac{7}{16}$, $\frac{28}{80}$ 　　　　35. $\frac{20}{12}$, $\frac{35}{21}$

36. **MULTIPLE CHOICE** What is the solution of $2n - 4 > -12$? *(Section 4.4)*

Ⓐ $n < -10$ 　　Ⓑ $n < -4$ 　　Ⓒ $n > -2$ 　　Ⓓ $n > -4$

# 6.3 The Percent Proportion

## Essential Question How can you use models to estimate percent questions?

The statement "25% of 12 is 3" has three numbers. In real-life problems, any one of these numbers can be unknown.

| Question | Which number is missing? | Type of Question |
|---|---|---|
| What is 25% of 12? | 3 | Find a part of a number. |
| 3 is what percent of 12? | 25% | Find a percent. |
| 3 is 25% of what? | 12 | Find the whole. |

### 1 ACTIVITY: Estimating a Part

**Work with a partner. Use a model to estimate the answer to each question.**

a. What number is 50% of 30?

⋮⟶ So, from the model, ▢ is 50% of 30.

b. What number is 75% of 30?

c. What number is 40% of 30?

d. What number is 6% of 30?

e. What number is 65% of 30?

### 2 ACTIVITY: Estimating a Percent

**Work with a partner. Use a model to estimate the answer to each question.**

a. 15 is what percent of 75?

⋮⟶ So, from the model, 15 is ▢ of 75.

b. 5 is what percent of 20?

c. 18 is what percent of 40?

d. 50 is what percent of 80?

e. 75 is what percent of 50?

COMMON CORE

**Percent Proportion**

In this lesson, you will

- use the percent proportion to find parts, wholes, and percents.

Learning Standard
7.RP.3

**3** **ACTIVITY: Estimating a Whole**

**Math Practice 4**

**Use a Model**
What quantities are given? How can you use the model to find the unknown quantity?

Work with a partner. Use a model to estimate the answer to each question.

**a.** 24 is $33\frac{1}{3}$% of what number?

So, from the model, 24 is $33\frac{1}{3}$% of ☐ .

**b.** 13 is 25% of what number?

**c.** 110 is 20% of what number?

**d.** 75 is 75% of what number?

**e.** 81 is 45% of what number?

**4** **ACTIVITY: Using Ratio Tables**

Work with a partner. Use a ratio table to answer each question. Then compare your answer to the estimate you found using the model.

**1d** **a.** What number is 6% of 30?

| Part | 6 | | |
|------|---|---|---|
| Whole | 100 | | 30 |

**1e** **b.** What number is 65% of 30?

| Part | 65 | | |
|------|----|---|---|
| Whole | 100 | | 30 |

**2c** **c.** 18 is what percent of 40?

| Part | 18 | | |
|------|----|---|-----|
| Whole | 40 | | 100 |

**3e** **d.** 81 is 45% of what number?

| Part | 45 | | 81 |
|------|----|---|----|
| Whole | 100 | | |

## What Is Your Answer?

**5.** **IN YOUR OWN WORDS** How can you use models to estimate percent questions? Give examples to support your answer.

**6.** Complete the proportion below using the given labels.

percent

whole

100

part

$$\frac{\phantom{xxx}}{\phantom{xxx}} = \frac{\phantom{xxx}}{\phantom{xxx}}$$

**Practice**

Use what you learned about estimating percent questions to complete Exercises 5–10 on page 230.

Key Idea

**The Percent Proportion**

**Words** You can represent "$a$ is $p$ percent of $w$" with the proportion

$$\frac{a}{w} = \frac{p}{100}$$

where $a$ is part of the whole $w$, and $p\%$, or $\frac{p}{100}$, is the percent.

**Study Tip**

In percent problems, the word *of* is usually followed by the whole.

**Numbers**　　　　　　　　3 out of 4 is 75%.

part → $\dfrac{3}{4} = \dfrac{75}{100}$ ← percent

whole →

---

**EXAMPLE 1 Finding a Percent**

**What percent of 15 is 12?**

$\dfrac{a}{w} = \dfrac{p}{100}$ 　　　Write the percent proportion.

$\dfrac{12}{15} = \dfrac{p}{100}$ 　　　Substitute 12 for $a$ and 15 for $w$.

$100 \cdot \dfrac{12}{15} = 100 \cdot \dfrac{p}{100}$ 　　　Multiplication Property of Equality

$80 = p$ 　　　Simplify.

∴ So, 80% of 15 is 12.

---

**EXAMPLE 2 Finding a Part**

**What number is 36% of 50?**

$\dfrac{a}{w} = \dfrac{p}{100}$ 　　　Write the percent proportion.

$\dfrac{a}{50} = \dfrac{36}{100}$ 　　　Substitute 50 for $w$ and 36 for $p$.

$50 \cdot \dfrac{a}{50} = 50 \cdot \dfrac{36}{100}$ 　　　Multiplication Property of Equality

$a = 18$ 　　　Simplify.

∴ So, 18 is 36% of 50.

### 150% of what number is 24?

$$\frac{a}{w} = \frac{p}{100}$$    Write the percent proportion.

$$\frac{24}{w} = \frac{150}{100}$$    Substitute 24 for $a$ and 150 for $p$.

$$24 \cdot 100 = w \cdot 150$$    Cross Products Property

$$2400 = 150w$$    Multiply.

$$16 = w$$    Divide each side by 150.

⋮ So, 150% of 16 is 24.

## On Your Own

Now You're Ready
Exercises 11–18

**Write and solve a proportion to answer the question.**

1. What percent of 5 is 3?

2. 25 is what percent of 20?

3. What number is 80% of 60?

4. 10% of 40.5 is what number?

5. 0.1% of what number is 4?

6. $\frac{1}{2}$ is 25% of what number?

---

2011 Alabama Tornadoes

**The bar graph shows the strengths of tornadoes that occurred in Alabama in 2011. What percent of the tornadoes were EF1s?**

The total number of tornadoes, 145, is the *whole*, and the number of EF1 tornadoes, 58, is the *part*.

$$\frac{a}{w} = \frac{p}{100}$$    Write the percent proportion.

$$\frac{58}{145} = \frac{p}{100}$$    Substitute 58 for $a$ and 145 for $w$.

$$100 \cdot \frac{58}{145} = 100 \cdot \frac{p}{100}$$    Multiplication Property of Equality

$$40 = p$$    Simplify.

⋮ So, 40% of the tornadoes were EF1s.

## On Your Own

7. Twenty percent of the tornadoes occurred in central Alabama on April 27. How many tornadoes does this represent?

**Check It Out**
Help with Homework
BigIdeasMath ✓com

 **Vocabulary and Concept Check**

1. **VOCABULARY** Write the percent proportion in words.

2. **WRITING** Explain how to use a proportion to find 30% of a number.

3. **NUMBER SENSE** Write and solve the percent proportion represented by the model.

   0%  20%  40%  60%  80%  100%

   0                              40

4. **WHICH ONE DOESN'T BELONG?** Which proportion does *not* belong with the other three? Explain your reasoning.

   $$\frac{15}{w} = \frac{50}{100}$$    $$\frac{12}{15} = \frac{40}{n}$$    $$\frac{15}{25} = \frac{p}{100}$$    $$\frac{a}{20} = \frac{35}{100}$$

 **Practice and Problem Solving**

**Use a model to estimate the answer to the question. Use a ratio table to check your answer.**

5. What number is 24% of 80?

6. 15 is what percent of 40?

7. 15 is 30% of what number?

8. What number is 120% of 70?

9. 20 is what percent of 52?

10. 48 is 75% of what number?

**Write and solve a proportion to answer the question.**

① 11. What percent of 25 is 12?

12. 14 is what percent of 56?

② 13. 25% of what number is 9?

14. 36 is 0.9% of what number?

③ 15. 75% of 124 is what number?

16. 110% of 90 is what number?

17. What number is 0.4% of 40?

18. 72 is what percent of 45?

$$\frac{a}{w} = \frac{p}{100}$$
$$\frac{a}{34} = \frac{40}{100}$$
$$a = 13.6$$

19. **ERROR ANALYSIS** Describe and correct the error in using the percent proportion to answer the question below.

    "40% of what number is 34?"

20. **FITNESS** Of 140 seventh-grade students, 15% earn the Presidential Physical Fitness Award. How many students earn the award?

21. **COMMISSION** A salesperson receives a 3% commission on sales. The salesperson receives $180 in commission. What is the amount of sales?

**Write and solve a proportion to answer the question.**

**22.** 0.5 is what percent of 20?

**23.** 14.2 is 35.5% of what number?

**24.** $\frac{3}{4}$ is 60% of what number?

**25.** What number is 25% of $\frac{7}{8}$?

**26.** **HOMEWORK** You are assigned 32 math exercises for homework. You complete 87.5% of these before dinner. How many do you have left to do after dinner?

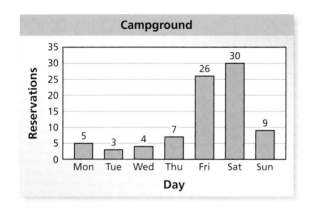

**27.** **HOURLY WAGE** Your friend earns $10.50 per hour. This is 125% of her hourly wage last year. How much did your friend earn per hour last year?

**28.** **CAMPSITE** The bar graph shows the numbers of reserved campsites at a campground for one week. What percent of the reservations were for Friday or Saturday?

**29.** **PROBLEM SOLVING** A classmate displays the results of a class president election in the bar graph shown.

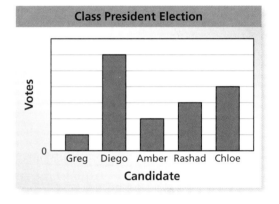

    **a.** What is missing from the bar graph?

    **b.** What percent of the votes does the last-place candidate receive? Explain your reasoning.

    **c.** There are 124 votes total. How many votes does Chloe receive?

**30.** **REASONING** 20% of a number is $x$. What is 100% of the number? Assume $x > 0$.

**31.** **Structure** Answer each question. Assume $x > 0$.

    **a.** What percent of $8x$ is $5x$?

    **b.** What is 65% of $80x$?

 **Fair Game Review** *What you learned in previous grades & lessons*

**Evaluate the expression when $a = -15$ and $b = -5$.** *(Section 1.5)*

**32.** $a \div b$

**33.** $\dfrac{b + 14}{a}$

**34.** $\dfrac{b^2}{a + 5}$

**35.** **MULTIPLE CHOICE** What is the solution of $9x = -1.8$? *(Section 3.4)*

    **A** $x = -5$      **B** $x = -0.2$      **C** $x = 0.2$      **D** $x = 5$

## 6.4 The Percent Equation

**Essential Question** How can you use an equivalent form of the percent proportion to solve a percent problem?

### 1 ACTIVITY: Solving Percent Problems Using Different Methods

**Work with a partner.** The circle graph shows the number of votes received by each candidate during a school election. So far, only half the students have voted.

**Votes Received by Each Candidate**

a. Complete the table.

| Candidate | Number of votes received / Total number of votes |
|-----------|------------------------------------------------|
| Sue | |
| Miguel | |
| Leon | |
| Hong | |

b. Find the percent of students who voted for each candidate. Explain the method you used to find your answers.

c. Compare the method you used in part (b) with the methods used by other students in your class. Which method do you prefer? Explain.

### 2 ACTIVITY: Finding Parts Using Different Methods

**Work with a partner.** The circle graph shows the final results of the election.

**Final Results**

a. Find the number of students who voted for each candidate. Explain the method you used to find your answers.

b. Compare the method you used in part (a) with the methods used by other students in your class. Which method do you prefer? Explain.

**COMMON CORE**

**Percent Equation**

In this lesson, you will
- use the percent equation to find parts, wholes, and percents.
- solve real-life problems.

Learning Standards
7.RP.3
7.EE.3

**ACTIVITY: Deriving the Percent Equation**

**Work with a partner. In Section 6.3, you used the percent proportion to find the missing percent, part, or whole. You can also use the *percent equation* to find these missing values.**

a.  Complete the steps below to find the percent equation.

$$\frac{\text{part}}{\text{whole}} = \text{percent}$$   Definition of percent

$$\frac{\text{part}}{\text{whole}} \cdot \boxed{\phantom{xx}} = \boxed{\phantom{xx}} \cdot \boxed{\phantom{xx}}$$   Multiply each side by the ☐.

$$\text{part} = \boxed{\phantom{xx}} \cdot \boxed{\phantom{xx}}$$   Divide out common factors.
This is the percent equation.

b.  Use the percent equation to find the number of students who voted for each candidate in Activity 2. How does this method compare to the percent proportion?

4  **ACTIVITY: Identifying Different Equations**

**Work with a partner. Without doing any calculations, choose the equation that you cannot use to answer each question.**

**Math Practice 3**

**Justify Conclusions**

How can you justify the equations that you chose?

a.  What number is 55% of 80?

$$a = 0.55 \cdot 80 \qquad a = \frac{11}{20} \cdot 80 \qquad 80a = 0.55 \qquad \frac{a}{80} = \frac{55}{100}$$

b.  24 is 60% of what number?

$$\frac{24}{w} = \frac{60}{100} \qquad 24 = 0.6 \cdot w \qquad \frac{24}{60} = w \qquad 24 = \frac{3}{5} \cdot w$$

## What Is Your Answer?

5.  **IN YOUR OWN WORDS** How can you use an equivalent form of the percent proportion to solve a percent problem?

6.  Write a percent proportion and a percent equation that you can use to answer the question below.

16 is what percent of 250?

**Practice**

Use what you learned about solving percent problems to complete Exercises 4–9 on page 236.

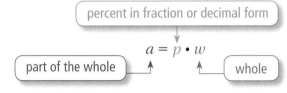 Key Idea

**The Percent Equation**

**Words**  To represent "$a$ is $p$ percent of $w$," use an equation.

percent in fraction or decimal form

$$a = p \cdot w$$

part of the whole          whole

**Numbers**          $15 = 0.5 \cdot 30$

---

**EXAMPLE** **1**  **Finding a Part of a Number**

**What number is 24% of 50?**          **Estimate**  0%  25%          100%

0          12.5          50

$a = p \cdot w$          Write percent equation.

$= \dfrac{24}{100} \cdot 50$          Substitute $\frac{24}{100}$ for $p$ and 50 for $w$.

$= 12$          Simplify.

∴ So, 12 is 24% of 50.          **Reasonable?**  $12 \approx 12.5$

---

**EXAMPLE** **2**  **Finding a Percent**

**9.5 is what percent of 25?**          **Estimate**  0%  40%          100%

0          10          25

$a = p \cdot w$          Write percent equation.

$9.5 = p \cdot 25$          Substitute 9.5 for $a$ and 25 for $w$.

$\dfrac{9.5}{25} = \dfrac{p \cdot 25}{25}$          Division Property of Equality

$0.38 = p$          Simplify.

∴ Because 0.38 equals 38%, 9.5 is 38% of 25.          **Reasonable?**  $38\% \approx 40\%$

**EXAMPLE 3** **Finding a Whole**

**39 is 52% of what number?**   **Estimate**

$$a = p \cdot w$$   Write percent equation.

$$39 = 0.52 \cdot w$$   Substitute 39 for $a$ and 0.52 for $p$.

$$75 = w$$   Divide each side by 0.52.

∴ So, 39 is 52% of 75.   **Reasonable?** $75 \approx 78$ ✓

## On Your Own

**Now You're Ready**
Exercises 10–17

**Write and solve an equation to answer the question.**

1. What number is 10% of 20?   2. What number is 150% of 40?

3. 3 is what percent of 600?   4. 18 is what percent of 20?

5. 8 is 80% of what number?   6. 90 is 18% of what number?

---

**EXAMPLE 4** **Real-Life Application**

### 8th Street Cafe

DATE:   MAY04'13 05:45PM
TABLE:   29
SERVER: JANE

**Food Total**      27.50
**Tax**          1.65
**Subtotal**       29.15

TIP: _____

TOTAL: _____

*Thank You*

**a. Find the percent of sales tax on the food total.**

Answer the question:  $1.65 is what percent of $27.50?

$$a = p \cdot w$$   Write percent equation.

$$1.65 = p \cdot 27.50$$   Substitute 1.65 for $a$ and 27.50 for $w$.

$$0.06 = p$$   Divide each side by 27.50.

∴ Because 0.06 equals 6%, the percent of sales tax is 6%.

**b. Find the amount of a 16% tip on the food total.**

Answer the question:  What tip amount is 16% of $27.50?

$$a = p \cdot w$$   Write percent equation.

$$= 0.16 \cdot 27.50$$   Substitute 0.16 for $p$ and 27.50 for $w$.

$$= 4.40$$   Multiply.

∴ So, the amount of the tip is $4.40.

## On Your Own

7. **WHAT IF?** Find the amount of a 20% tip on the food total.

## ✓ Vocabulary and Concept Check

1. **VOCABULARY** Write the percent equation in words.

2. **REASONING** A number $n$ is 150% of number $m$. Is $n$ *greater than, less than,* or *equal to m*? Explain your reasoning.

3. **DIFFERENT WORDS, SAME QUESTION** Which is different? Find "both" answers.

| | |
|---|---|
| What number is 20% of 55? | 55 is 20% of what number? |
| 20% of 55 is what number? | 0.2 • 55 is what number? |

## Practice and Problem Solving

**Answer the question. Explain the method you chose.**

4. What number is 24% of 80?

5. 15 is what percent of 40?

6. 15 is 30% of what number?

7. What number is 120% of 70?

8. 20 is what percent of 52?

9. 48 is 75% of what number?

**Write and solve an equation to answer the question.**

10. 20% of 150 is what number?

11. 45 is what percent of 60?

12. 35% of what number is 35?

13. 0.8% of 150 is what number?

14. 29 is what percent of 20?

15. 0.5% of what number is 12?

16. What percent of 300 is 51?

17. 120% of what number is 102?

**ERROR ANALYSIS** Describe and correct the error in using the percent equation.

18. What number is 35% of 20?

$$✗ \quad a = p \cdot w$$
$$= 35 \cdot 20$$
$$= 700$$

19. 30 is 60% of what number?

$$✗ \quad a = p \cdot w$$
$$= 0.6 \cdot 30$$
$$= 18$$

20. **COMMISSION** A salesperson receives a 2.5% commission on sales. What commission does the salesperson receive for $8000 in sales?

21. **FUNDRAISING** Your school raised 125% of its fundraising goal. The school raised $6750. What was the goal?

22. **SURFBOARD** The sales tax on a surfboard is $12. What is the percent of sales tax?

SALE
$240

**PUZZLE** There were $w$ signers of the Declaration of Independence. The youngest was Edward Rutledge, who was $x$ years old. The oldest was Benjamin Franklin, who was $y$ years old.

**23.** $x$ is 25% of 104. What was Rutledge's age?

**24.** 7 is 10% of $y$. What was Franklin's age?

**25.** $w$ is 80% of $y$. How many signers were there?

**26.** $y$ is what percent of $(w + y - x)$?

**Favorite Sport**

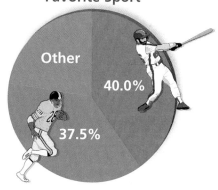

Other

40.0%

37.5%

**27.** **LOGIC** How can you tell whether the percent of a number will be *greater than*, *less than*, or *equal to* the number? Give examples to support your answer.

**28.** **SURVEY** In a survey, a group of students were asked their favorite sport. Eighteen students chose "other" sports.

    **a.** How many students participated?

    **b.** How many chose football?

**29.** **WATER TANK** Water tank $A$ has a capacity of 550 gallons and is 66% full. Water tank $B$ is 53% full. The ratio of the capacity of Tank $A$ to Tank $B$ is 11 : 15.

    **a.** How much water is in Tank $A$?

    **b.** What is the capacity of Tank $B$?

    **c.** How much water is in Tank $B$?

**30.** **TRUE OR FALSE?** Tell whether the statement is *true* or *false*. Explain your reasoning.

    If $W$ is 25% of $Z$, then $Z : W$ is 75 : 25.

**31.** **Reasoning** The table shows your test results for math class. What test score do you need on the last exam to earn 90% of the total points?

| Test Score | Point Value |
|---|---|
| 83% | 100 |
| 91.6% | 250 |
| 88% | 150 |
| ? | 300 |

**Fair Game Review** *What you learned in previous grades & lessons*

**Simplify. Write the answer as a decimal.** *(Skills Review Handbook)*

**32.** $\dfrac{10 - 4}{10}$

**33.** $\dfrac{25 - 3}{25}$

**34.** $\dfrac{105 - 84}{84}$

**35.** $\dfrac{170 - 125}{125}$

**36.** **MULTIPLE CHOICE** There are 160 people in a grade. The ratio of boys to girls is 3 to 5. Which proportion can you use to find the number $x$ of boys? *(Section 5.3)*

    Ⓐ $\dfrac{3}{8} = \dfrac{x}{160}$      Ⓑ $\dfrac{3}{5} = \dfrac{x}{160}$      Ⓒ $\dfrac{5}{8} = \dfrac{x}{160}$      Ⓓ $\dfrac{3}{5} = \dfrac{160}{x}$

**Check It Out**
Graphic Organizer
BigIdeasMath.com

You can use a **summary triangle** to explain a concept. Here is an example of a summary triangle for writing a percent as a decimal.

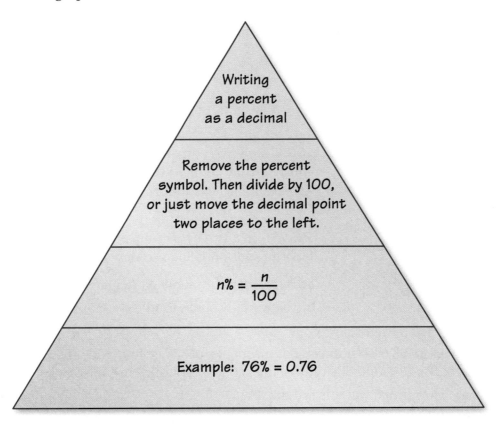

Writing a percent as a decimal

Remove the percent symbol. Then divide by 100, or just move the decimal point two places to the left.

$n\% = \dfrac{n}{100}$

Example: $76\% = 0.76$

## On Your Own

**Make summary triangles to help you study these topics.**

1. writing a decimal as a percent

2. comparing and ordering fractions, decimals, and percents

3. the percent proportion

4. the percent equation

**After you complete this chapter, make summary triangles for the following topics.**

5. percent of change
6. discount

7. markup
8. simple interest

I'm writing to Classy Calicos about this one.

Lonely?
Buy a pet.
Dogs are cool!
Example: Beagles

**"I found this great summary triangle in my *Beautiful Beagle Magazine*."**

**Write the percent as a decimal.**    *(Section 6.1)*

**1.** 34%

**2.** 0.12%

**3.** 62.5%

**Write the decimal as a percent.**    *(Section 6.1)*

**4.** 0.67

**5.** 5.35

**6.** 0.685

**Tell which number is greater.**    *(Section 6.2)*

**7.** $\frac{11}{15}$, 74%

**8.** 3%, 0.3

**Use a number line to order the numbers from least to greatest.**    *(Section 6.2)*

**9.** 125%, $\frac{6}{5}$, 1.22

**10.** 42%, 0.43, $\frac{17}{40}$

**Write and solve a proportion to answer the question.**    *(Section 6.3)*

**11.** What percent of 15 is 6?

**12.** 35 is what percent of 25?

**13.** What number is 40% of 50?

**14.** 0.5% of what number is 5?

**Write and solve an equation to answer the question.**    *(Section 6.4)*

**15.** What number is 28% of 75?

**16.** 42 is 21% of what number?

**17. FISHING** On a fishing trip, 38% of the fish that you catch are perch. Write this percent as a decimal.    *(Section 6.1)*

**18. SCAVENGER HUNT** The table shows the results of 8 teams competing in a scavenger hunt. Which team collected the most items? Which team collected the fewest items?    *(Section 6.2)*

| Team | 1 | 2 | 3 | 4 | 5 | 6 | 7 | 8 |
|---|---|---|---|---|---|---|---|---|
| Portion Collected | $\frac{3}{4}$ | 0.8 | 77.5% | 0.825 | $\frac{29}{40}$ | 76.25% | $\frac{63}{80}$ | 81.25% |

**19. COMPLETIONS** A quarterback completed 68% of his passes in a game. He threw 25 passes. How many passes did the quarterback complete?    *(Section 6.3)*

**20. TEXT MESSAGES** You have 44 text messages in your inbox. How many messages can your cell phone hold?    *(Section 6.4)*

## 6.5 Percents of Increase and Decrease

**Essential Question** What is a percent of decrease? What is a percent of increase?

1 **ACTIVITY: Percent of Decrease**

**Work with a partner.**

Each year in the Columbia River Basin, adult salmon swim upriver to streams to lay eggs and hatch their young.

To go up the river, the adult salmon use fish ladders. But to go down the river, the young salmon must pass through several dams.

At one time, there were electric turbines at each of the eight dams on the main stem of the Columbia and Snake Rivers. About 88% of the young salmon passed through these turbines unharmed.

**COMMON CORE**

**Percents**

In this lesson, you will
- find percents of increase.
- find percents of decrease.

Learning Standard
7.RP.3

**a.** Copy and complete the table to show the number of young salmon that made it through the dams.

| Dam | 0 | 1 | 2 | 3 | 4 | 5 | 6 | 7 | 8 |
|---|---|---|---|---|---|---|---|---|---|
| Salmon | 1000 | 880 | 774 | | | | | | |

88% of 1000 = 0.88 • 1000
= 880

88% of 880 = 0.88 • 880
= 774.4
≈ 774

**b.** Display the data in a bar graph.

**c.** By what percent did the number of young salmon decrease when passing through each dam?

**Math Practice 1**

**Consider Similar Problems**

How is this activity similar to the previous activity?

**Work with a partner. In 2013, the population of a city was 18,000 people.**

a. An organization projects that the population will increase by 2% each year for the next 7 years. Copy and complete the table to find the populations of the city for 2014 through 2020. Then display the data in a bar graph.

For 2014:

$$2\% \text{ of } 18{,}000 = 0.02 \cdot 18{,}000$$
$$= 360$$

$$\underset{\substack{\uparrow \\ \text{2013} \\ \text{Population}}}{18{,}000} + \underset{\substack{\uparrow \\ \text{Increase}}}{360} = \underset{\substack{\uparrow \\ \text{2014} \\ \text{Population}}}{18{,}360}$$

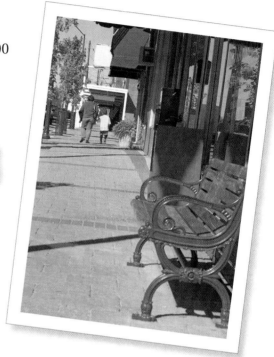

| Year | Population |
|------|-----------|
| 2013 | 18,000 |
| 2014 | 18,360 |
| 2015 | |
| 2016 | |
| 2017 | |
| 2018 | |
| 2019 | |
| 2020 | |

b. Another organization projects that the population will increase by 3% each year for the next 7 years. Repeat part (a) using this percent.

c. Which organization projects the larger populations? How many more people do they project for 2020?

## What Is Your Answer?

3. **IN YOUR OWN WORDS** What is a percent of decrease? What is a percent of increase?

4. Describe real-life examples of a percent of decrease and a percent of increase.

**Practice**  Use what you learned about percent of increase and percent of decrease to complete Exercises 4–7 on page 244.

## 6.5 Lesson

Check It Out
Lesson Tutorials
BigIdeasMath.com

**Key Vocabulary**
percent of change,
  p. 242
percent of increase,
  p. 242
percent of decrease,
  p. 242
percent error, p. 243

A **percent of change** is the percent that a quantity changes from the original amount.

$$\text{percent of change} = \frac{\text{amount of change}}{\text{original amount}}$$

 **Key Idea**

**Percents of Increase and Decrease**

When the original amount increases, the percent of change is called a **percent of increase**.

$$\text{percent of increase} = \frac{\text{new amount} - \text{original amount}}{\text{original amount}}$$

When the original amount decreases, the percent of change is called a **percent of decrease**.

$$\text{percent of decrease} = \frac{\text{original amount} - \text{new amount}}{\text{original amount}}$$

**EXAMPLE 1**  **Finding a Percent of Increase**

The table shows the numbers of hours you spent online last weekend. What is the percent of change in your online time from Saturday to Sunday?

| Day | Hours Online |
|---|---|
| Saturday | 2 |
| Sunday | 4.5 |

The number of hours on Sunday is greater than the number of hours on Saturday. So, the percent of change is a percent of increase.

$$\text{percent of increase} = \frac{\text{new amount} - \text{original amount}}{\text{original amount}}$$

$$= \frac{4.5 - 2}{2} \qquad \text{Substitute.}$$

$$= \frac{2.5}{2} \qquad \text{Subtract.}$$

$$= 1.25, \text{ or } 125\% \qquad \text{Write as a percent.}$$

⁘ So, your online time increased 125% from Saturday to Sunday.

● **On Your Own**

Find the percent of change. Round to the nearest tenth of a percent if necessary.

**1.** 10 inches to 25 inches    **2.** 57 people to 65 people

🔊 Multi-Language Glossary at BigIdeasMath✓com

EXAMPLE (2) **Finding a Percent of Decrease**

**The bar graph shows a softball player's home run totals. What was the percent of change from 2012 to 2013?**

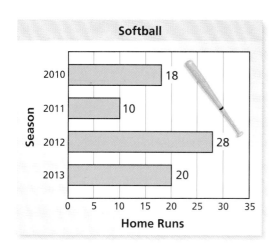

The number of home runs decreased from 2012 to 2013. So, the percent of change is a percent of decrease.

$$\text{percent of decrease} = \frac{\text{original amount} - \text{new amount}}{\text{original amount}}$$

$$= \frac{28 - 20}{28} \qquad \text{Substitute.}$$

$$= \frac{8}{28} \qquad \text{Subtract.}$$

$$\approx 0.286, \text{ or } 28.6\% \qquad \text{Write as a percent.}$$

∴ So, the number of home runs decreased about 28.6%.

 **Key Idea**

**Study Tip**

The amount of error is always positive.

**Percent Error**

A **percent error** is the percent that an estimated quantity differs from the actual amount.

$$\text{percent error} = \frac{\text{amount of error}}{\text{actual amount}}$$

**EXAMPLE (3)** **Finding a Percent Error**

**You estimate that the length of your classroom is 16 feet. The actual length is 21 feet. Find the percent error.**

The amount of error is $21 - 16 = 5$ feet.

$$\text{percent error} = \frac{\text{amount of error}}{\text{actual amount}} \qquad \text{Write percent error equation.}$$

$$= \frac{5}{21} \qquad \text{Substitute.}$$

$$\approx 0.238, \text{ or } 23.8\% \qquad \text{Write as a percent.}$$

∴ The percent error is about 23.8%.

**On Your Own**

Exercises 8–15 and 18

3. In Example 2, what was the percent of change from 2010 to 2011?

4. **WHAT IF?** In Example 3, your friend estimates that the length of the classroom is 23 feet. Who has the greater percent error? Explain.

Vocabulary and Concept Check

1. **VOCABULARY** How do you know whether a percent of change is a *percent of increase* or a *percent of decrease*?

2. **NUMBER SENSE** Without calculating, which has a greater percent of increase?
   - 5 bonus points on a 50-point exam
   - 5 bonus points on a 100-point exam

3. **WRITING** What does it mean to have a 100% decrease?

 Practice and Problem Solving

**Find the new amount.**

4. 8 meters increased by 25%

5. 15 liters increased by 60%

6. 50 points decreased by 26%

7. 25 penalties decreased by 32%

**Identify the percent of change as an *increase* or a *decrease*. Then find the percent of change. Round to the nearest tenth of a percent if necessary.**

 8. 12 inches to 36 inches

9. 75 people to 25 people

10. 50 pounds to 35 pounds

11. 24 songs to 78 songs

12. 10 gallons to 24 gallons

13. 72 paper clips to 63 paper clips

14. 16 centimeters to 44.2 centimeters

15. 68 miles to 42.5 miles

16. **ERROR ANALYSIS** Describe and correct the error in finding the percent increase from 18 to 26.

$$\frac{26 - 18}{26} \approx 0.31 = 31\%$$

17. **VIDEO GAME** Last week, you finished Level 2 of a video game in 32 minutes. Today, you finish Level 2 in 28 minutes. What is your percent of change?

18. **PIG** You estimate that a baby pig weighs 20 pounds. The actual weight of the baby pig is 16 pounds. Find the percent error.

19. **CONCERT** You estimate that 200 people attended a school concert. The actual attendance was 240 people.

   a. Find the percent error.

   b. What other estimate gives the same percent error? Explain your reasoning.

**Identify the percent of change as an *increase* or a *decrease*. Then find the percent of change. Round to the nearest tenth of a percent if necessary.**

**20.** $\frac{1}{4}$ to $\frac{1}{2}$      **21.** $\frac{4}{5}$ to $\frac{3}{5}$      **22.** $\frac{3}{8}$ to $\frac{7}{8}$      **23.** $\frac{5}{4}$ to $\frac{3}{8}$

**24. CRITICAL THINKING** Explain why a change from 20 to 40 is a 100% increase, but a change from 40 to 20 is a 50% decrease.

**25. POPULATION** The table shows population data for a community.

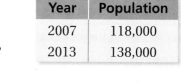

| Year | Population |
|------|------------|
| 2007 | 118,000 |
| 2013 | 138,000 |

    **a.** What is the percent of change from 2007 to 2013?

    **b.** Use this percent of change to predict the population in 2019.

**26. GEOMETRY** Suppose the length and the width of the sandbox are doubled.

    **a.** Find the percent of change in the perimeter.

    **b.** Find the percent of change in the area.

6 ft      10 ft

**27. CEREAL** A cereal company fills boxes with 16 ounces of cereal. The acceptable percent error in filling a box is 2.5%. Find the least and the greatest acceptable weights.

7:45 June      5:51 September

**28. PRECISION** Find the percent of change from June to September in the time to run a mile.

**29. CRITICAL THINKING** A number increases by 10%, and then decreases by 10%. Will the result be *greater than*, *less than*, or *equal to* the original number? Explain.

**30. DONATIONS** Donations to an annual fundraiser are 15% greater this year than last year. Last year, donations were 10% greater than the year before. The amount raised this year is $10,120. How much was raised 2 years ago?

**31. Reasoning** Forty students are in the science club. Of those, 45% are girls. This percent increases to 56% after new girls join the club. How many new girls join?

## Fair Game Review   What you learned in previous grades & lessons

**Write and solve an equation to answer the question.** *(Section 6.4)*

**32.** What number is 25% of 64?

**33.** 39.2 is what percent of 112?

**34.** 5 is 5% of what number?

**35.** 18 is 32% of what number?

**36. MULTIPLE CHOICE** Which set of ratios does *not* form a proportion? *(Section 5.2)*

    Ⓐ $\frac{1}{4}, \frac{6}{24}$      Ⓑ $\frac{4}{7}, \frac{7}{10}$      Ⓒ $\frac{16}{24}, \frac{2}{3}$      Ⓓ $\frac{36}{10}, \frac{18}{5}$

**Essential Question** How can you find discounts and selling prices?

### 1 ACTIVITY: Comparing Discounts

**Work with a partner. The same pair of sneakers is on sale at three stores. Which one is the best buy? Explain.**

**a.** Regular Price: $45    **b.** Regular Price: $49    **c.** Regular Price: $39

**a.**

$0    $9    $18    $27    $36    $45

**b.**

$0    $9.80    $19.60    $29.40    $39.20    $49

**c.**

$0    $7.80    $15.60    $23.40    $31.20    $39

### 2 ACTIVITY: Finding the Original Price

**Work with a partner.**

**COMMON CORE**

**Percents**

In this lesson, you will
- use percent of discounts to find prices of items.
- use percent of markups to find selling prices of items.

Learning Standard
7.RP.3

**a.** You buy a shirt that is on sale for 30% off. You pay $22.40. Your friend wants to know the original price of the shirt. Show how you can use the model below to find the original price.

**b.** Explain how you can use the percent proportion to find the original price.

$0    $22.40    Original Price

**Math Practice 2**

**Make Sense of Quantities**

What do the quantities represent? What is the relationship between the quantities?

You own a small jewelry store. You increase the price of the jewelry by 125%.

Work with a partner. Use a model to estimate the selling price of the jewelry. Then use a calculator to find the selling price.

**a.** Your cost is $250.

**b.** Your cost is $50.

**c.** Your cost is $170.

## What Is Your Answer?

**4. IN YOUR OWN WORDS** How can you find discounts and selling prices? Give examples of each.

**Practice**

Use what you learned about discounts to complete Exercises 4, 9, and 14 on page 250.

**Check It Out**
Lesson Tutorials
BigIdeasMath  com

**Key Vocabulary** ◀))
discount, *p. 248*
markup, *p. 248*

## Key Ideas

### Discounts

A **discount** is a decrease in the original price of an item.

### Markups

To make a profit, stores charge more than what they pay. The increase from what the store pays to the selling price is called a **markup**.

---

**EXAMPLE** **1** **Finding a Sale Price**

**The original price of the shorts is $35. What is the sale price?**

**Method 1:** First, find the discount. The discount is 25% of $35.

$$a = p \cdot w$$      Write percent equation.

$$= 0.25 \cdot 35$$      Substitute 0.25 for $p$ and 35 for $w$.

$$= 8.75$$      Multiply.

Next, find the sale price.

| sale price | = | original price | − | discount |
|---|---|---|---|---|
| | = | 35 | − | 8.75 |
| | = 26.25 | | | |

⋮∙ So, the sale price is $26.25.

**Method 2:** First, find the percent of the original price.

$$100\% - 25\% = 75\%$$

Next, find the sale price.

sale price = 75% of $35

$$= 0.75 \cdot 35$$

$$= 26.25$$

**Study Tip**

A 25% discount is the same as paying 75% of the original price.

⋮∙ So, the sale price is $26.25.

**Check**

0%   25%        75%  100%

✓

0    8.75      26.25  35

## On Your Own

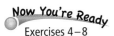
**Now You're Ready**
Exercises 4–8

**1.** The original price of a skateboard is $50. The sale price includes a 20% discount. What is the sale price?

---

◀)) Multi-Language Glossary at BigIdeasMath  com

EXAMPLE **2** **Finding an Original Price**

**What is the original price of the shoes?**

The sale price is
100% − 40% = 60%
of the original price.

Answer the question: 33 is 60% of what number?

| | |
|---|---|
| $a = p \cdot w$ | Write percent equation. |
| $33 = 0.6 \cdot w$ | Substitute 33 for $a$ and 0.6 for $p$. |
| $55 = w$ | Divide each side by 0.6. |

So, the original price
of the shoes is $55.

**Check**

EXAMPLE **3** **Finding a Selling Price**

**A store pays $70 for a bicycle. The percent of markup is 20%. What is the selling price?**

**Method 1:** First, find the markup.
The markup is 20% of $70.

$a = p \cdot w$

$\quad = 0.20 \cdot 70$

$\quad = 14$

Next, find the selling price.

$$\begin{array}{ccc} \text{selling} \\ \text{price} \end{array} = \begin{array}{c} \text{cost to} \\ \text{store} \end{array} + \text{markup}$$

$$\quad\quad - \quad 70 \quad | \quad 14$$

$$\quad = 84$$

So, the selling
price is $84.

**Method 2:** Use a ratio table.
The selling price is 120% of
the cost to the store.

| Percent | Dollars |
|---------|---------|
| 100% | $70 |
| 20% | $14 |
| 120% | $84 |

÷ 5        ÷ 5
× 6        × 6

So, the selling price is $84.

**Check**

**On Your Own**

Now You're Ready
Exercises 9–13
and 17–19

**2.** The discount on a DVD is 50%. It is on sale for $10. What is the original price of the DVD?

**3.** A store pays $75 for an aquarium. The markup is 20%. What is the selling price?

Vocabulary and Concept Check

1. **WRITING** Describe how to find the sale price of an item that has been discounted 25%.

2. **WRITING** Describe how to find the selling price of an item that has been marked up 110%.

3. **REASONING** Which would you rather pay? Explain your reasoning.

   a. | 6% tax on a discounted price | or | 6% tax on the original price |

   b. | 30% markup on a $30 shirt | or | $30 markup on a $30 shirt |

 Practice and Problem Solving

**Copy and complete the table.**

| | | Original Price | Percent of Discount | Sale Price |
|---|---|---|---|---|
| ① | 4. | $80 | 20% | |
| | 5. | $42 | 15% | |
| | 6. | $120 | 80% | |
| | 7. | $112 | 32% | |
| | 8. | $69.80 | 60% | |
| ② | 9. | | 25% | $40 |
| | 10. | | 5% | $57 |
| | 11. | | 80% | $90 |
| | 12. | | 64% | $72 |
| | 13. | | 15% | $146.54 |
| | 14. | $60 | | $45 |
| | 15. | $82 | | $65.60 |
| | 16. | $95 | | $61.75 |

**Find the selling price.**

③ 17. Cost to store: $50
   Markup: 10%

18. Cost to store: $80
   Markup: 60%

19. Cost to store: $140
   Markup: 25%

**20. YOU BE THE TEACHER** The cost to a store for an MP3 player is $60. The selling price is $105. A classmate says that the markup is 175% because $\frac{\$105}{\$60} = 1.75$. Is your classmate correct? If not, explain how to find the correct percent of markup.

**21. SCOOTER** The scooter is on sale for 90% off the original price. Which of the methods can you use to find the sale price? Which method do you prefer? Explain.

| Multiply $45.85 by 0.9. | Multiply $45.85 by 0.1. |
|---|---|
| Multiply $45.85 by 0.9, then add to $45.85. | Multiply $45.85 by 0.9, then subtract from $45.85. |

**22. GAMING** You are shopping for a video game system.

a. At which store should you buy the system?

b. Store A has a weekend sale. What discount must Store A offer for you to buy the system there?

| Store | Cost to Store | Markup |
|---|---|---|
| A | $162 | 40% |
| B | $155 | 30% |
| C | $160 | 25% |

**23. STEREO** A $129.50 stereo is discounted 40%. The next month, the sale price is discounted 60%. Is the stereo now "free"? If not, what is the sale price?

**24. CLOTHING** You buy a pair of jeans at a department store.

a. What is the percent of discount to the nearest percent?

b. What is the percent of sales tax to the nearest tenth of a percent?

c. The price of the jeans includes a 60% markup. After the discount, what is the percent of markup to the nearest percent?

Department Store

| Jeans | 39.99 |
|---|---|
| Discount | -10.00 |
| Subtotal | 29.99 |
| Sales Tax | 1.95 |
| Total | 31.94 |

*Thank You*

**25.**  **Critical Thinking** You buy a bicycle helmet for $22.26, which includes 6% sales tax. The helmet is discounted 30% off the selling price. What is the original price?

**Fair Game Review** What you learned in previous grades & lessons

**Evaluate.** *(Skills Review Handbook)*

**26.** 2000(0.085)  **27.** 1500(0.04)(3)  **28.** 3200(0.045)(8)

**29. MULTIPLE CHOICE** Which measurement is greater than 1 meter? *(Skills Review Handbook)*

(A) 38 inches  (B) 1 yard  (C) 3.4 feet  (D) 98 centimeters

## 6.7 Simple Interest

**Essential Question** How can you find the amount of simple interest earned on a savings account? How can you find the amount of interest owed on a loan?

**Simple interest** is money earned on a savings account or an investment. It can also be money you pay for borrowing money.

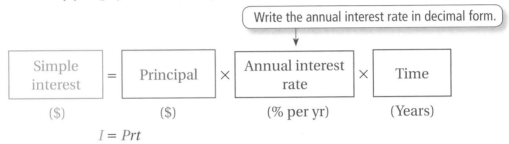

Write the annual interest rate in decimal form.

$$\underset{(\$)}{\boxed{\begin{array}{c}\text{Simple}\\\text{interest}\end{array}}} = \underset{(\$)}{\boxed{\text{Principal}}} \times \underset{(\% \text{ per yr})}{\boxed{\begin{array}{c}\text{Annual interest}\\\text{rate}\end{array}}} \times \underset{(\text{Years})}{\boxed{\text{Time}}}$$

$$I = Prt$$

### 1 ACTIVITY: Finding Simple Interest

Work with a partner. You put $100 in a savings account. The account earns 6% simple interest per year. **(a)** Find the interest earned and the balance at the end of 6 months. **(b)** Copy and complete the table. Then make a bar graph that shows how the balance grows in 6 months.

a.    $I = Prt$                  Write simple interest formula.

      = ▭                        Substitute values.

      = ▭                        Multiply.

      At the end of 6 months, you earn $▭ in interest. So, your balance is $▭ .

b.

| Time | Interest | Balance |
|---|---|---|
| 0 month | $0 | $100 |
| 1 month | | |
| 2 months | | |
| 3 months | | |
| 4 months | | |
| 5 months | | |
| 6 months | | |

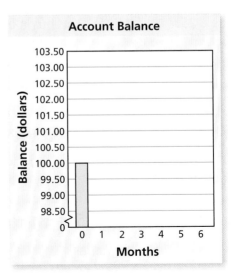

**COMMON CORE**

**Percents**

In this lesson, you will

- use the simple interest formula to find interest earned or paid, annual interest rates, and amounts paid on loans.

Learning Standard
7.RP.3

## 2 ACTIVITY: Financial Literacy

**Work with a partner. Use the following information to write a report about credit cards. In the report, describe how a credit card works. Include examples that show the amount of interest paid each month on a credit card.**

**Math Practice 5**

**Use Other Resources**

What resources can you use to find more information about credit cards?

### U.S. Credit Card Data

- A typical household with credit card debt in the United States owes about $16,000 to credit card companies.

- A typical credit card interest rate is 14% to 16% per year. This is called the annual percentage rate.

## 3 ACTIVITY: The National Debt

**Work with a partner. In 2012, the United States owed about $16 trillion in debt. The interest rate on the national debt is about 1% per year.**

**a.** Write $16 trillion in decimal form. How many zeros does this number have?

**b.** How much interest does the United States pay each year on its national debt?

**c.** How much interest does the United States pay each day on its national debt?

**d.** The United States has a population of about 314 million people. Estimate the amount of interest that each person pays per year toward interest on the national debt.

$16 Trillion in Debt

## What Is Your Answer?

**4. IN YOUR OWN WORDS** How can you find the amount of simple interest earned on a savings account? How can you find the amount of interest owed on a loan? Give examples with your answer.

**Practice** Use what you learned about simple interest to complete Exercises 4–7 on page 256.

Interest is money paid or earned for the use of money. The **principal** is the amount of money borrowed or deposited.

**Key Vocabulary** 🔊
interest, *p. 254*
principal, *p. 254*
simple interest,
*p. 254*

 **Key Idea**

**Simple Interest**

**Words** **Simple interest** is money paid or earned only on the principal.

**Algebra**

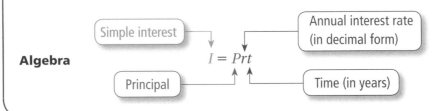

**EXAMPLE 1** **Finding Interest Earned**

You put $500 in a savings account. The account earns 3% simple interest per year. (a) What is the interest earned after 3 years? (b) What is the balance after 3 years?

**a.** $I = Prt$          Write simple interest formula.

    $= 500(0.03)(3)$     Substitute 500 for *P*, 0.03 for *r*, and 3 for *t*.

    $= 45$              Multiply.

⋮ So, the interest earned is $45 after 3 years.

**b.** To find the balance, add the interest to the principal.

⋮ So, the balance is $500 + $45 = $545 after 3 years.

**EXAMPLE 2** **Finding an Annual Interest Rate**

You put $1000 in an account. The account earns $100 simple interest in 4 years. What is the annual interest rate?

       $I = Prt$          Write simple interest formula.

   $100 = 1000(r)(4)$     Substitute 100 for *I*, 1000 for *P*, and 4 for *t*.

   $100 = 4000r$        Simplify.

  $0.025 = r$         Divide each side by 4000.

⋮ So, the annual interest rate of the account is 0.025, or 2.5%.

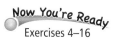
Now You're Ready
Exercises 4–16

## On Your Own

**1.** In Example 1, what is the balance of the account after 9 months?

**2.** You put $350 in an account. The account earns $17.50 simple interest in 2.5 years. What is the annual interest rate?

---

**EXAMPLE** ③ **Finding an Amount of Time**

A bank offers three savings accounts. The simple interest rate is determined by the principal. How long does it take an account with a principal of $800 to earn $100 in interest?

3.0%
More than $5000

2.0%
$500-$5000

1.5%
Less than $500

The pictogram shows that the interest rate for a principal of $800 is 2%.

| $I = Prt$ | Write simple interest formula. |
| $100 = 800(0.02)(t)$ | Substitute 100 for $I$, 800 for $P$, and 0.02 for $r$. |
| $100 = 16t$ | Simplify. |
| $6.25 = t$ | Divide each side by 16. |

So, the account earns $100 in interest in 6.25 years.

---

**EXAMPLE** ④ **Finding an Amount Paid on a Loan**

You borrow $600 to buy a violin. The simple interest rate is 15%. You pay off the loan after 5 years. How much do you pay for the loan?

| $I = Prt$ | Write simple interest formula. |
| $= 600(0.15)(5)$ | Substitute 600 for $P$, 0.15 for $r$, and 5 for $t$. |
| $= 450$ | Multiply. |

To find the amount you pay, add the interest to the loan amount.

So, you pay $600 + $450 = $1050 for the loan.

## On Your Own

Now You're Ready
Exercises 17–20
and 24–27

**3.** In Example 3, how long does it take an account with a principal of $10,000 to earn $750 in interest?

**4.** **WHAT IF?** In Example 4, you pay off the loan after 2 years. How much money do you save?

**Vocabulary and Concept Check**

1. **VOCABULARY** Define each variable in $I = Prt$.

2. **WRITING** In each situation, tell whether you would want a *higher* or *lower* interest rate. Explain your reasoning.

   **a.** you borrow money        **b.** you open a savings account

3. **REASONING** An account earns 6% simple interest. You want to find the interest earned on $200 after 8 months. What conversions do you need to make before you can use the formula $I = Prt$?

 **Practice and Problem Solving**

**An account earns simple interest. (a) Find the interest earned. (b) Find the balance of the account.**

**4.** $600 at 5% for 2 years              **5.** $1500 at 4% for 5 years

**6.** $350 at 3% for 10 years            **7.** $1800 at 6.5% for 30 months

**8.** $700 at 8% for 6 years             **9.** $1675 at 4.6% for 4 years

**10.** $925 at 2% for 2.4 years          **11.** $5200 at 7.36% for 54 months

12. **ERROR ANALYSIS** Describe and correct the error in finding the simple interest earned on $500 at 6% for 18 months.

$I = (500)(0.06)(18)$
$= \$540$

**Find the annual interest rate.**

**13.** $I = \$24$, $P = \$400$, $t = 2$ years        **14.** $I = \$562.50$, $P = \$1500$, $t = 5$ years

**15.** $I = \$54$, $P = \$900$, $t = 18$ months       **16.** $I = \$160.67$, $P = \$2000$, $t = 8$ months

**Find the amount of time.**

**17.** $I = \$30$, $P = \$500$, $r = 3\%$          **18.** $I = \$720$, $P = \$1000$, $r = 9\%$

**19.** $I = \$54$, $P = \$800$, $r = 4.5\%$        **20.** $I = \$450$, $P = \$2400$, $r = 7.5\%$

21. **BANKING** A savings account earns 5% simple interest per year. The principal is $1200. What is the balance after 4 years?

22. **SAVINGS** You put $400 in an account. The account earns $18 simple interest in 9 months. What is the annual interest rate?

23. **CD** You put $3000 in a CD (certificate of deposit) at the promotional rate. How long will it take to earn $336 in interest?

*Certificate of Deposit*

This certificate is the original Specimen and valid document from the treasury and Security department of this trust true financial group & associates. The agreement herein construed are thorough, correct and binding on the parties. The agreement herein construed after it has been legally transferred.

**Promotional Rate 5.6%**
**Simple Interest**

DIRECTOR'S SIGNATURE

**Find the amount paid for the loan.**

④ **24.** $1500 at 9% for 2 years

**25.** $2000 at 12% for 3 years

**26.** $2400 at 10.5% for 5 years

**27.** $4800 at 9.9% for 4 years

**Copy and complete the table.**

| | Principal | Interest Rate | Time | Simple Interest |
|---|---|---|---|---|
| **28.** | $12,000 | 4.25% | 5 years | |
| **29.** | | 6.5% | 18 months | $828.75 |
| **30.** | $15,500 | 8.75% | | $5425.00 |
| **31.** | $18,000 | | 54 months | $4252.50 |

**32. ZOO** A family charges a trip to the zoo on a credit card. The simple interest rate is 12%. The charges are paid after 3 months. What is the total amount paid for the trip?

| Zoo Trip | |
|---|---|
| Tickets | 67.70 |
| Food | 62.34 |
| Gas | 45.50 |
| Total Cost | ? |

**33. MONEY MARKET** You deposit $5000 in an account earning 7.5% simple interest. How long will it take for the balance of the account to be $6500?

**11.8% Simple Interest
Equal monthly
payments for 2 years**

**34. LOANS** A music company offers a loan to buy a drum set for $1500. What is the monthly payment?

**35. REASONING** How many years will it take for $2000 to double at a simple interest rate of 8%? Explain how you found your answer.

**36. PROBLEM SOLVING** You have two loans, for 2 years each. The total interest for the two loans is $138. On the first loan, you pay 7.5% simple interest on a principal of $800. On the second loan, you pay 3% simple interest. What is the principal for the second loan?

**37.**  *Critical Thinking* You put $500 in an account that earns 4% annual interest. The interest earned each year is added to the principal to create a new principal. Find the total amount in your account after each year for 3 years.

**Fair Game Review** What you learned in previous grades & lessons

**Solve the inequality. Graph the solution.** *(Section 4.2)*

**38.** $x + 5 < 2$

**39.** $b - 2 \geq -1$

**40.** $w + 6 \leq -3$

**41. MULTIPLE CHOICE** What is the solution of $4x + 5 = -11$? *(Section 3.5)*

Ⓐ $x = -4$     Ⓑ $x = -1.5$     Ⓒ $x = 1.5$     Ⓓ $x = 4$

**Identify the percent of change as an *increase* or a *decrease*. Then find the percent of change. Round to the nearest tenth of a percent if necessary.**   *(Section 6.5)*

1. 8 inches to 24 inches

2. 300 miles to 210 miles

**Find the original price, discount, sale price, or selling price.**   *(Section 6.6)*

3. Original price: $30
   Discount: 10%
   Sale price: ?

4. Original price: $55
   Discount: ?
   Sale price: $46.75

5. Original price: ?
   Discount: 75%
   Sale price: $74.75

6. Cost to store: $152
   Markup: 50%
   Selling price: ?

**An account earns simple interest. Find the interest earned, principal, interest rate, or time.**   *(Section 6.7)*

7. Interest earned: ?
   Principal: $1200
   Interest rate: 2%
   Time: 5 years

8. Interest earned: $25
   Principal: $500
   Interest rate: 5%
   Time: ?

9. Interest earned: $76
   Principal: $800
   Interest rate: ?
   Time: 2 years

10. Interest earned: $119.88
    Principal: ?
    Interest rate: 3.6%
    Time: 3 years

11. **HEIGHT** You estimate that your friend is 50 inches tall. The actual height of your friend is 54 inches. Find the percent error.   *(Section 6.5)*

12. **DIGITAL CAMERA** A digital camera costs $230. The camera is on sale for 30% off, and you have a coupon for an additional 15% off the sale price. What is the final price?   *(Section 6.6)*

13. **WATER SKIS** The original price of the water skis was $200. What is the percent of discount?   *(Section 6.6)*

2 **Ways to Own:**
1. $75 cash back with 3.5% simple interest
2. No interest for 2 years

14. **SAXOPHONE** A saxophone costs $1200. A store offers two loan options. Which option saves more money if you pay the loan in 2 years?   *(Section 6.7)*

15. **LOAN** You borrow $200. The simple interest rate is 12%. You pay off the loan after 2 years. How much do you pay for the loan?   *(Section 6.7)*

**Check It Out**
Vocabulary Help
BigIdeasMath ✓com

## Review Key Vocabulary

percent of change, *p. 242*     percent error, *p. 243*     interest, *p. 254*
percent of increase, *p. 242*     discount, *p. 248*     principal, *p. 254*
percent of decrease, *p. 242*     markup, *p. 248*     simple interest, *p. 254*

## Review Examples and Exercises

 **Percents and Decimals**  *(pp. 214–219)*

**a.** Write 64% as a decimal.

$$64\% = 64.\% = 0.64$$

**b.** Write 0.023 as a percent.

$$0.023 = 0.023 = 2.3\%$$

### Exercises

**Write the percent as a decimal. Use a model to check your answer.**

**1.** 76%     **2.** 6%     **3.** 334%

**Write the decimal as a percent. Use a model to check your answer.**

**4.** 0.15     **5.** 1.24     **6.** 0.097

 **Comparing and Ordering Fractions, Decimals, and Percents**
*(pp. 220–225)*

Which is greater, $\dfrac{9}{10}$ or 88%?

Write $\dfrac{9}{10}$ as a percent: $\dfrac{9}{10} = \dfrac{90}{100} = 90\%$

∴ 88% is less than 90%. So, $\dfrac{9}{10}$ is the greater number.

### Exercises

**Tell which number is greater.**

**7.** $\dfrac{1}{2}$, 52%     **8.** $\dfrac{12}{5}$, 245%

**9.** 0.46, 43%     **10.** 0.023, 22%

**Use a number line to order the numbers from least to greatest.**

**11.** $\dfrac{41}{50}$, 0.83, 80%     **12.** $\dfrac{9}{4}$, 220%, 2.15

**13.** 0.67, 66%, $\dfrac{2}{3}$     **14.** 0.88, $\dfrac{7}{8}$, 90%

**The Percent Proportion** *(pp. 226–231)*

**a. What percent of 24 is 9?**

$$\frac{a}{w} = \frac{p}{100}$$    Write the percent proportion.

$$\frac{9}{24} = \frac{p}{100}$$    Substitute 9 for $a$ and 24 for $w$.

$$100 \cdot \frac{9}{24} = 100 \cdot \frac{p}{100}$$    Multiplication Property of Equality

$$37.5 = p$$    Simplify.

∴ So, 37.5% of 24 is 9.

**b. What number is 15% of 80?**

$$\frac{a}{w} = \frac{p}{100}$$    Write the percent proportion.

$$\frac{a}{80} = \frac{15}{100}$$    Substitute 80 for $w$ and 15 for $p$.

$$80 \cdot \frac{a}{80} = 80 \cdot \frac{15}{100}$$    Multiplication Property of Equality

$$a = 12$$    Simplify.

∴ So, 12 is 15% of 80.

**c. 120% of what number is 54?**

$$\frac{a}{w} = \frac{p}{100}$$    Write the percent proportion.

$$\frac{54}{w} = \frac{120}{100}$$    Substitute 54 for $a$ and 120 for $p$.

$$54 \cdot 100 = w \cdot 120$$    Cross Products Property

$$5400 = 120w$$    Multiply.

$$45 = w$$    Divide each side by 120.

∴ So, 120% of 45 is 54.

## Exercises

**Write and solve a proportion to answer the question.**

**15.** What percent of 60 is 18?

**16.** 40 is what percent of 32?

**17.** What number is 70% of 70?

**18.** $\frac{3}{4}$ is 75% of what number?

## 6.4 The Percent Equation *(pp. 232–237)*

**a. What number is 72% of 25?**

$$a = p \cdot w \qquad \text{Write percent equation.}$$
$$= 0.72 \cdot 25 \qquad \text{Substitute 0.72 for } p \text{ and 25 for } w.$$
$$= 18 \qquad \text{Multiply.}$$

⋮ So, 72% of 25 is 18.

**b. 28 is what percent of 70?**

$$a = p \cdot w \qquad \text{Write percent equation.}$$
$$28 = p \cdot 70 \qquad \text{Substitute 28 for } a \text{ and 70 for } w.$$
$$\frac{28}{70} = \frac{p \cdot 70}{70} \qquad \text{Division Property of Equality}$$
$$0.4 = p \qquad \text{Simplify.}$$

⋮ Because 0.4 equals 40%, 28 is 40% of 70.

**c. 22.1 is 26% of what number?**

$$a = p \cdot w \qquad \text{Write percent equation.}$$
$$22.1 = 0.26 \cdot w \qquad \text{Substitute 22.1 for } a \text{ and 0.26 for } p.$$
$$85 = w \qquad \text{Divide each side by 0.26.}$$

⋮ So, 22.1 is 26% of 85.

### Exercises

**Write and solve an equation to answer the question.**

**19.** What number is 24% of 25?

**20.** 9 is what percent of 20?

**21.** 60.8 is what percent of 32?

**22.** 91 is 130% of what number?

**23.** 85% of what number is 10.2?

**24.** 83% of 20 is what number?

**25. PARKING** 15% of the school parking spaces are handicap spaces. The school has 18 handicap spaces. How many parking spaces are there?

**26. FIELD TRIP** Of the 25 students on a field trip, 16 students bring cameras. What percent of the students bring cameras?

**6.5** **Percents of Increase and Decrease** *(pp. 240–245)*

**The table shows the numbers of skim boarders at a beach on Saturday and Sunday. What was the percent of change in boarders from Saturday to Sunday?**

The number of skim boarders on Sunday is less than the number of skim boarders on Saturday. So, the percent of change is a percent of decrease.

percent of decrease = $\dfrac{\text{original amount} - \text{new amount}}{\text{original amount}}$

| Day | Number of Skim Boarders |
|---|---|
| Saturday | 12 |
| Sunday | 9 |

$= \dfrac{12 - 9}{12}$   Substitute.

$= \dfrac{3}{12}$   Subtract.

$= 0.25 = 25\%$   Write as a percent.

⋮ So, the number of skim boarders decreased by 25% from Saturday to Sunday.

### Exercises

**Identify the percent of change as an *increase* or a *decrease*. Then find the percent of change. Round to the nearest tenth of a percent if necessary.**

**27.** 6 yards to 36 yards

**28.** 120 meals to 52 meals

**29.** **MARBLES** You estimate that a jar contains 68 marbles. The actual number of marbles is 60. Find the percent error.

**6.6** **Discounts and Markups** *(pp. 246–251)*

**What is the original price of the tennis racquet?**

The sale price is 100% − 30% = 70% of the original price.

Answer the question: 21 is 70% of what number?

$a = p \cdot w$   Write percent equation.

$21 = 0.7 \cdot w$   Substitute 21 for *a* and 0.7 for *p*.

$30 = w$   Divide each side by 0.7.

SALE 30% off Now $21

⋮ So, the original price of the tennis racquet is $30.

### Exercises

**Find the sale price or original price.**

**30.** Original price: $50
Discount: 15%
Sale price: ?

**31.** Original price: ?
Discount: 20%
Sale price: $75

**Simple Interest** *(pp. 252–257)*

You put $200 in a savings account. The account earns 2% simple interest per year.

a. **What is the interest earned after 4 years?**

b. **What is the balance after 4 years?**

a. $I = Prt$           Write simple interest formula.

    $= 200(0.02)(4)$     Substitute 200 for *P*, 0.02 for *r*, and 4 for *t*.

    $= 16$            Multiply.

      So, the interest earned is $16 after 4 years.

b. To find the balance, add the interest to the principal.

      So, the balance is $200 + $16 = $216 after 4 years.

You put $500 in an account. The account earns $55 simple interest in 5 years. What is the annual interest rate?

         $I = Prt$         Write simple interest formula.

     $55 = 500(r)(5)$     Substitute 55 for *I*, 500 for *P*, and 5 for *t*.

     $55 = 2500r$       Simplify.

   $0.022 = r$        Divide each side by 2500.

      So, the annual interest rate of the account is 0.022, or 2.2%.

## Exercises

**An account earns simple interest.**

a. **Find the interest earned.**

b. **Find the balance of the account.**

**32.** $300 at 4% for 3 years          **33.** $2000 at 3.5% for 4 years

**Find the annual simple interest rate.**

**34.** $I = \$17, P = \$500, t = 2$ years      **35.** $I = \$426, P = \$1200, t = 5$ years

**Find the amount of time.**

**36.** $I = \$60, P = \$400, r - 5\%$        **37.** $I = \$237.90, P - \$1525, r = 2.6\%$

**38. SAVINGS** You put $100 in an account. The account earns $2 simple interest in 6 months. What is the annual interest rate?

**Write the percent as a decimal.**

**1.** 0.96%

**2.** 65%

**3.** 25.7%

**Write the decimal as a percent.**

**4.** 0.42

**5.** 7.88

**6.** 0.5854

**Tell which number is greater.**

**7.** $\frac{16}{25}$, 65%

**8.** 56%, 5.6

**Use a number line to order the numbers from least to greatest.**

**9.** 85%, $\frac{15}{18}$, 0.84

**10.** 58.3%, 0.58, $\frac{7}{12}$

**Answer the question.**

**11.** What percent of 28 is 21?

**12.** 64 is what percent of 40?

**13.** What number is 80% of 45?

**14.** 0.8% of what number is 6?

**Identify the percent of change as an *increase* or a *decrease*. Then find the percent of change. Round to the nearest tenth of a percent if necessary.**

**15.** 4 strikeouts to 10 strikeouts

**16.** $24 to $18

**Find the sale price or selling price.**

**17.** Original price: $15
Discount: 5%
Sale price: ?

**18.** Cost to store: $5.50
Markup: 75%
Selling price: ?

**An account earns simple interest. Find the interest earned or the principal.**

**19.** Interest earned: ?
Principal: $450
Interest rate: 6%
Time: 8 years

**20.** Interest earned: $27
Principal: ?
Interest rate: 1.5%
Time: 2 years

**21.** **BASKETBALL** You, your cousin, and a friend each take the same number of free throws at a basketball hoop. Who made the most free throws?

**22.** **PARKING LOT** You estimate that there are 66 cars in a parking lot. The actual number of cars is 75.

| Player | Made |
|--------|------|
| You | $\frac{17}{20}$ |
| Cousin | 0.8 |
| Friend | 87.5% |

  **a.** Find the percent error.

  **b.** What other estimate gives the same percent error? Explain your reasoning.

**23.** **INVESTMENT** You put $800 in an account that earns 4% simple interest. Find the total amount in your account after each year for 3 years.

1. A movie theater offers 30% off the price of a movie ticket to students from your school. The regular price of a movie ticket is $8.50. What is the discounted price that you would pay for a ticket? *(7.RP.3)*

    **A.** $2.55

    **B.** $5.50

    **C.** $5.95

    **D.** $8.20

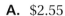
2. You are comparing the prices of four boxes of cereal. Two of the boxes contain free extra cereal.

    - Box F costs $3.59 and contains 16 ounces.

    - Box G costs $3.79 and contains 16 ounces, plus an additional 10% for free.

    - Box H costs $4.00 and contains 500 grams.

    - Box I costs $4.69 and contains 500 grams, plus an additional 20% for free.

    Which box has the least unit cost? (1 ounce = 28.35 grams)  *(7.RP.3)*

    **F.** Box F

    **G.** Box G

    **H.** Box H

    **I.** Box I

3. What value makes the equation $11 - 3x = -7$ true?  *(7.EE.4a)*

4. Which proportion represents the problem below?  *(7.RP.3)*

    "17% of a number is 43. What is the number?"

    **A.** $\dfrac{17}{43} = \dfrac{n}{100}$

    **B.** $\dfrac{n}{17} = \dfrac{43}{100}$

    **C.** $\dfrac{n}{43} = \dfrac{17}{100}$

    **D.** $\dfrac{43}{n} = \dfrac{17}{100}$

**5.** Which list of numbers is in order from least to greatest? *(7.EE.3)*

**F.** 0.8, $\frac{5}{8}$, 70%, 0.09

**H.** $\frac{5}{8}$, 70%, 0.8, 0.09

**G.** 0.09, $\frac{5}{8}$, 0.8, 70%

**I.** 0.09, $\frac{5}{8}$, 70%, 0.8

**6.** What is the value of $\frac{9}{8} \div \left(-\frac{11}{4}\right)$? *(7.NS.2b)*

**7.** A pair of running shoes is on sale for 25% off the original price.

ORIGINAL PRICE
$123.75

Which price is closest to the sale price of the running shoes? *(7.RP.3)*

**A.** $93

**C.** $124

**B.** $99

**D.** $149

**8.** What is the slope of the line? *(7.RP.2b)*

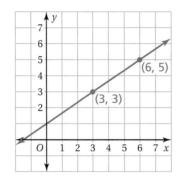

**F.** $\frac{2}{3}$

**H.** 2

**G.** $\frac{3}{2}$

**I.** 3

**9.** Brad solved the equation in the box shown.

What should Brad do to correct the error that he made?  *(7.EE.4a)*

$$-3(2 + w) = -45$$
$$2 + w = -15$$
$$w = -17$$

**A.** Multiply $-45$ by $-3$ to get $2 + w = 135$.

**B.** Add 3 to $-45$ to get $2 + w = -42$.

**C.** Add 2 to $-15$ to get $w = -13$.

**D.** Divide $-45$ by $-3$ to get 15.

**10.** You are comparing the costs of a certain model of ladder at a hardware store and at an online store.  *(7.RP.3)*

*Part A*   What is the cost of the ladder at each of the stores? Show your work and explain your reasoning.

*Part B*   Suppose that the hardware store is offering 10% off the price of the ladder and that the online store is offering free shipping and handling. Which store offers the better final cost? by how much? Show your work and explain your reasoning.

**11.** Which graph represents the inequality below?  *(7.EE.4b)*

$$-5 - 3x \geq -11$$

**F.**

**G.**

**H.**

**I.**

# 7 Constructions and Scale Drawings

"Move 4 of the lines to make 3 equilateral triangles."

"Well done, Descartes!"

"I love puzzles!"

"I'm at 3rd base. You are running to 1st base, and Fluffy is running to 2nd base."

"Should I throw the ball to 2nd to get Fluffy out or throw it to 1st to get you out?"

"Throw to the closer angle. Heh, heh, heh."

# What You Learned Before

"Look at this baby crocodile! Isn't it cute?"

*Yes, it's very acute.*

## Measuring Angles (4.MD.6)

**Example 1** Use a protractor to find the measure of each angle. Then classify the angle as *acute*, *obtuse*, *right*, or *straight*.

**a.**

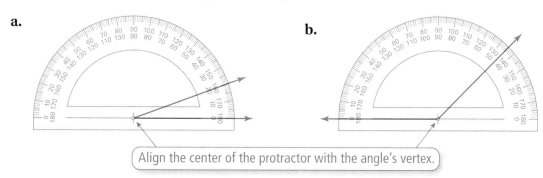

Align the center of the protractor with the angle's vertex.

**b.**

⋰ The angle measure is 20°.
  So, the angle is acute.

⋰ The angle measure is 135°.
  So, the angle is obtuse.

## Drawing Angles (4.G.1)

**Example 2** Use a protractor to draw a 45° angle.

Draw a ray. Place the center of the protractor on the endpoint of the ray and align the protractor so the ray passes through the 0° mark. Make a mark at 45°. Then draw a ray from the endpoint at the center of the protractor through the mark at 45°.

45°

### Try It Yourself

Use a protractor to find the measure of the angle. Then classify the angle as *acute*, *obtuse*, *right*, or *straight*.

**1.**        **2.**        **3.**

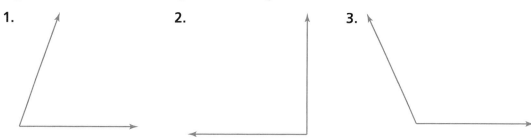

Use a protractor to draw an angle with the given measure.

**4.** 55°        **5.** 160°        **6.** 85°        **7.** 180°

## Essential Question

What can you conclude about the angles formed by two intersecting lines?

### Classification of Angles

**Acute:**
Less than 90°

**Right:**
Equal to 90°

**Obtuse:**
Greater than 90° and less than 180°

**Straight:**
Equal to 180°

### 1 ACTIVITY: Drawing Angles

**Work with a partner.**

a. Draw the hands of the clock to represent the given type of angle.

| Acute | Straight | Right | Obtuse |
|---|---|---|---|

b. What is the measure of the angle formed by the hands of the clock at the given time?

9:00          6:00          12:00

**COMMON CORE**

**Geometry**
In this lesson, you will
- identify adjacent and vertical angles.
- find angle measures using adjacent and vertical angles.

Learning Standard
7.G.5

## The Meaning of a Word ● Adjacent

When two states are **adjacent**,

they are next to each other and they share a common border.

Maine

New Hampshire

Maine

New Hampshire

## ② ACTIVITY: Naming Angles

Work with a partner. Some angles, such as ∠A, can be named by a single letter. When this does not clearly identify an angle, you should use three letters, as shown.

**Math Practice 3**

**Justify Conclusions**

When you name an angle, does the order in which you write the letters matter? Explain.

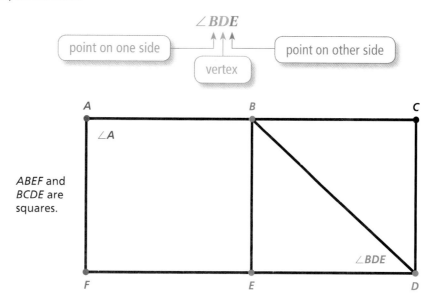

∠BDE

point on one side — vertex — point on other side

*ABEF* and *BCDE* are squares.

**a.** Name all the right angles, acute angles, and obtuse angles.

**b.** Which pairs of angles do you think are *adjacent*? Explain.

## ③ ACTIVITY: Measuring Angles

Work with a partner.

**a.** How many angles are formed by the intersecting roads? Number the angles.

**b.** **CHOOSE TOOLS** Measure each angle formed by the intersecting roads. What do you notice?

## What Is Your Answer?

**4.** **IN YOUR OWN WORDS** What can you conclude about the angles formed by two intersecting lines?

**5.** Draw two acute angles that are adjacent.

**Practice**

Use what you learned about angles and intersecting lines to complete Exercises 3 and 4 on page 274.

**Check It Out**
Lesson Tutorials
BigIdeasMath√com

**Key Vocabulary** ◀))
adjacent angles,
   p. 272
vertical angles, p. 272
congruent angles,
   p. 272

 **Key Ideas**

### Adjacent Angles

**Words**   Two angles are **adjacent angles** when they share a common side and have the same vertex.

**Examples**

∠1 and ∠2 are adjacent.

∠2 and ∠4 are not adjacent.

### Vertical Angles

**Words**   Two angles are **vertical angles** when they are opposite angles formed by the intersection of two lines. Vertical angles are **congruent angles**, meaning they have the same measure.

**Examples**

∠1 and ∠3 are vertical angles.

∠2 and ∠4 are vertical angles.

---

**EXAMPLE** ① **Naming Angles**

**Use the figure shown.**

**a.   Name a pair of adjacent angles.**

∠ABC and ∠ABF share a common side and have the same vertex B.

⋮• So, ∠ABC and ∠ABF are adjacent angles.

**b.   Name a pair of vertical angles.**

∠ABF and ∠CBD are opposite angles formed by the intersection of two lines.

⋮• So, ∠ABF and ∠CBD are vertical angles.

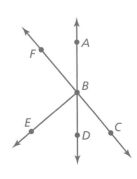

### On Your Own

**Now You're Ready**
Exercises 5 and 6

**Name two pairs of adjacent angles and two pairs of vertical angles in the figure.**

**1.**

**2.**

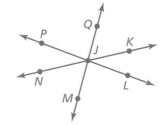

**EXAMPLE** 2 **Using Adjacent and Vertical Angles**

**Tell whether the angles are *adjacent* or *vertical*. Then find the value of $x$.**

**a.**

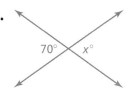

The angles are vertical angles. Because vertical angles are congruent, the angles have the same measure.

⋮∴ So, the value of $x$ is 70.

**Remember**

You can add angle measures. When two or more adjacent angles form a larger angle, the sum of the measures of the smaller angles is equal to the measure of the larger angle.

**b.**

The angles are adjacent angles. Because the angles make up a right angle, the sum of their measures is 90°.

$$(x + 4) + 31 = 90 \qquad \text{Write equation.}$$
$$x + 35 = 90 \qquad \text{Combine like terms.}$$
$$x = 55 \qquad \text{Subtract 35 from each side.}$$

⋮∴ So, the value of $x$ is 55.

**EXAMPLE** 3 **Constructing Angles**

**Draw a pair of vertical angles with a measure of 40°.**

**Step 1:** Use a protractor to draw a 40° angle.

**Step 2:** Use a straightedge to extend the sides to form two intersecting lines.

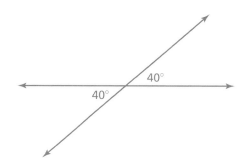

**On Your Own**

Now You're Ready
Exercises 8–17

**Tell whether the angles are *adjacent* or *vertical*. Then find the value of $x$.**

**3.**

**4.**

**5.**

**6.** Draw a pair of vertical angles with a measure of 75°.

# 7.1 Exercises

Check It Out
Help with Homework
BigIdeasMath.com

## ✓ Vocabulary and Concept Check

1. **VOCABULARY** When two lines intersect, how many pairs of vertical angles are formed? How many pairs of adjacent angles are formed?

2. **REASONING** Identify the congruent angles in the figure. Explain your reasoning.

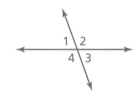

## Practice and Problem Solving

**Use the figure at the right.**

3. Measure each angle formed by the intersecting lines.

4. Name two angles that are adjacent to ∠ABC.

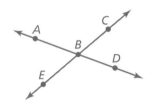

**Name two pairs of adjacent angles and two pairs of vertical angles in the figure.**

**5.**

**6.**

7. **ERROR ANALYSIS** Describe and correct the error in naming a pair of vertical angles.

∠ACB and ∠BCD
are vertical angles.

**Tell whether the angles are *adjacent* or *vertical*. Then find the value of x.**

**8.**

**9.**

**10.**

**11.**

**12.**

**13.**

**Draw a pair of vertical angles with the given measure.**

**③ 14.** 25°          **15.** 85°          **16.** 110°          **17.** 135°

**18. IRON CROSS** The iron cross is a skiing trick in which the tips of the skis are crossed while the skier is airborne. Find the value of $x$ in the iron cross shown.

127°

$(2x + 41)°$

**19. OPEN-ENDED** Draw a pair of adjacent angles with the given description.

    **a.** Both angles are acute.

    **b.** One angle is acute, and one is obtuse.

    **c.** The sum of the angle measures is 135°.

**20. PRECISION** Explain two procedures that you can use to draw adjacent angles with given measures.

**Determine whether the statement is *always, sometimes,* or *never* true.**

**21.** When the measure of ∠1 is 70°, the measure of ∠3 is 110°.

**22.** When the measure of ∠4 is 120°, the measure of ∠1 is 60°.

**23.** ∠2 and ∠3 are congruent.

**24.** The measure of ∠1 plus the measure of ∠2 equals the measure of ∠3 plus the measure of ∠4.

**25. REASONING** Draw a figure in which ∠1 and ∠2 are acute vertical angles, ∠3 is a right angle adjacent to ∠2, and the sum of the measure of ∠1 and the measure of ∠4 is 180°.

**26. Structure** For safety reasons, a ladder should make a 15° angle with a wall. Is the ladder shown leaning at a safe angle? Explain.

120°

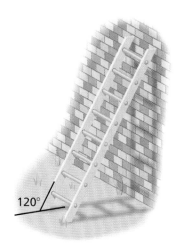

![Fair Game Review] **Fair Game Review** What you learned in previous grades & lessons

**Solve the inequality. Graph the solution.** *(Section 4.3)*

**27.** $-6n > 54$          **28.** $-\dfrac{1}{2}x \le 17$          **29.** $-1.6 < \dfrac{m}{-2.5}$

**30. MULTIPLE CHOICE** What is the slope of the line that passes through the points (2, 3) and (6, 8)? *(Section 5.5)*

  **Ⓐ** $\dfrac{4}{5}$          **Ⓑ** $\dfrac{5}{4}$          **Ⓒ** $\dfrac{4}{3}$          **Ⓓ** $\dfrac{3}{2}$

**Essential Question** How can you classify two angles as complementary or supplementary?

## 1 ACTIVITY: Complementary and Supplementary Angles

**Work with a partner.**

**a.** The graph represents the measures of *complementary angles*. Use the graph to complete the table.

| x |  | 20° |  | 30° | 45° |  | 75° |
|---|---|---|---|---|---|---|---|
| y | 80° |  | 65° | 60° |  | 40° |  |

Angle measure (degrees) vs. Angle measure (degrees)

**b.** How do you know when two angles are complementary? Explain.

**c.** The graph represents the measures of *supplementary angles*. Use the graph to complete the table.

| x | 20° |  | 60° | 90° |  | 140° |  |
|---|---|---|---|---|---|---|---|
| y |  | 150° |  | 90° | 50° |  | 30° |

Angle measure (degrees) vs. Angle measure (degrees)

**d.** How do you know when two angles are supplementary? Explain.

## 2 ACTIVITY: Exploring Rules About Angles

**COMMON CORE**

**Geometry**

In this lesson, you will
- classify pairs of angles as complementary, supplementary, or neither.
- find angle measures using complementary and supplementary angles.

Learning Standard
7.G.5

**Work with a partner. Copy and complete each sentence with *always*, *sometimes*, or *never*.**

**a.** If x and y are complementary angles, then both x and y are _____ acute.

**b.** If x and y are supplementary angles, then x is _____ acute.

**c.** If x is a right angle, then x is _____ acute.

**d.** If x and y are complementary angles, then x and y are _____ adjacent.

**e.** If x and y are supplementary angles, then x and y are _____ vertical.

**Work with a partner. Tell whether the two angles shown on the clocks are** *complementary*, *supplementary*, **or** *neither*. **Explain your reasoning.**

a.

b.

c.

d.

④ **ACTIVITY: Identifying Angles**

**Work with a partner. Use a protractor and the figure shown.**

a. Name four pairs of complementary angles and four pairs of supplementary angles.

**Math Practice** ③

**Use Definitions**
How can you use the definitions of *complementary*, *supplementary*, and *vertical angles* to answer the questions?

b. Name two pairs of vertical angles.

## What Is Your Answer?

5. **IN YOUR OWN WORDS** How can you classify two angles as complementary or supplementary? Give examples of each type.

**Practice**

Use what you learned about complementary and supplementary angles to complete Exercises 3–5 on page 280.

**Key Vocabulary**
complementary
angles, *p. 278*
supplementary
angles, *p. 278*

## Key Ideas

**Complementary Angles**

**Words** Two angles are **complementary angles** when the sum of their measures is 90°.

**Examples**

∠1 and ∠2 are complementary angles.

**Supplementary Angles**

**Words** Two angles are **supplementary angles** when the sum of their measures is 180°.

**Examples**

∠3 and ∠4 are supplementary angles.

---

**EXAMPLE 1** **Classifying Pairs of Angles**

Tell whether the angles are *complementary*, *supplementary*, or *neither*.

**a.**

70° + 110° = 180°

∴ So, the angles are supplementary.

**b.**

41° + 49° = 90°

∴ So, the angles are complementary.

**c.**

128° + 62° = 190°

∴ So, the angles are *neither* complementary nor supplementary.

## On Your Own

**Now You're Ready**
Exercises 6–11

Tell whether the angles are *complementary*, *supplementary*, or *neither*.

**1.** 26° 64°

**2.** 136° 44°

**3.** 70° 19°

---

EXAMPLE **2** Using Complementary and Supplementary Angles

**Tell whether the angles are *complementary* or *supplementary*. Then find the value of *x*.**

**a.**

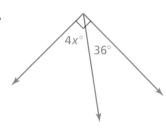

The two angles make up a right angle. So, the angles are complementary angles, and the sum of their measures is 90°.

$$4x + 36 = 90 \quad \text{Write equation.}$$
$$4x = 54 \quad \text{Subtract 36 from each side.}$$
$$x = 13.5 \quad \text{Divide each side by 4.}$$

**b.**

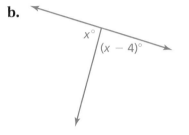

The two angles make up a straight angle. So, the angles are supplementary angles, and the sum of their measures is 180°.

$$x + (x - 4) = 180 \quad \text{Write equation.}$$
$$2x - 4 = 180 \quad \text{Combine like terms.}$$
$$2x = 184 \quad \text{Add 4 to each side.}$$
$$x = 92 \quad \text{Divide each side by 2.}$$

---

EXAMPLE **3** Constructing Angles

**Draw a pair of adjacent supplementary angles so that one angle has a measure of 60°.**

**Step 1:** Use a protractor to draw a 60° angle.

**Step 2:** Extend one of the sides to form a line.

## On Your Own

Now You're Ready
Exercises 12–14
and 17–20

**Tell whether the angles are *complementary* or *supplementary*. Then find the value of *x*.**

**4.**

**5.**

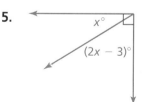

**6.** Draw a pair of adjacent supplementary angles so that one angle has a measure of 15°.

---

## Vocabulary and Concept Check

1. **VOCABULARY** Explain how complementary angles and supplementary angles are different.

2. **REASONING** Can adjacent angles be supplementary? complementary? neither? Explain.

 ## Practice and Problem Solving

**Tell whether the statement is *always*, *sometimes*, or *never* true. Explain.**

3. If *x* and *y* are supplementary angles, then *x* is obtuse.

4. If *x* and *y* are right angles, then *x* and *y* are supplementary angles.

5. If *x* and *y* are complementary angles, then *y* is a right angle.

**Tell whether the angles are *complementary*, *supplementary*, or *neither*.**

① 6.

7.

8.

9.

10.

11.

**Tell whether the angles are *complementary* or *supplementary*. Then find the value of *x*.**

② 12.

13.

14.

15. **INTERSECTION** What are the measures of the other three angles formed by the intersection?

16. **TRIBUTARY** A tributary joins a river at an angle. Find the value of *x*.

**Draw a pair of adjacent supplementary angles so that one angle has the given measure.**

③ **17.** 20°  **18.** 35°  **19.** 80°  **20.** 130°

**21. PRECISION** Explain two procedures that you can use to draw two adjacent complementary angles. Then draw a pair of adjacent complementary angles so that one angle has a measure of 30°.

**22. OPEN-ENDED** Give an example of an angle that can be a supplementary angle but cannot be a complementary angle. Explain.

**23. VANISHING POINT** The vanishing point of the picture is represented by point *B*.

   **a.** The measure of ∠*ABD* is 6.2 times greater than the measure of ∠*CBD*. Find the measure of ∠*CBD*.

   **b.** ∠*FBE* and ∠*EBD* are congruent. Find the measure of ∠*FBE*.

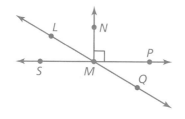

**24. LOGIC** Your friend says that ∠*LMN* and ∠*PMQ* are complementary angles. Is she correct? Explain.

**25. RATIO** The measures of two complementary angles have a ratio of 3 : 2. What is the measure of the larger angle?

**26. REASONING** Two angles are vertical angles. What are their measures if they are also complementary angles? supplementary angles?

**27.** ⚡*Problem Solving*⚡ Find the values of *x* and *y*.

 **Fair Game Review** *What you learned in previous grades & lessons*

**Solve the equation. Check your solution.** *(Section 3.3)*

**28.** $x + 7 = -8$  **29.** $\dfrac{1}{3} = n + \dfrac{3}{4}$  **30.** $-12.7 = y - 3.4$

**31. MULTIPLE CHOICE** Which decimal is equal to 3.7%? *(Section 6.1)*

  Ⓐ 0.0037  Ⓑ 0.037  Ⓒ 0.37  Ⓓ 3.7

# 7.3 Triangles

**Essential Question** How can you construct triangles?

## 1 ACTIVITY: Constructing Triangles Using Side Lengths

Work with a partner. Cut different-colored straws to the lengths shown. Then construct a triangle with the specified straws if possible. Compare your results with those of others in your class.

2 cm

4 cm

6 cm

7 cm

**a.** blue, green, purple

**b.** red, green, purple

**c.** red, blue, purple

**d.** red, blue, green

## 2 ACTIVITY: Using Technology to Draw Triangles (Side Lengths)

Work with a partner. Use geometry software to draw a triangle with the two given side lengths. What is the length of the third side of your triangle? Compare your results with those of others in your class.

**a.** 4 units, 7 units

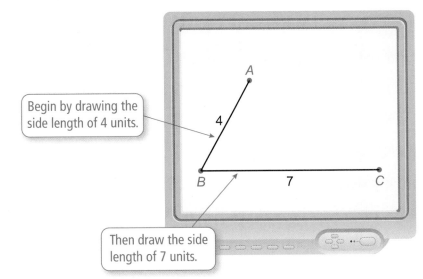

Begin by drawing the side length of 4 units.

Then draw the side length of 7 units.

**COMMON CORE**

**Geometry**

In this lesson, you will
- construct triangles with given angle measures.
- construct triangles with given side lengths.

Learning Standard
7.G.2

**b.** 3 units, 5 units

**c.** 2 units, 8 units

**d.** 1 unit, 1 unit

## 3  ACTIVITY: Constructing Triangles Using Angle Measures

**Work with a partner. Two angle measures of a triangle are given. Draw the triangle. What is the measure of the third angle? Compare your results with those of others in your class.**

**a.** 40°, 70°

Begin by drawing the angle measure of 40°.

40°

**b.** 60°, 75°     **c.** 90°, 30°     **d.** 100°, 40°

## 4  ACTIVITY: Using Technology to Draw Triangles (Angle Measures)

**Math Practice 5**

**Recognize Usefulness of Tools**

What are some advantages and disadvantages of using geometry software to draw a triangle?

**Work with a partner. Use geometry software to draw a triangle with the two given angle measures. What is the measure of the third angle? Compare your results with those of others in your class.**

**a.** 45°, 55°

**b.** 50°, 40°

**c.** 110°, 35°

Begin by drawing the angle measure of 45°.

45°

## What Is Your Answer?

5. **IN YOUR OWN WORDS** How can you construct triangles?

6. **REASONING** Complete the table below for each set of side lengths in Activity 2. Write a rule that compares the sum of any two side lengths to the third side length.

| Side Length | | | |
|---|---|---|---|
| Sum of Other Two Side Lengths | | | |

7. **REASONING** Use a table to organize the angle measures of each triangle you formed in Activity 3. Include the sum of the angle measures. Then describe the pattern in the table and write a conclusion based on the pattern.

**Practice**

Use what you learned about constructing triangles to complete Exercises 3–5 on page 286.

**Key Vocabulary**
congruent sides,
  p. 284

You can use side lengths and angle measures to classify triangles.

## Key Ideas

**Classifying Triangles Using Angles**

| *acute* triangle | *obtuse* triangle | *right* triangle | *equiangular* triangle |

all acute angles    1 obtuse angle    1 right angle    3 congruent angles

**Classifying Triangles Using Sides**

**Congruent sides** have the same length.

*scalene* triangle     *isosceles* triangle     *equilateral* triangle

no congruent sides    at least 2 congruent sides    3 congruent sides

**Reading**

Red arcs indicate congruent angles.
Red tick marks indicate congruent sides.

---

**EXAMPLE** **1** **Classifying Triangles**

Classify each triangle.

a.

b.

The triangle has one obtuse angle and no congruent sides.

∴ So, the triangle is an obtuse scalene triangle.

The triangle has all acute angles and two congruent sides.

∴ So, the triangle is an acute isosceles triangle.

## On Your Own

**Now You're Ready**
Exercises 6–11

Classify the triangle.

1.

2.

EXAMPLE 2 **Constructing a Triangle Using Angle Measures**

**Draw a triangle with angle measures of 30°, 60°, and 90°. Then classify the triangle.**

**Step 1:** Use a protractor to draw the 30° angle.

**Step 2:** Use a protractor to draw the 60° angle.

**Step 3:** The protractor shows that the measure of the remaining angle is 90°.

**Study Tip**

After drawing the first two angles, make sure you check the remaining angle.

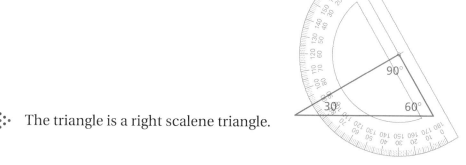

∴ The triangle is a right scalene triangle.

---

EXAMPLE 3 **Constructing a Triangle Using Side Lengths**

**Draw a triangle with a 3-centimeter side and a 4-centimeter side that meet at a 20° angle. Then classify the triangle.**

**Step 1:** Use a protractor to draw a 20° angle.

**Step 2:** Use a ruler to mark 3 centimeters on one ray and 4 centimeters on the other ray.

**Step 3:** Draw the third side to form the triangle.

∴ The triangle is an obtuse scalene triangle.

4 cm

20°

3 cm

**On Your Own**

**Now You're Ready**
Exercises 14–19

**3.** Draw a triangle with angle measures of 45°, 45°, and 90°. Then classify the triangle.

**4.** Draw a triangle with a 1-inch side and a 2-inch side that meet at a 60° angle. Then classify the triangle.

**Vocabulary and Concept Check**

1. **WRITING** How can you classify triangles using angles? using sides?

2. **DIFFERENT WORDS, SAME QUESTION** Which is different? Find "both" answers.

Construct an equilateral triangle.

Construct a triangle with 3 congruent sides.

Construct an equiangular triangle.

Construct a triangle with no congruent sides.

 **Practice and Problem Solving**

**Construct a triangle with the given description.**

3. side lengths: 4 cm, 6 cm

4. side lengths: 5 cm, 12 cm

5. angles: 65°, 55°

**Classify the triangle.**

① 6.

7.

8.

9.

10.

11.

12. **ERROR ANALYSIS** Describe and correct the error in classifying the triangle.

The triangle is acute and scalene because it has two acute angles and no congruent sides.

13. **MOSAIC TILE** A mosaic is a pattern or picture made of small pieces of colored material. Classify the yellow triangle used in the mosaic.

**Draw a triangle with the given angle measures. Then classify the triangle.**

**2** **14.** 15°, 75°, 90°　　　　　**15.** 20°, 60°, 100°　　　　　**16.** 30°, 30°, 120°

**Draw a triangle with the given description.**

**3** **17.** a triangle with a 2-inch side and a 3-inch side that meet at a 40° angle

**18.** a triangle with a 45° angle connected to a 60° angle by an 8-centimeter side

**19.** an acute scalene triangle

**20.** **LOGIC** You are constructing a triangle. You draw the first angle, as shown. Your friend says that you must be constructing an acute triangle. Is your friend correct? Explain your reasoning.

**Determine whether you can construct *many, one,* or *no* triangle(s) with the given description. Explain your reasoning.**

**21.** a triangle with angle measures of 50°, 70°, and 100°

**22.** a triangle with one angle measure of 60° and one 4-centimeter side

**23.** a scalene triangle with a 3-centimeter side and a 7-centimeter side

**24.** an isosceles triangle with two 4-inch sides that meet at an 80° angle

**25.** an isosceles triangle with two 2-inch sides and one 5-inch side

**26.** a right triangle with three congruent sides

**27.** **Critical Thinking** Consider the three isosceles triangles.

　　**a.** Find the value of *x* for each triangle.

　　**b.** What do you notice about the angle measures of each triangle?

　　**c.** Write a rule about the angle measures of an isosceles triangle.

 **Fair Game Review** What you learned in previous grades & lessons

**Tell whether *x* and *y* show direct variation. Explain your reasoning. If so, find the constant of proportionality.** *(Section 5.6)*

**28.** $x = 2y$　　　　　**29.** $y \quad x = 6$　　　　　**30.** $xy = 5$

**31.** **MULTIPLE CHOICE** A savings account earns 6% simple interest per year. The principal is $800. What is the balance after 18 months? *(Section 6.7)*

　　**Ⓐ** $864　　　　**Ⓑ** $872　　　　**Ⓒ** $1664　　　　**Ⓓ** $7200

**Check It Out**
Lesson Tutorials
BigIdeasMath ✓com

## Key Idea

**Sum of the Angle Measures of a Triangle**

**Words** The sum of the angle measures of a triangle is 180°.

**Algebra** $x + y + z = 180$

**EXAMPLE 1** **Finding Angle Measures**

Find each value of $x$. Then classify each triangle.

**COMMON CORE**

**Geometry**

In this extension, you will

- understand that the sum of the angle measures of any triangle is 180°.
- find missing angle measures in triangles.

Learning Standard
7.G.5

**a.**

$$x + 28 + 50 = 180$$
$$x + 78 = 180$$
$$x = 102$$

∴ The value of $x$ is 102. The triangle has one obtuse angle and no congruent sides. So, it is an obtuse scalene triangle.

**b.**

$$x + 45 + 90 = 180$$
$$x + 135 = 180$$
$$x = 45$$

∴ The value of $x$ is 45. The triangle has a right angle and two congruent sides. So, it is a right isosceles triangle.

## Practice

Find the value of $x$. Then classify the triangle.

**1.**

**2.**

**3.**

**4.**

**5.**

**6.**

Tell whether a triangle can have the given angle measures. If not, change the first angle measure so that the angle measures form a triangle.

**7.** 76.2°, 81.7°, 22.1°

**8.** 115.1°, 47.5°, 93°

**9.** $5\frac{2}{3}°$, $64\frac{1}{3}°$, 87°

**10.** $31\frac{3}{4}°$, $53\frac{1}{2}°$, $94\frac{3}{4}°$

EXAMPLE 2 Finding Angle Measures

**Analyze Givens**

Math Practice 1

What information is given in the problem? How can you use this information to answer the question?

**Find each value of x. Then classify each triangle.**

**a.** Flag of Jamaica

$x + x + 128 = 180$
$2x + 128 = 180$
$2x = 52$
$x = 26$

∴ The value of x is 26. The triangle has one obtuse angle and two congruent sides. So, it is an obtuse isosceles triangle.

**b.** Flag of Cuba

$x + x + 60 = 180$
$2x + 60 = 180$
$2x = 120$
$x = 60$

∴ The value of x is 60. All three angles are congruent. So, it is an equilateral and equiangular triangle.

## Practice

**Find the value of x. Then classify the triangle.**

**11.**

**12.**

**13.**

**14.**

**15.**

**16. REASONING** Explain why all triangles have at least two acute angles.

**17. CARDS** One method of stacking cards is shown.

**a.** Find the value of x.

**b.** Describe how to stack the cards with different angles. Is the value of x limited? If so, what are the limitations? Explain your reasoning.

You can use an **example and non-example chart** to list examples and non-examples of a vocabulary word or item. Here is an example and non-example chart for complementary angles.

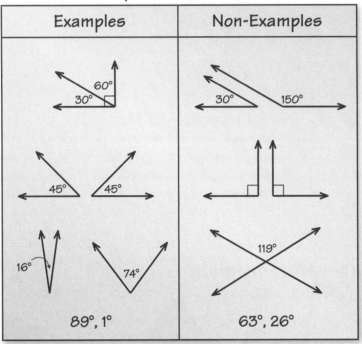

**On Your Own**

Make example and non-example charts to help you study these topics.

1. adjacent angles

2. vertical angles

3. supplementary angles

After you complete this chapter, make example and non-example charts for the following topics.

4. quadrilaterals

5. scale factor

"What do you think of my example & non-example chart for popular cat toys?"

**Name two pairs of adjacent angles and two pairs of vertical angles in the figure.** *(Section 7.1)*

**1.**

**2.**

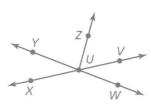

**Tell whether the angles are *adjacent* or *vertical*. Then find the value of *x*.** *(Section 7.1)*

**3.**

**4.**

**5.**

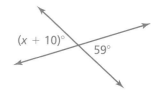

**Tell whether the angles are *complementary* or *supplementary*. Then find the value of *x*.** *(Section 7.2)*

**6.**

**7.**

**Draw a triangle with the given description.** *(Section 7.3)*

**8.** a triangle with angle measures of 35°, 65°, and 80°

**9.** a triangle with a 5-centimeter side and a 7-centimeter side that meet at a 70° angle

**10.** an obtuse scalene triangle

**Find the value of *x*. Then classify the triangle.** *(Section 7.3)*

**11.**

**12.**

**13.**

**14. RAILROAD CROSSING** Describe two ways to find the measure of ∠2. *(Section 7.1 and Section 7.2)*

# 7.4 Quadrilaterals

## Essential Question How can you classify quadrilaterals?

*Quad* means *four* and *lateral* means *side*. So, *quadrilateral* means a polygon with *four sides*.

**Quadrilaterals**

### 1 ACTIVITY: Using Descriptions to Form Quadrilaterals

**Work with a partner. Use a geoboard to form a quadrilateral that fits the given description. Record your results on geoboard dot paper.**

**a.** Form a quadrilateral with exactly one pair of parallel sides.

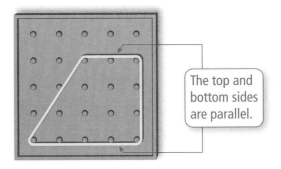

The top and bottom sides are parallel.

**b.** Form a quadrilateral with four congruent sides and four right angles.

**c.** Form a quadrilateral with four right angles that is *not* a square.

**d.** Form a quadrilateral with four congruent sides that is *not* a square.

**e.** Form a quadrilateral with two pairs of congruent adjacent sides and whose opposite sides are *not* congruent.

**f.** Form a quadrilateral with congruent and parallel opposite sides that is *not* a rectangle.

### 2 ACTIVITY: Naming Quadrilaterals

**Work with a partner. Match the names *square, rectangle, rhombus, parallelogram, trapezoid,* and *kite* with your 6 drawings in Activity 1.**

COMMON CORE

**Geometry**

In this lesson, you will
- understand that the sum of the angle measures of any quadrilateral is 360°.
- find missing angle measures in quadrilaterals.
- construct quadrilaterals.

Learning Standard
7.G.2

**3** **ACTIVITY: Forming Quadrilaterals**

Work with a partner. Form each quadrilateral on your geoboard. Then move *only one* vertex to create the new type of quadrilateral. Record your results on geoboard dot paper.

**a.** Trapezoid ⟹ Kite

**b.** Kite ⟹ Rhombus (*not* a square)

**4** **ACTIVITY: Using Technology to Draw Quadrilaterals**

**Math Practice 5**

**Use Technology to Explore**

How does geometry software help you learn about the characteristics of a quadrilateral?

Work with a partner. Use geometry software to draw a quadrilateral that fits the given description.

**a.** a square with a side length of 3 units

**b.** a rectangle with a width of 2 units and a length of 5 units

**c.** a parallelogram with side lengths of 6 units and 1 unit

**d.** a rhombus with a side length of 4 units

## What Is Your Answer?

**5.** **REASONING** Measure the angles of each quadrilateral you formed in Activity 1. Record your results in a table. Include the sum of the angle measures. Then describe the pattern in the table and write a conclusion based on the pattern.

**6.** **IN YOUR OWN WORDS** How can you classify quadrilaterals? Explain using properties of sides and angles.

**Practice** Use what you learned about quadrilaterals to complete Exercises 4–6 on page 296.

**Key Vocabulary** 🔊
kite, *p. 294*

A quadrilateral is a polygon with four sides. The diagram shows properties of different types of quadrilaterals and how they are related. When identifying a quadrilateral, use the name that is most specific.

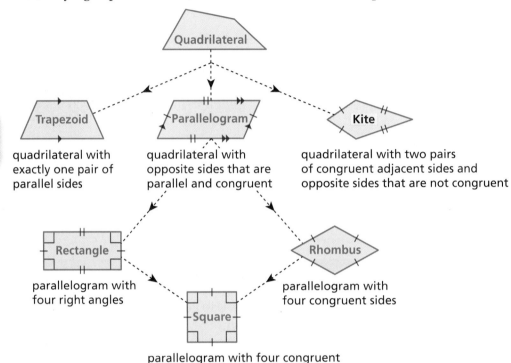

**Reading**

Red arrows indicate parallel sides.

**Trapezoid**
quadrilateral with exactly one pair of parallel sides

**Parallelogram**
quadrilateral with opposite sides that are parallel and congruent

**Kite**
quadrilateral with two pairs of congruent adjacent sides and opposite sides that are not congruent

**Rectangle**
parallelogram with four right angles

**Rhombus**
parallelogram with four congruent sides

**Square**
parallelogram with four congruent sides and four right angles

---

**EXAMPLE** 1 **Classifying Quadrilaterals**

**Classify the quadrilateral.**

**Study Tip**

In Example 1(a), the square is also a parallelogram, a rectangle, and a rhombus. Square is the most specific name.

a.

The quadrilateral has four congruent sides and four right angles.

⋮ So, the quadrilateral is a square.

b.

The quadrilateral has two pairs of congruent adjacent sides and opposite sides that are not congruent.

⋮ So, the quadrilateral is a kite.

🔵 **On Your Own**

Now You're Ready
Exercises 4–9

**Classify the quadrilateral.**

1.

2.

3.

**Key Idea**

**Sum of the Angle Measures of a Quadrilateral**

**Words** The sum of the angle measures of a quadrilateral is 360°.

**Algebra** $w + x + y + z = 360$

---

**EXAMPLE** 2 **Finding an Angle Measure of a Quadrilateral**

**Find the value of $x$.**

| | | | |
|---|---|---|---|
| $70 + 75 + 115 + x =$ | $360$ | Write an equation. |
| $260 + x =$ | $360$ | Combine like terms. |
| $-\ 260$ | $-\ 260$ | Subtraction Property of Equality |
| $x =$ | $100$ | Simplify. |

∴ The value of $x$ is 100.

---

**EXAMPLE** 3 **Constructing a Quadrilateral**

**Draw a parallelogram with a 60° angle and a 120° angle.**

**Step 1:** Draw a line.

**Step 2:** Draw a 60° angle and a 120° angle that each have one side on the line.

**Step 3:** Draw the remaining side. Make sure that both pairs of opposite sides are parallel and congruent.

---

⬤ **On Your Own**

*Now You're Ready*
Exercises 10–12 and 14–17

**Find the value of $x$.**

4.

5.

6. Draw a right trapezoid whose parallel sides have lengths of 3 centimeters and 5 centimeters.

---

## Vocabulary and Concept Check

1. **VOCABULARY** Which statements are true?

   a. All squares are rectangles.    b. All squares are parallelograms.

   c. All rectangles are parallelograms.    d. All squares are rhombuses.

   e. All rhombuses are parallelograms.

2. **REASONING** Name two types of quadrilaterals with four right angles.

3. **WHICH ONE DOESN'T BELONG?** Which type of quadrilateral does *not* belong with the other three? Explain your reasoning.

   | rectangle | parallelogram | square | kite |

 ## Practice and Problem Solving

**Classify the quadrilateral.**

① **4.**

**5.**

**6.**

**7.**

**8.**

**9.**

**Find the value of x.**

② **10.**

65°   115°

115°   x°

**11.**

128°   x°

82°

40°

**12.**

x°   52°

13. **KITE MAKING** What is the measure of the angle at the tail end of the kite?

122°

**Draw a quadrilateral with the given description.**

**3** **14.** a trapezoid with a pair of congruent, nonparallel sides

**15.** a rhombus with 3-centimeter sides and two 100° angles

**16.** a parallelogram with a 45° angle and a 135° angle

**17.** a parallelogram with a 75° angle and a 4-centimeter side

**Copy and complete using *always*, *sometimes*, or *never*.**

**18.** A square is __?__ a rectangle.    **19.** A square is __?__ a rhombus.

**20.** A rhombus is __?__ a square.    **21.** A parallelogram is __?__ a trapezoid.

**22.** A trapezoid is __?__ a kite.    **23.** A rhombus is __?__ a rectangle.

**24.** **DOOR** The dashed line shows how you cut the bottom of a rectangular door so it opens more easily.

   **a.** Identify the new shape of the door. Explain.

   **b.** What is the new angle at the bottom left side of the door? Explain.

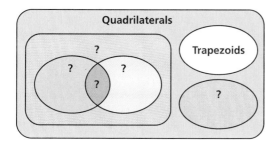

**25.** **VENN DIAGRAM** The diagram shows that some quadrilaterals are trapezoids, and all trapezoids are quadrilaterals. Copy the diagram. Fill in the names of the types of quadrilaterals to show their relationships.

**26.** **Structure** Consider the parallelogram.

   **a.** Find the values of $x$ and $y$.

   **b.** Make a conjecture about opposite angles in a parallelogram.

   **c.** In polygons, consecutive interior angles share a common side. Make a conjecture about consecutive interior angles in a parallelogram.

 **Fair Game Review** What you learned in previous grades & lessons

**Write the ratio as a fraction in simplest form.** *(Section 5.1)*

**27.** 3 turnovers : 12 assists    **28.** 18 girls to 27 boys    **29.** 42 pens : 35 pencils

**30.** **MULTIPLE CHOICE** Computer sales decreased from 40 to 32. What is the percent of decrease? *(Section 6.5)*

   **A** 8%    **B** 20%    **C** 25%    **D** 80%

**Essential Question** How can you enlarge or reduce a drawing proportionally?

### 1 ACTIVITY: Comparing Measurements

**Work with a partner. The diagram shows a food court at a shopping mall. Each centimeter in the diagram represents 40 meters.**

**a.** Find the length and the width of the drawing of the food court.

length: [ ] cm     width: [ ] cm

**b.** Find the actual length and width of the food court. Explain how you found your answers.

length: [ ] m     width: [ ] m

**c.** Find the ratios $\dfrac{\text{drawing length}}{\text{actual length}}$ and $\dfrac{\text{drawing width}}{\text{actual width}}$. What do you notice?

### 2 ACTIVITY: Recreating a Drawing

**Work with a partner. Draw the food court in Activity 1 on the grid paper so that each centimeter represents 20 meters.**

**a.** What happens to the size of the drawing?

**b.** Find the length and the width of your drawing. Compare these dimensions to the dimensions of the original drawing in Activity 1.

**COMMON CORE**

**Geometry**

In this lesson, you will

- use scale drawings to find actual distances.
- find scale factors.
- use scale drawings to find actual perimeters and areas.
- recreate scale drawings at a different scale.

Learning Standard
7.G.1

Work with a partner. The diagram shows a sketch of a painting.
Each unit in the sketch represents 8 inches.

**a.** Find the length and the width of the sketch.

length: ▭ units          width: ▭ units

**b.** Find the actual length and width of the painting.
Explain how you found your answers.

length: ▭ in.    width: ▭ in.

**c.** Find the ratios $\dfrac{\text{sketch length}}{\text{actual length}}$ and $\dfrac{\text{sketch width}}{\text{actual width}}$.
What do you notice?

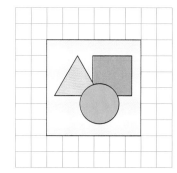

**Math Practice 6**

**Specify Units**
How do you know whether to use feet or units for each measurement?

Work with a partner. Let each unit in the grid paper represent 2 feet. Now sketch the painting in Activity 3 onto the grid paper.

**a.** What happens to the size of the sketch?

**b.** Find the length and the width of your sketch. Compare these dimensions to the dimensions of the original sketch in Activity 3.

## What Is Your Answer?

5. **IN YOUR OWN WORDS** How can you enlarge or reduce a drawing proportionally?

6. Complete the table for both the food court and the painting.

|           | Actual Object | Original Drawing | Your Drawing |
|-----------|---------------|------------------|--------------|
| Perimeter |               |                  |              |
| Area      |               |                  |              |

Compare the measurements in each table. What conclusions can you make?

7. **RESEARCH** Look at some maps in your school library or on the Internet. Make a list of the different scales used on the maps.

8. When you view a map on the Internet, how does the scale change when you zoom out? How does the scale change when you zoom in?

**Practice**

Use what you learned about enlarging or reducing drawings to complete Exercises 4–7 on page 303.

**Check It Out**
Lesson Tutorials
BigIdeasMath✓com

## 🔑 Key Ideas

**Scale Drawings and Models**

A **scale drawing** is a proportional, two-dimensional drawing of an object.
A **scale model** is a proportional, three-dimensional model of an object.

**Scale**

The measurements in scale drawings and models are proportional to the measurements of the actual object. The **scale** gives the ratio that compares the measurements of the drawing or model with the actual measurements.

$$\frac{1 \text{ in.}}{10 \text{ mi}}$$ ← drawing distance
← actual distance

$$1 \text{ in.} : 10 \text{ mi}$$
↑ drawing   ↑ actual

**Study Tip**

Scales are written so that the drawing distance comes first in the ratio.

---

**EXAMPLE** **1** **Finding an Actual Distance**

**What is the actual distance *d* between Cadillac and Detroit?**

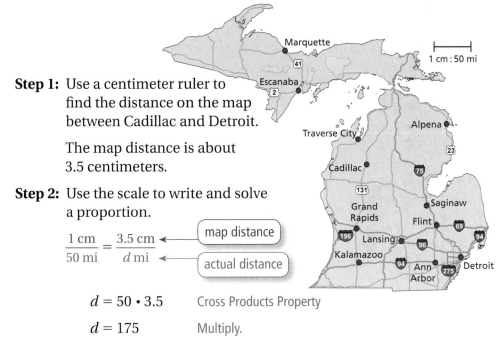

**Step 1:** Use a centimeter ruler to find the distance on the map between Cadillac and Detroit.

The map distance is about 3.5 centimeters.

**Step 2:** Use the scale to write and solve a proportion.

$$\frac{1 \text{ cm}}{50 \text{ mi}} = \frac{3.5 \text{ cm}}{d \text{ mi}}$$ ← map distance
← actual distance

$$d = 50 \cdot 3.5$$     Cross Products Property
$$d = 175$$             Multiply.

∴ So, the distance between Cadillac and Detroit is about 175 miles.

## On Your Own

**Now You're Ready**
Exercises 8–11

**1.** What is the actual distance between Traverse City and Marquette?

---

◀)) Multi-Language Glossary at BigIdeasMath✓com

EXAMPLE **2** Finding a Distance in a Model

The liquid outer core of Earth is 2300 kilometers thick. A scale model of the layers of Earth has a scale of 1 in. : 500 km. How thick is the liquid outer core of the model?

(A) 0.2 in.     (B) 4.6 in.     (C) 0.2 km     (D) 4.6 km

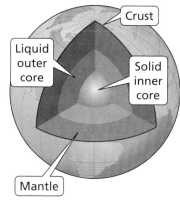

Crust
Liquid outer core
Solid inner core
Mantle

$$\frac{1 \text{ in.}}{500 \text{ km}} = \frac{x \text{ in.}}{2300 \text{ km}}$$ ← model thickness
← actual thickness

$$\frac{1 \text{ in.}}{500 \text{ km}} \cdot 2300 \text{ km} = \frac{x \text{ in.}}{2300 \text{ km}} \cdot 2300 \text{ km}$$  Multiplication Property of Equality

$$4.6 = x$$  Simplify.

⋮∙ So, the liquid outer core of the model is 4.6 inches thick. The correct answer is (B).

● **On Your Own**

2. The mantle of Earth is 2900 kilometers thick. How thick is the mantle of the model?

A scale can be written without units when the units are the same. A scale without units is called a **scale factor**.

EXAMPLE **3** Finding a Scale Factor

A scale model of the Sergeant Floyd Monument is 10 inches tall. The actual monument is 100 feet tall.

a. What is the scale of the model?

$$\frac{\text{model height}}{\text{actual height}} = \frac{10 \text{ in.}}{100 \text{ ft}} = \frac{1 \text{ in.}}{10 \text{ ft}}$$

⋮∙ The scale is 1 in. : 10 ft.

b. What is the scale factor of the model?

Write the scale with the same units. Use the fact that 1 ft = 12 in.

$$\text{scale factor} = \frac{1 \text{ in.}}{10 \text{ ft}} = \frac{1 \text{ in.}}{120 \text{ in.}} = \frac{1}{120}$$

⋮∙ The scale factor is 1 : 120.

● **On Your Own**

Now You're Ready
Exercises 12–16

3. A drawing has a scale of 1 mm : 20 cm. What is the scale factor of the drawing?

EXAMPLE 4   **Finding an Actual Perimeter and Area**

1 cm : 2 mm

The scale drawing of a computer chip helps you see the individual components on the chip.

a. **Find the perimeter and the area of the computer chip in the scale drawing.**

When measured using a centimeter ruler, the scale drawing of the computer chip has a side length of 4 centimeters.

∴ So, the perimeter of the computer chip in the scale drawing is $4(4) = 16$ centimeters, and the area is $4^2 = 16$ square centimeters.

b. **Find the actual perimeter and area of the computer chip.**

$$\frac{1 \text{ cm}}{2 \text{ mm}} = \frac{4 \text{ cm}}{s \text{ mm}}$$ ← drawing distance ← actual distance

$$s = 2 \cdot 4$$      Cross Products Property

$$s = 8$$      Multiply.

The side length of the actual computer chip is 8 millimeters.

∴ So, the actual perimeter of the computer chip is $4(8) = 32$ millimeters, and the actual area is $8^2 = 64$ square millimeters.

c. **Compare the ratios $\dfrac{\text{drawing perimeter}}{\text{actual perimeter}}$ and $\dfrac{\text{drawing area}}{\text{actual area}}$ to the scale factor.**

Use the fact that 1 cm = 10 mm.

$$\text{scale factor} = \frac{1 \text{ cm}}{2 \text{ mm}} = \frac{10 \text{ mm}}{2 \text{ mm}} = \frac{5}{1}$$

$$\frac{\text{drawing perimeter}}{\text{actual perimeter}} = \frac{16 \text{ cm}}{32 \text{ mm}} = \frac{1 \text{ cm}}{2 \text{ mm}} = \frac{5}{1}$$

$$\frac{\text{drawing area}}{\text{actual area}} = \frac{16 \text{ cm}^2}{64 \text{ mm}^2} = \frac{1 \text{ cm}^2}{4 \text{ mm}^2} = \left(\frac{1 \text{ cm}}{2 \text{ mm}}\right)^2 = \left(\frac{5}{1}\right)^2$$

∴ So, the ratio of the perimeters is equal to the scale factor, and the ratio of the areas is equal to the square of the scale factor.

**Study Tip**

The ratios tell you that the perimeter of the drawing is 5 times the actual perimeter, and the area of the drawing is $5^2 = 25$ times the actual area.

**● On Your Own**

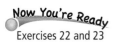

Now You're Ready
Exercises 22 and 23

4. **WHAT IF?** The scale of the drawing of the computer chip is 1 cm : 3 mm. How do the answers in parts (a)–(c) change? Justify your answer.

### ✓ Vocabulary and Concept Check

1. **VOCABULARY** Compare and contrast the terms *scale* and *scale factor*.

2. **CRITICAL THINKING** The scale of a drawing is 2 cm : 1 mm. Is the scale drawing *larger* or *smaller* than the actual object? Explain.

3. **REASONING** How would you find the scale factor of a drawing that shows a length of 4 inches when the actual object is 8 feet long?

### Practice and Problem Solving

**Use the drawing and a centimeter ruler. Each centimeter in the drawing represents 5 feet.**

4. What is the actual length of the flower garden?

5. What are the actual dimensions of the rose bed?

6. What are the actual perimeters of the perennial beds?

7. The area of the tulip bed is what percent of the area of the rose bed?

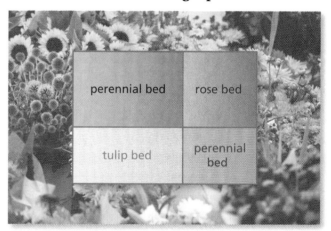

**Use the map in Example 1 to find the actual distance between the cities.**

① 8. Kalamazoo and Ann Arbor

9. Lansing and Flint

10. Grand Rapids and Escanaba

11. Saginaw and Alpena

**Find the missing dimension. Use the scale factor 1 : 12.**

| Item | Model | Actual |
|---|---|---|
| ②③ 12. Mattress | Length: 6.25 in. | Length:          in. |
| 13. Corvette | Length:          in. | Length: 15 ft |
| 14. Water tower | Depth: 32 cm | Depth:          m |
| 15. Wingspan | Width: 5.4 ft | Width:          yd |
| 16. Football helmet | Diameter:          mm | Diameter: 21 cm |

17. **ERROR ANALYSIS** A scale is 1 cm : 20 m. Describe and correct the error in finding the actual distance that corresponds to 5 centimeters.

$$\bcancel{\quad} \frac{1\,cm}{20\,m} = \frac{x\,m}{5\,cm}$$

$$x = 0.25\,m$$

**Use a centimeter ruler to measure the segment shown. Find the scale of the drawing.**

**18.** ⊢———— 120 m ————⊣

**19.**

**20.** **REASONING** You know the length and the width of a scale model. What additional information do you need to know to find the scale of the model?

**21.** **OPEN-ENDED** You are in charge of creating a billboard advertisement with the dimensions shown.

    **a.** Choose a product. Then design the billboard using words and a picture.

    **b.** What is the scale factor of your design?

**④ 22.** **CENTRAL PARK** Central Park is a rectangular park in New York City.

1 cm : 320 m

    **a.** Find the perimeter and the area of Central Park in the scale drawing.

    **b.** Find the actual perimeter and area of Central Park.

**23.** **ICON** You are designing an icon for a mobile app.

    **a.** Find the perimeter and the area of the icon in the scale drawing.

    **b.** Find the actual perimeter and area of the icon.

1 cm : 2.5 mm

**24.** **CRITICAL THINKING** Use the results of Exercises 22 and 23 to make a conjecture about the relationship between the scale factor of a drawing and the ratios $\dfrac{\text{drawing perimeter}}{\text{actual perimeter}}$ and $\dfrac{\text{drawing area}}{\text{actual area}}$.

**Recreate the scale drawing so that it has a scale of 1 cm : 4 m.**

25.

1 cm : 8 m

26.

1 cm : 2 m

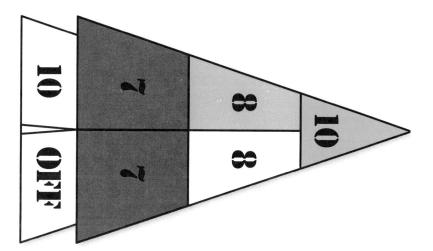

**The shuffleboard diagram has a scale of 1 cm : 1 ft. Find the actual area of the region.**

27. red region

28. blue region

29. green region

30. **BLUEPRINT** In a blueprint, each square has a side length of $\frac{1}{4}$ inch.

a. Ceramic tile costs $5 per square foot. How much would it cost to tile the bathroom?

b. Carpet costs $18 per square yard. How much would it cost to carpet the bedroom and living room?

c. Which has a greater unit cost, the tile or the carpet? Explain.

Reduced Drawing of Blueprint

31. **Modeling** You are making a scale model of the solar system. The radius of Earth is 6378 kilometers. The radius of the Sun is 695,500 kilometers. Is it reasonable to choose a baseball as a model of Earth? Explain your reasoning.

 **Fair Game Review** *What you learned in previous grades & lessons*

**Plot and label the ordered pair in a coordinate plane.** *(Skills Review Handbook)*

32. $A(-4, 3)$    33. $B(2, -6)$    34. $C(5, 1)$    35. $D(-3, -7)$

36. **MULTIPLE CHOICE** Which set of numbers is ordered from least to greatest? *(Section 6.2)*

Ⓐ $\frac{7}{20}$, 32%, 0.45    Ⓑ 17%, 0.21, $\frac{3}{25}$    Ⓒ 0.88, $\frac{7}{8}$, 93%    Ⓓ 57%, $\frac{11}{16}$, 5.7

**Classify the quadrilateral.** *(Section 7.4)*

1.

2.

**Find the value of *x*.** *(Section 7.4)*

3.

4.

**Draw a quadrilateral with the given description.** *(Section 7.4)*

5. a rhombus with 2-centimeter sides and two 50° angles

6. a parallelogram with a 65° angle and a 5-centimeter side

**Find the missing dimension. Use the scale factor 1 : 20.** *(Section 7.5)*

| | Item | Model | Actual |
|---|---|---|---|
| 7. | Basketball player | Height:       in. | Height: 90 in. |
| 8. | Dinosaur | Length: 3.75 ft | Length:       ft |

9. **SHED** The side of the storage shed is in the shape of a trapezoid. Find the value of *x*.   *(Section 7.4)*

10. **DOLPHIN** A dolphin in an aquarium is 12 feet long. A scale model of the dolphin is $3\frac{1}{2}$ inches long. What is the scale factor of the model?   *(Section 7.5)*

11. **SOCCER** A scale drawing of a soccer field is shown. The actual soccer field is 300 feet long. *(Section 7.5)*

    a. What is the scale of the drawing?

    b. What is the scale factor of the drawing?

**Check It Out**
Vocabulary Help
BigIdeasMath √com

## Review Key Vocabulary

adjacent angles, *p. 272*
vertical angles, *p. 272*
congruent angles, *p. 272*
complementary angles,
  *p. 278*

supplementary angles,
  *p. 278*
congruent sides, *p. 284*
kite, *p. 294*
scale drawing, *p. 300*

scale model, *p. 300*
scale, *p. 300*
scale factor, *p. 301*

## Review Examples and Exercises

 **Adjacent and Vertical Angles** *(pp. 270–275)*

**Tell whether the angles are *adjacent* or *vertical*. Then find the value of *x*.**

The angles are vertical angles. Because vertical angles are congruent, the angles have the same measure.

So, the value of *x* is 123.

### Exercises

**Tell whether the angles are *adjacent* or *vertical*. Then find the value of *x*.**

**1.**

**2.**

7.2 **Complementary and Supplementary Angles** *(pp. 276–281)*

**Tell whether the angles are *complementary* or *supplementary*. Then find the value of *x*.**

The two angles make up a right angle. So, the angles are complementary angles, and the sum of their measures is 90°.

| | |
|---|---|
| $(2x - 8) + 42 = 90$ | Write equation. |
| $2x + 34 = 90$ | Combine like terms. |
| $2x = 56$ | Subtract 34 from each side. |
| $x = 28$ | Divide each side by 2. |

So, the value of *x* is 28.

## Exercises

Tell whether the angles are *complementary* or *supplementary*. Then find the value of *x*.

3.

4.

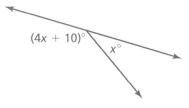

(7.3) **Triangles** *(pp. 282–289)*

Draw a triangle with a 3.5-centimeter side and a 4-centimeter side that meet at a 25° angle. Then classify the triangle.

**Step 1:** Use a protractor to draw a 25° angle.

**Step 2:** Use a ruler to mark 3.5 centimeters on one ray and 4 centimeters on the other ray.

**Step 3:** Draw the third side to form the triangle.

∴ The triangle is an obtuse scalene triangle.

## Exercises

Draw a triangle with the given description.

5. a triangle with angle measures of 40°, 50°, and 90°

6. a triangle with a 3-inch side and a 4-inch side that meet at a 30° angle

Find the value of *x*. Then classify the triangle.

7.

8.

## 7.4 Quadrilaterals (pp. 292–297)

**Draw a parallelogram with a 50° angle and a 130° angle.**

**Step 1:** Draw a line.

**Step 2:** Draw a 50° angle and a 130° angle that each have one side on the line.

**Step 3:** Draw the remaining side. Make sure that both pairs of opposite sides are parallel and congruent.

### Exercises

**Find the value of x.**

9.
$x°$
$128°$

10. $80°$ $x°$
$95°$ $38°$

11. Draw a rhombus with 5-centimeter sides and two 120° angles.

## 7.5 Scale Drawings (pp. 298–305)

**A lighthouse is 160 feet tall. A scale model of the lighthouse has a scale of 1 in. : 8 ft. How tall is the model of the lighthouse?**

$$\frac{1 \text{ in.}}{8 \text{ ft}} = \frac{x \text{ in.}}{160 \text{ ft}}$$   ← model height
← actual height

$$\frac{1 \text{ in.}}{8 \text{ ft}} \cdot 160 \text{ ft} = \frac{x \text{ in.}}{160 \text{ ft}} \cdot 160 \text{ ft}$$   Multiplication Property of Equality

$$20 = x$$   Simplify.

So, the model of the lighthouse is 20 inches tall.

### Exercises

**Use a centimeter ruler to measure the segment shown. Find the scale of the drawing.**

12. |———— 30 in. ————|

13. |—7.5 in.—|

**Check It Out**
Test Practice
BigIdeasMath⎷com

**Tell whether the angles are *adjacent* or *vertical*. Then find the value of *x*.**

**1.**
113°
*x*°

**2.**
(*x* + 6)°
56°

**Tell whether the angles are *complementary* or *supplementary*. Then find the value of *x*.**

**3.**
(8*x* + 2)° 74°

**4.**
15°
(4*x* − 5)°

**Draw a triangle with the given angle measures. Then classify the triangle.**

**5.** 10°, 80°, 90°

**6.** 30°, 40°, 110°

**Draw a triangle with the given description.**

**7.** a triangle with a 5-inch side and a 6-inch side that meet at a 50° angle

**8.** a right isosceles triangle

**Find the value of *x*. Then classify the triangle.**

**9.**
23° 129° *x*°

**10.**
*x*°
68° *x*°

**11.**
*x*°
*x*° *x*°

**Find the value of *x*.**

**12.**
*x*°

**13.**
95° *x*°
95°

**14.**
*x*° 84°
110° 96°

**Draw a quadrilateral with the given description.**

**15.** a rhombus with 6-centimeter sides and two 80° angles

**16.** a parallelogram with a 20° angle and a 160° angle

**17.** **FISH** Use a centimeter ruler to measure the fish. Find the scale factor of the drawing.

15 mm

**18.** **CAD** An engineer is using computer-aided design (CAD) software to design a component for a space shuttle. The scale of the drawing is 1 cm : 60 in. The actual length of the component is 12.5 feet. What is the length of the component in the drawing?

**Test-Taking Strategy**
**Solve Problem Before Looking at Choices**

"Solve the problem before looking at the choices. You know area increases as the square of the scale. So, it's 8 in.²"

1. The number of calories you burn by playing basketball is proportional to the number of minutes you play. Which of the following is a valid interpretation of the graph below? *(7.RP.2d)*

**Basketball**

A. The unit rate is $\frac{1}{9}$ calorie per minute.

B. You burn 5 calories by playing basketball for 45 minutes.

C. You do not burn any calories if you do not play basketball for at least 1 minute.

D. You burn an additional 9 calories for each minute of basketball you play.

2. A lighting store is holding a clearance sale. The store is offering discounts on all the lamps it sells. As the sale progresses, the store will increase the percent of discount it is offering.

   You want to buy a lamp that has an original price of $40. You will buy the lamp when its price is marked down to $10. What percent discount will you have received? *(7.RP.3)*

3. What is the value of the expression below? *(7.NS.1c)*

$$2 - 6 - (-9)$$

   F. $-13$

   G. $-5$

   H. 5

   I. 13

**4.** What is the solution to the proportion below?  *(7.RP.2c)*

$$\frac{8}{12} = \frac{x}{18}$$

**5.** Which graph represents the inequality below?  *(7.EE.4b)*

$$-5 - 6x \le -23$$

**A.**

**B.**

**C.**

**D.**

**6.** You are building a scale model of a park that is planned for a city. The model uses the scale below.

$$1 \text{ centimeter} = 2 \text{ meters}$$

The park will have a rectangular reflecting pool with a length of 20 meters and a width of 12 meters. In your scale model, what will be the area of the reflecting pool?  *(7.G.1)*

**F.** $60 \text{ cm}^2$            **H.** $480 \text{ cm}^2$

**G.** $120 \text{ cm}^2$           **I.** $960 \text{ cm}^2$

**7.** The quantities $x$ and $y$ are proportional. What is the missing value in the table?  *(7.RP.2a)*

| x | y |
|---|---|
| $\frac{5}{7}$ | 10 |
| $\frac{9}{7}$ | 18 |
| $\frac{15}{7}$ | 30 |
| 4 | |

**A.** 38            **C.** 46

**B.** 42            **D.** 56

**8.** $\angle 1$ and $\angle 2$ form a straight angle. $\angle 1$ has a measure of $28°$. What is the measure of $\angle 2$? *(7.G.5)*

   **F.** $62°$                         **H.** $152°$

   **G.** $118°$                      **I.** $208°$

**9.** Brett solved the equation in the box below. *(7.EE.4a)*

$$\frac{c}{5} - (-15) = -35$$

$$\frac{c}{5} + 15 = -35$$

$$\frac{c}{5} + 15 - 15 = -35 - 15$$

$$\frac{c}{5} = -50$$

$$\frac{c}{5} = \frac{-50}{5}$$

$$c = -10$$

What should Brett do to correct the error that he made?

   **A.** Subtract 15 from $-35$ to get $-20$.

   **B.** Rewrite $\frac{c}{5} - (-15)$ as $\frac{c}{5} - 15$.

   **C.** Multiply each side of the equation by 5 to get $c = -250$.

   **D.** Multiply each side of the equation by $-5$ to get $c = 250$.

**10.** A map of the state where Donna lives has the scale shown below. *(7.G.1)*

$$\frac{1}{2} \text{ inch} = 10 \text{ miles}$$

*Part A*   Donna measured the distance between her town and the state capital on the map. Her measurement was $4\frac{1}{2}$ inches. Based on Donna's measurement, what is the actual distance, in miles, between her town and the state capital? Show your work and explain your reasoning.

*Part B*   Donna wants to mark her favorite campsite on the map. She knows that the campsite is 65 miles north of her town. What distance on the map, in inches, represents an actual distance of 65 miles? Show your work and explain your reasoning.

# 8 Circles and Area

"Think of any number between 1 and 9."

"Okay, now add 4 to the number, multiply by 3, subtract 12, and divide by your original number."

"You end up with 3, don't you?"

"What do you get when you divide the circumference of a jack-o-lantern by its diameter?"

"Pumpkin pi, HE HE HE."

# What You Learned Before

"The area of the circle is pi $r$ squared. The area of the triangle is one-half $bh$."

## Classifying Figures (4.G.2)

**Identify the basic shapes in the figure.**

**Example 1**

∴ Rectangle, right triangle

**Example 2**

Semicircle, square, and triangle

### Try It Yourself
**Identify the basic shapes in the figure.**

**1.**

**2.**

**3.**

**4.**

**5.**

**6.**

## Squaring Numbers and Using Order of Operations (6.EE.1)

**Example 3    Evaluate $4^2$.**

$$4^2 = 4 \cdot 4 = 16$$

$4^2$ means to multiply 4 by itself.

**Example 4    Evaluate $3 \cdot 6^2$.**

$$3 \cdot 6^2 = 3 \cdot (6 \cdot 6) = 3 \cdot 36 = 108$$

Use order of operations. Evaluate the exponent, and then multiply.

### Try It Yourself
**Evaluate the expression.**

**7.** $5^2$

**8.** $12^2$

**9.** $3 \cdot 2^2$

**10.** $4 \cdot 7^2$

**11.** $3(1 + 8)^2$

**12.** $2(3 + 7)^2 - 3 \cdot 4$

**Essential Question** How can you find the circumference of a circle?

Archimedes was a Greek mathematician, physicist, engineer, and astronomer.

Archimedes discovered that in any circle the ratio of circumference to diameter is always the same. Archimedes called this ratio pi, or $\pi$ (a letter from the Greek alphabet).

$$\pi = \frac{\text{circumference}}{\text{diameter}}$$

In Activities 1 and 2, you will use the same strategy Archimedes used to approximate $\pi$.

### 1 ACTIVITY: Approximating Pi

**Work with a partner. Copy the table. Record your results in the table.**

- **Measure the perimeter of the large square in millimeters.**

- **Measure the diameter of the circle in millimeters.**

- **Measure the perimeter of the small square in millimeters.**

- **Calculate the ratios of the two perimeters to the diameter.**

- **The average of these two ratios is an approximation of $\pi$.**

Large Square

Small Square

**COMMON CORE**

**Geometry**

In this lesson, you will

- describe a circle in terms of radius and diameter.
- understand the concept of pi.
- find circumferences of circles and perimeters of semicircles.

Learning Standard
7.G.4

| Sides | Large Perimeter | Diameter of Circle | Small Perimeter | $\dfrac{\text{Large Perimeter}}{\text{Diameter}}$ | $\dfrac{\text{Small Perimeter}}{\text{Diameter}}$ | Average of Ratios |
|---|---|---|---|---|---|---|
| 4 | | | | | | |
| 6 | | | | | | |
| 8 | | | | | | |
| 10 | | | | | | |

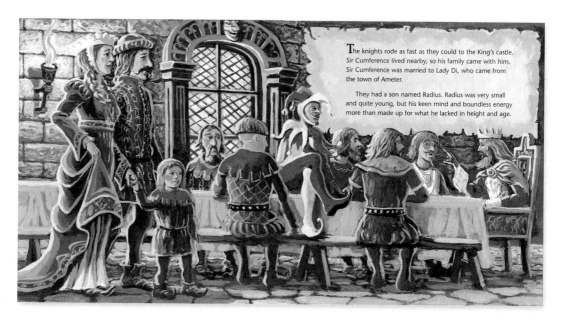

The knights rode as fast as they could to the King's castle. Sir Cumference lived nearby, so his family came with him. Sir Cumference was married to Lady Di, who came from the town of Ameter.

They had a son named Radius. Radius was very small and quite young, but his keen mind and boundless energy more than made up for what he lacked in height and age.

A page from *Sir Cumference and the First Round Table* by Cindy Neuschwander

## 2  ACTIVITY: Approximating Pi

**Math Practice 3**

**Make Conjectures**

How can you use the results of the activity to find an approximation of pi?

Continue your approximation of pi. Complete the table from Activity 1 using a hexagon (6 sides), an octagon (8 sides), and a decagon (10 sides).

**a.**  Large Hexagon    Small Hexagon

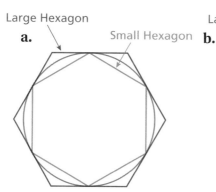

**b.**  Large Octagon    Small Octagon

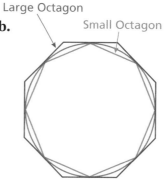

**c.**  Large Decagon    Small Decagon

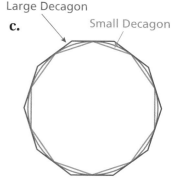

**d.**  From the table, what can you conclude about the value of $\pi$? Explain your reasoning.

**e.**  Archimedes calculated the value of $\pi$ using polygons with 96 sides. Do you think his calculations were more or less accurate than yours?

## What Is Your Answer?

**3. IN YOUR OWN WORDS** Now that you know an approximation for pi, explain how you can use it to find the circumference of a circle. Write a formula for the circumference $C$ of a circle whose diameter is $d$.

**4. CONSTRUCTION** Use a compass to draw three circles. Use your formula from Question 3 to find the circumference of each circle.

**Practice**

Use what you learned about circles and circumference to complete Exercises 9–11 on page 321.

**Key Vocabulary** ◀))
circle, *p. 318*
center, *p. 318*
radius, *p. 318*
diameter, *p. 318*
circumference, *p. 319*
pi, *p. 319*
semicircle, *p. 320*

A **circle** is the set of all points in a plane that are the same distance from a point called the **center**.

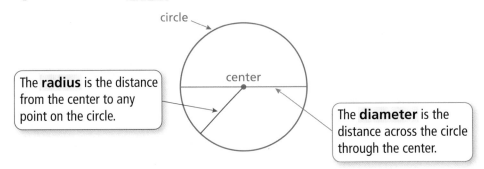

The **radius** is the distance from the center to any point on the circle.

The **diameter** is the distance across the circle through the center.

 **Key Idea**

**Radius and Diameter**

**Words** The diameter $d$ of a circle is twice the radius $r$. The radius $r$ of a circle is one-half the diameter $d$.

**Algebra** **Diameter:** $d = 2r$     **Radius:** $r = \dfrac{d}{2}$

**EXAMPLE** 1 **Finding a Radius and a Diameter**

a. **The diameter of a circle is 20 feet. Find the radius.**

20 ft

$$r = \frac{d}{2}$$     Radius of a circle

$$= \frac{20}{2}$$     Substitute 20 for $d$.

$$= 10$$     Divide.

∴ The radius is 10 feet.

b. **The radius of a circle is 7 meters. Find the diameter.**

7 m

$$d = 2r$$     Diameter of a circle

$$= 2(7)$$     Substitute 7 for $r$.

$$= 14$$     Multiply.

∴ The diameter is 14 meters.

**On Your Own**

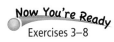
Now You're Ready
Exercises 3–8

1. The diameter of a circle is 16 centimeters. Find the radius.

2. The radius of a circle is 9 yards. Find the diameter.

The distance around a circle is called the **circumference**. The ratio $\dfrac{\text{circumference}}{\text{diameter}}$ is the same for *every* circle and is represented by the Greek letter $\pi$, called **pi**. The value of $\pi$ can be approximated as 3.14 or $\dfrac{22}{7}$.

**Study Tip**

When the radius or diameter is a multiple of 7, it is easier to use $\dfrac{22}{7}$ as the estimate of $\pi$.

 **Key Idea**

**Circumference of a Circle**

**Words** The circumference $C$ of a circle is equal to $\pi$ times the diameter $d$ or $\pi$ times twice the radius $r$.

**Algebra** $C = \pi d$ or $C = 2\pi r$

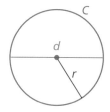

**EXAMPLE 2    Finding Circumferences of Circles**

5 in.

**a.** **Find the circumference of the flying disc. Use 3.14 for $\pi$.**

$$C = 2\pi r \qquad\qquad \text{Write formula for circumference.}$$
$$\approx 2 \cdot 3.14 \cdot 5 \qquad \text{Substitute 3.14 for } \pi \text{ and 5 for } r.$$
$$= 31.4 \qquad\qquad \text{Multiply.}$$

:•: The circumference is about 31.4 inches.

28 mm

**b.** **Find the circumference of the watch face. Use $\dfrac{22}{7}$ for $\pi$.**

$$C = \pi d \qquad\qquad \text{Write formula for circumference.}$$
$$\approx \frac{22}{7} \cdot 28 \qquad \text{Substitute } \frac{22}{7} \text{ for } \pi \text{ and 28 for } d.$$
$$= 88 \qquad\qquad \text{Multiply.}$$

:•: The circumference is about 88 millimeters.

⬤ **On Your Own**

**Now You're Ready**
Exercises 9–11

**Find the circumference of the object. Use 3.14 or $\dfrac{22}{7}$ for $\pi$.**

3.

2 cm.

4.
14 ft

5.

9 in.

EXAMPLE 3    **Estimating a Diameter**

**The circumference of the roll of caution tape decreases 10.5 inches after a construction worker uses some of the tape. Which is the best estimate of the diameter of the roll after the decrease?**

   **A** 5 inches    **B** 7 inches    **C** 10 inches    **D** 12 inches

C = 31.4 in.

After the decrease, the circumference of the roll is
$31.4 - 10.5 = 20.9$ inches.

| | |
|---|---|
| $C = \pi d$ | Write formula for circumference. |
| $20.9 \approx 3.14 \cdot d$ | Substitute 20.9 for C and 3.14 for $\pi$. |
| $21 \approx 3d$ | Round 20.9 up to 21. Round 3.14 down to 3. |
| $7 = d$ | Divide each side by 3. |

⁘ The correct answer is **B**.

● **On Your Own**

6. **WHAT IF?** The circumference of the roll of tape decreases 5.25 inches. Estimate the diameter of the roll after the decrease.

EXAMPLE 4    **Finding the Perimeter of a Semicircular Region**

**A semicircle is one-half of a circle. Find the perimeter of the semicircular region.**

The straight side is 6 meters long. The distance around the curved part is one-half the circumference of a circle with a diameter of 6 meters.

6 m

| | |
|---|---|
| $\dfrac{C}{2} = \dfrac{\pi d}{2}$ | Divide the circumference by 2. |
| $\approx \dfrac{3.14 \cdot 6}{2}$ | Substitute 3.14 for $\pi$ and 6 for $d$. |
| $= 9.42$ | Simplify. |

⁘ So, the perimeter is about $6 + 9.42 = 15.42$ meters.

● **On Your Own**

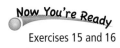
*Now You're Ready*
Exercises 15 and 16

**Find the perimeter of the semicircular region.**

7.
2 ft

8.
7 cm

9.
⊢15 in.⊣

## 8.1 Exercises

 **Vocabulary and Concept Check**

1. **VOCABULARY** What is the relationship between the radius and the diameter of a circle?

2. **WHICH ONE DOESN'T BELONG?** Which phrase does *not* belong with the other three? Explain your reasoning.

| the distance around a circle | $\pi$ times twice the radius |

| $\pi$ times the diameter | the distance from the center to any point on the circle |

 **Practice and Problem Solving**

**Find the radius of the button.**

① 3.

5 cm

4.

28 mm

5.

$3\frac{1}{2}$ in.

**Find the diameter of the object.**

6.

6 cm

7.

2 in.

8.

0.8 ft

**Find the circumference of the pizza. Use 3.14 or $\frac{22}{7}$ for $\pi$.**

② 9.

10 in.

10.

7 in.

11.

18 in.

12. **CHOOSE TOOLS** Choose a real-life circular object. Explain why you might need to know its circumference. Then find the circumference.

13. **SINKHOLE** A circular sinkhole has a circumference of 75.36 meters. A week later, it has a circumference of 150.42 meters.

    a. Estimate the diameter of the sinkhole each week.

    b. How many times greater is the diameter of the sinkhole now compared to the previous week?

14. **REASONING** Consider the circles *A*, *B*, *C*, and *D*.

    a. Without calculating, which circle has the greatest circumference?

    b. Without calculating, which circle has the least circumference?

**Find the perimeter of the window.**

④ **15.**

**16.**

**Find the circumferences of both circles.**

**17.**

**18.**

**19.**

20. **STRUCTURE** Because the ratio $\dfrac{\text{circumference}}{\text{diameter}}$ is the same for every circle, is the ratio $\dfrac{\text{circumference}}{\text{radius}}$ the same for every circle? Explain.

21. **WIRE** A wire is bent to form four semicircles. How long is the wire?

22. **CRITICAL THINKING** Explain how to draw a circle with a circumference of $\pi^2$ inches. Then draw the circle.

**23. AROUND THE WORLD** "Lines" of latitude on Earth are actually circles. The Tropic of Cancer is the northernmost line of latitude at which the Sun appears directly overhead at noon. The Tropic of Cancer has a radius of 5854 kilometers.

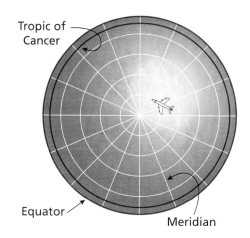

To qualify for an around-the-world speed record, a pilot must cover a distance no less than the circumference of the Tropic of Cancer, cross all meridians, and land on the same airfield where he started.

  **a.** What is the minimum distance that a pilot must fly to qualify for an around-the-world speed record?

  **b. RESEARCH** Estimate the time it would take for a pilot to qualify for the speed record.

9 in.    60 in.

**24. PROBLEM SOLVING** Bicycles in the late 1800s looked very different than they do today.

  **a.** How many rotations does each tire make after traveling 600 feet? Round your answers to the nearest whole number.

  **b.** Would you rather ride a bicycle made with two large wheels or two small wheels? Explain.

**25.** **Logic** The length of the minute hand is 150% of the length of the hour hand.

36 mm

  **a.** What distance will the tip of the minute hand move in 45 minutes? Explain how you found your answer.

  **b.** In 1 hour, how much farther does the tip of the minute hand move than the tip of the hour hand? Explain how you found your answer.

## Fair Game Review *What you learned in previous grades & lessons*

**Find the perimeter of the polygon.** *(Skills Review Handbook)*

**26.**

4 ft

7 ft

**27.**

6 m   5 m

9 m

**28.**

16 in.

12 in.       12 in.

25 in.

**29. MULTIPLE CHOICE** What is the median of the data set? *(Skills Review Handbook)*

    12, 25, 16, 9, 5, 22, 27, 20

  **Ⓐ** 7        **Ⓑ** 16        **Ⓒ** 17        **Ⓓ** 18

**Essential Question** How can you find the perimeter of a composite figure?

### 1 ACTIVITY: Finding a Pattern

**Work with a partner. Describe the pattern of the perimeters. Use your pattern to find the perimeter of the tenth figure in the sequence. (Each small square has a perimeter of 4.)**

a.

b.

c.

### 2 ACTIVITY: Combining Figures

**Work with a partner.**

a. A rancher is constructing a rectangular corral and a trapezoidal corral, as shown. How much fencing does the rancher need to construct both corrals?

b. Another rancher is constructing one corral by combining the two corrals above, as shown. Does this rancher need more or less fencing? Explain your reasoning.

c. How can the rancher in part (b) combine the two corrals to use even less fencing?

**COMMON CORE**

**Geometry**

In this lesson, you will
- find perimeters of composite figures.

Applying Standard
7.G.4

**Work with a partner. You want to bid on a tiling contract. You will be supplying and installing the brown tile that borders the swimming pool. In the figure, each grid square represents 1 square foot.**

- **Your cost for the tile is $4 per linear foot.**
- **It takes about 15 minutes to prepare, install, and clean each foot of tile.**

**a.** How many brown tiles do you need for the border?

**b.** Write a bid for how much you will charge to supply and install the tile. Include what you want to charge as an hourly wage. Estimate what you think your profit will be.

**Math Practice 6**

**Communicate Precisely**

What do you need to include to create an accurate bid? Explain.

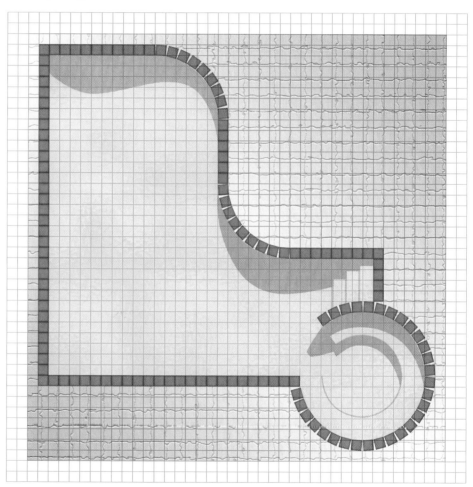

## What Is Your Answer?

**4. IN YOUR OWN WORDS** How can you find the perimeter of a composite figure? Use a semicircle, a triangle, and a parallelogram to draw a composite figure. Label the dimensions. Find the perimeter of the figure.

**Practice** ➤ Use what you learned about perimeters of composite figures to complete Exercises 3–5 on page 328.

**Key Vocabulary**
composite figure,
p. 326

A **composite figure** is made up of triangles, squares, rectangles, semicircles, and other two-dimensional figures. Here are two examples.

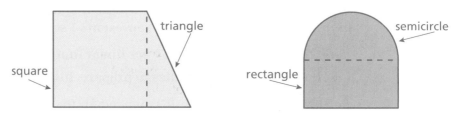

To find the perimeter of a composite figure, find the distance around the figure.

---

**EXAMPLE** **1** **Estimating a Perimeter Using Grid Paper**

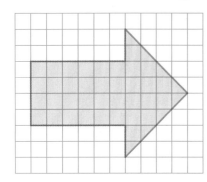

**Estimate the perimeter of the arrow.**

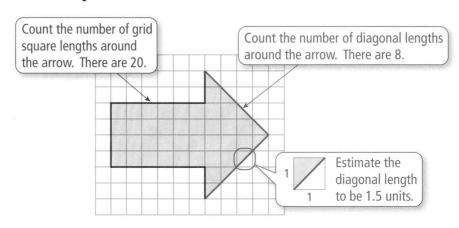

Count the number of grid square lengths around the arrow. There are 20.

Count the number of diagonal lengths around the arrow. There are 8.

Estimate the diagonal length to be 1.5 units.

Length of 20 grid square lengths: $20 \times 1 = 20$ units

Length of 8 diagonal lengths: $8 \times 1.5 = 12$ units

∴ So, the perimeter is about $20 + 12 = 32$ units.

**On Your Own**

Now You're Ready
Exercises 3–8

**Estimate the perimeter of the figure.**

1.

2.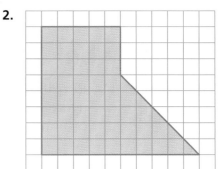

---

EXAMPLE 2 **Finding a Perimeter**

**The figure is made up of a semicircle and a triangle. Find the perimeter.**

The distance around the triangular part of the figure is $6 + 8 = 14$ feet.

The distance around the semicircle is one-half the circumference of a circle with a diameter of 10 feet.

$$\frac{C}{2} = \frac{\pi d}{2} \qquad \text{Divide the circumference by 2.}$$

$$\approx \frac{3.14 \cdot 10}{2} \qquad \text{Substitute 3.14 for } \pi \text{ and 10 for } d.$$

$$= 15.7 \qquad \text{Simplify.}$$

So, the perimeter is about $14 + 15.7 = 29.7$ feet.

EXAMPLE 3 **Finding a Perimeter**

**The running track is made up of a rectangle and two semicircles. Find the perimeter.**

The semicircular ends of the track form a circle with a radius of 32 meters. Find its circumference.

$$C = 2\pi r \qquad \text{Write formula for circumference.}$$

$$\approx 2 \cdot 3.14 \cdot 32 \qquad \text{Substitute 3.14 for } \pi \text{ and 32 for } r.$$

$$= 200.96 \qquad \text{Multiply.}$$

So, the perimeter is about $100 + 100 + 200.96 = 400.96$ meters.

## On Your Own

**Now You're Ready**
Exercises 9–11

**3.** The figure is made up of a semicircle and a triangle. Find the perimeter.

**4.** The figure is made up of a square and two semicircles. Find the perimeter.

**Vocabulary and Concept Check**

1. **REASONING** Is the perimeter of the composite figure equal to the sum of the perimeters of the individual figures? Explain.

2. **OPEN-ENDED** Draw a composite figure formed by a parallelogram and a trapezoid.

 **Practice and Problem Solving**

**Estimate the perimeter of the figure.**

① 3.

4.

5.

6.

7.

8.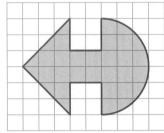

**Find the perimeter of the figure.**

② 9.

10.

11.

12. **ERROR ANALYSIS** Describe and correct the error in finding the perimeter of the figure.

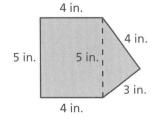

Perimeter = 4 + 3 + 4 + 5 + 4 + 5
         = 25 in.

**Find the perimeter of the figure.**

**13.**

7 in.

5 in.

7 in.    5 in.

**14.**

12 in.

5 in.    5 in.

9 in.

**15.**

3 ft    3 ft

3 ft

3 ft    3 ft

|— 12 ft —|

240 ft    285 ft

450 ft    450 ft

450 ft

**16. PASTURE** A farmer wants to fence a section of land for a horse pasture. Fencing costs $27 per yard. How much will it cost to fence the pasture?

40% of a circle of radius 225 ft

225 ft

300 ft

**17. BASEBALL** You run around the perimeter of the baseball field at a rate of 9 feet per second. How long does it take you to run around the baseball field?

**18. TRACK** In Example 3, the running track has six lanes. Explain why the starting points for the six runners are staggered. Draw a diagram as part of your explanation.

**19.** *Critical Thinking* How can you add a figure to a composite figure without increasing its perimeter? Draw a diagram to support your answer.

## Fair Game Review  What you learned in previous grades & lessons

**Evaluate the expression.** *(Skills Review Handbook)*

**20.** $2.15(3)^2$

**21.** $4.37(8)^2$

**22.** $3.14(7)^2$

**23.** $8.2(5)^2$

**24. MULTIPLE CHOICE** Which expression is equivalent to $(5y + 4) - 2(7 - 2y)$? *(Section 3.2)*

Ⓐ $y - 10$       Ⓑ $9y + 18$       Ⓒ $3y - 10$       Ⓓ $9y - 10$

You can use a **word magnet** to organize formulas or phrases that are associated with a vocabulary word or term. Here is an example of a word magnet for circle.

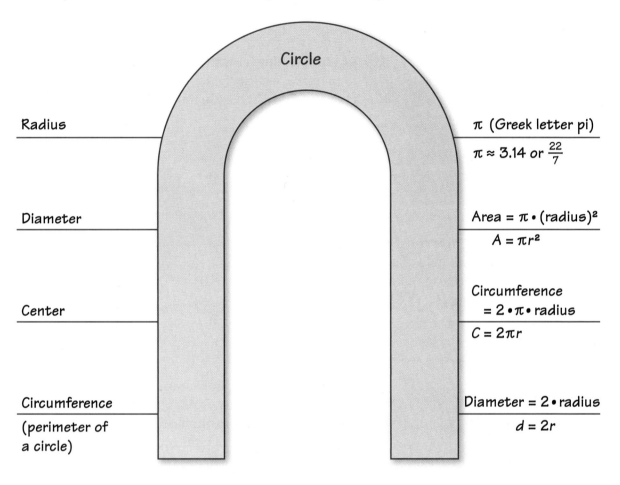

Circle

Radius

Diameter

Center

Circumference
(perimeter of a circle)

$\pi$ (Greek letter pi)

$\pi \approx 3.14$ or $\frac{22}{7}$

Area = $\pi \cdot$ (radius)²

$A = \pi r^2$

Circumference
= 2 • $\pi$ • radius

$C = 2\pi r$

Diameter = 2 • radius

$d = 2r$

## On Your Own

**Make word magnets to help you study these topics.**

1. semicircle

2. composite figure

3. perimeter

**After you complete this chapter, make word magnets for the following topics.**

4. area of a circle

5. area of a composite figure

"I'm trying to make a word magnet for happiness, but I can only think of two words."

1. The diameter of a circle is 36 centimeters. Find the radius. *(Section 8.1)*

2. The radius of a circle is 11 inches. Find the diameter. *(Section 8.1)*

**Estimate the perimeter of the figure.** *(Section 8.2)*

3.

4.

5.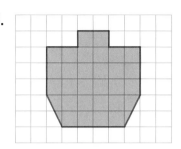

**Find the circumference of the circle. Use 3.14 or $\frac{22}{7}$ for $\pi$.** *(Section 8.1)*

6.
6 mm

7.
1.5 ft

8.
7 cm

**Find the perimeter of the figure.** *(Section 8.1 and Section 8.2)*

9.
8 in.
20 in.
12 in.
24 in.

10.
8 ft   6 ft
8 ft
10 ft

11.
3 ft

12. **BUTTON** What is the circumference of a circular button with a diameter of 8 millimeters? *(Section 8.1)*

12 ft
14 ft
8 ft
10 ft
18 ft

13. **GARDEN** You want to fence part of a yard to make a vegetable garden. How many feet of fencing do you need to surround the garden? *(Section 8.2)*

14. **BAKING** A baker is using two circular pans. The larger pan has a diameter of 12 inches. The smaller pan has a diameter of 7 inches. How much greater is the circumference of the larger pan than that of the smaller pan? *(Section 8.1)*

**Essential Question** How can you find the area of a circle?

---

**1** **ACTIVITY: Estimating the Area of a Circle**

**Work with a partner. Each square in the grid is 1 unit by 1 unit.**

a. Find the area of the large 10-by-10 square.

b. Copy and complete the table.

| Region | 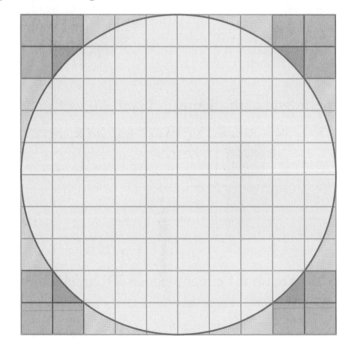 square | triangle | triangle |
|---|---|---|---|
| Area (square units) | | | |

c. Use your results to estimate the area of the circle. Explain your reasoning.

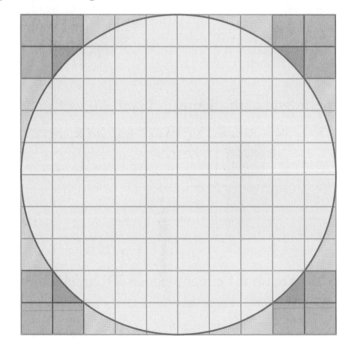

**COMMON CORE**

**Geometry**

In this lesson, you will

• find areas of circles and semicircles.

Learning Standard
7.G.4

d. Fill in the blanks. Explain your reasoning.

$$\text{Area of large square} = \boxed{\phantom{0}} \cdot 5^2 \text{ square units}$$

$$\text{Area of circle} \approx \boxed{\phantom{0}} \cdot 5^2 \text{ square units}$$

e. What dimension of the circle does 5 represent? What can you conclude?

**ACTIVITY: Approximating the Area of a Circle**

**Work with a partner.**

**a.** Draw a circle. Label the radius as *r*.

**b.** Divide the circle into 24 equal sections.

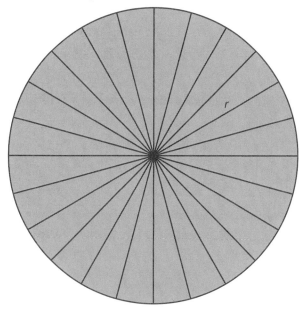

**c.** Cut the sections apart. Then arrange them to approximate a parallelogram.

> **Math Practice 1**
>
> **Interpret a Solution**
>
> What does the area of the parallelogram represent? Explain.

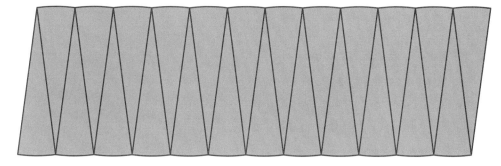

**d.** What is the approximate height and base of the parallelogram?

**e.** Find the area of the parallelogram. What can you conclude?

## What Is Your Answer?

**3.** **IN YOUR OWN WORDS** How can you find the area of a circle?

**4.** Write a formula for the area of a circle with radius *r*. Find an object that is circular. Use your formula to find the area.

**Practice** Use what you learned about areas of circles to complete Exercises 3–5 on page 336.

## Key Idea

**Area of a Circle**

**Words**   The area $A$ of a circle is the product of $\pi$ and the square of the radius.

**Algebra**   $A = \pi r^2$

**EXAMPLE  1   Finding Areas of Circles**

7 cm

**a.   Find the area of the circle. Use $\dfrac{22}{7}$ for $\pi$.**

**Estimate**   $3 \times 7^2 \approx 3 \times 50 = 150$

$$A = \pi r^2 \qquad \text{Write formula for area.}$$

$$\approx \frac{22}{7} \cdot 7^2 \qquad \text{Substitute } \frac{22}{7} \text{ for } \pi \text{ and 7 for } r.$$

$$= \frac{22}{\cancel{7}_{1}} \cdot \cancel{49}^{7} \qquad \text{Evaluate } 7^2. \text{ Divide out the common factor.}$$

$$= 154 \qquad \text{Multiply.}$$

∴  The area is about 154 square centimeters.

**Reasonable?**   $154 \approx 150$ ✓

**b.   Find the area of the circle. Use 3.14 for $\pi$.**

26 in.

The radius is $26 \div 2 = 13$ inches.

**Estimate**   $3 \times 13^2 \approx 3 \times 170 = 510$

$$A = \pi r^2 \qquad \text{Write formula for area.}$$

$$\approx 3.14 \cdot 13^2 \qquad \text{Substitute 3.14 for } \pi \text{ and 13 for } r.$$

$$= 3.14 \cdot 169 \qquad \text{Evaluate } 13^2.$$

$$= 530.66 \qquad \text{Multiply.}$$

∴  The area is about 530.66 square inches.

**Reasonable?**   $530.66 \approx 510$ ✓

## On Your Own

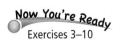
Now You're Ready
Exercises 3–10

**1.**   Find the area of a circle with a radius of 6 feet. Use 3.14 for $\pi$.

**2.**   Find the area of a circle with a diameter of 28 meters. Use $\dfrac{22}{7}$ for $\pi$.

**You want to find the distance the monster truck travels when the tires make one 360-degree rotation. Which best describes this distance?**

- (A) the radius of the tire
- (B) the diameter of the tire
- (C) the circumference of the tire
- (D) the area of the tire

The distance the truck travels after one rotation is the same as the distance *around* the tire. So, the circumference of the tire best describes the distance in one rotation.

⋮ The correct answer is Ⓒ.

● **On Your Own**

3. You want to find the height of one of the tires. Which measurement would best describe the height?

---

EXAMPLE ③ **Finding the Area of a Semicircle**

**Find the area of the semicircular orchestra pit.**

The area of the orchestra pit is one-half the area of a circle with a diameter of 30 feet.

The radius of the circle is 30 ÷ 2 = 15 feet.

$$\frac{A}{2} = \frac{\pi r^2}{2}$$   Divide the area by 2.

$$\approx \frac{3.14 \cdot 15^2}{2}$$   Substitute 3.14 for $\pi$ and 15 for $r$.

$$= \frac{3.14 \cdot 225}{2}$$   Evaluate $15^2$.

$$= 353.25$$   Simplify.

⋮ So, the area of the orchestra pit is about 353.25 square feet.

● **On Your Own**

*Now You're Ready*
Exercises 13–15

**Find the area of the semicircle.**

4.

8 m

5.   5 yd

6.

11 cm

**Vocabulary and Concept Check**

1. **VOCABULARY** Explain how to find the area of a circle given its diameter.

2. **DIFFERENT WORDS, SAME QUESTION** Which is different? Find "both" answers.

> What is the area of a circle with a diameter of 1 m?

> What is the area of a circle with a diameter of 100 cm?

> What is the area of a circle with a radius of 100 cm?

> What is the area of a circle with a radius of 500 mm?

 **Practice and Problem Solving**

**Find the area of the circle. Use 3.14 or $\frac{22}{7}$ for $\pi$.**

① 3.
9 mm

4.
14 cm

5.
10 in.

6.
3 in.

7.
2 cm

8. 1.5 ft

9. Find the area of a circle with a diameter of 56 millimeters.

10. Find the area of a circle with a radius of 5 feet.

11. **TORTILLA** The diameter of a flour tortilla is 12 inches. What is the area?

12. **LIGHTHOUSE** The Hillsboro Inlet Lighthouse lights up how much more area than the Jupiter Inlet Lighthouse?

Jupiter Inlet Lighthouse    18 mi

PALM BEACH

Hillsboro Inlet Lighthouse    28 mi

BROWARD

**Find the area of the semicircle.**

③ **13.**

|— 40 cm —|

**14.**

|— 24 in. —|

**15.**

|— 2 ft —|

**16. REPEATED REASONING** Consider five circles with radii of 1, 2, 4, 8, and 16 inches.

   **a.** Copy and complete the table. Write your answers in terms of $\pi$.

   **b.** Compare the areas and circumferences. What happens to the circumference of a circle when you double the radius? What happens to the area?

   **c.** What happens when you triple the radius?

| Radius | Circumference | Area |
|--------|---------------|------|
| 1 | $2\pi$ in. | $\pi$ in.$^2$ |
| 2 | | |
| 4 | | |
| 8 | | |
| 16 | | |

20 ft

**17. DOG** A dog is leashed to the corner of a house. How much running area does the dog have? Explain how you found your answer.

**18. CRITICAL THINKING** Is the area of a semicircle with a diameter of $x$ *greater than*, *less than*, or *equal to* the area of a circle with a diameter of $\frac{1}{2}x$? Explain.

**Reasoning** **Find the area of the shaded region. Explain how you found your answer.**

**19.**

5 in.

**20.**

9 m

9 m

**21.**

4 ft

4 ft

**Fair Game Review** What you learned in previous grades & lessons

**Evaluate the expression.**   (*Skills Review Handbook*)

**22.** $\frac{1}{2}(7)(4) + 6(5)$

**23.** $\frac{1}{2} \cdot 8^2 + 3(7)$

**24.** $12(6) + \frac{1}{4} \cdot 2^2$

**25. MULTIPLE CHOICE** What is the product of $-8\frac{1}{3}$ and $3\frac{2}{5}$?   (*Section 2.4*)

   Ⓐ $-28\frac{1}{3}$       Ⓑ $-24\frac{2}{15}$       Ⓒ $24\frac{2}{15}$       Ⓓ $28\frac{1}{3}$

**Essential Question** How can you find the area of a composite figure?

### 1 ACTIVITY: Estimating Area

**Work with a partner.**

a. Choose a state. On grid paper, draw a larger outline of the state.

b. Use your drawing to estimate the area (in square miles) of the state.

c. Which state areas are easy to find? Which are difficult? Why?

COMMON CORE

**Geometry**

In this lesson, you will

- find areas of composite figures by separating them into familiar figures.
- solve real-life problems.

Learning Standard
7.G.6

**ACTIVITY: Estimating Areas**

**Work with a partner. The completed puzzle has an area of 150 square centimeters.**

**a.** Estimate the area of each puzzle piece.

**b.** Check your work by adding the six areas. Why is this a check?

3 **ACTIVITY: Filling a Square with Circles**

**Math Practice 1**

**Make a Plan**

What steps will you use to solve this problem?

**Work with a partner. Which pattern fills more of the square with circles? Explain.**

**a.**

8

8

**b.**

8

8

**c.**

8

8

**d.**

8

8

# What Is Your Answer?

**4. IN YOUR OWN WORDS** How can you find the area of a composite figure?

**5.** Summarize the area formulas for all the basic figures you have studied. Draw a single composite figure that has each type of basic figure. Label the dimensions and find the total area.

Use what you learned about areas of composite figures to complete Exercises 3–5 on page 342.

## 8.4 Lesson

To find the area of a composite figure, separate it into figures with areas you know how to find. Then find the sum of the areas of those figures.

**EXAMPLE 1 Finding an Area Using Grid Paper**

**Find the area of the yellow figure.**

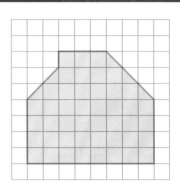

Count the number of squares that lie entirely in the figure. There are 45.

Count the number of half squares in the figure. There are 5.

The area of a half square is $1 \div 2 = 0.5$ square unit.

Area of 45 squares: $45 \times 1 = 45$ square units

Area of 5 half squares: $5 \times 0.5 = 2.5$ square units

So, the area is $45 + 2.5 = 47.5$ square units.

**On Your Own**

**Now You're Ready**
Exercises 3–8

**Find the area of the shaded figure.**

1.

2.
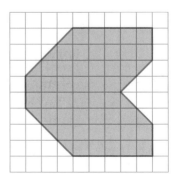

**EXAMPLE 2 Finding an Area**

**Find the area of the portion of the basketball court shown.**

The figure is made up of a rectangle and a semicircle. Find the area of each figure.

**Area of Rectangle**

$$A = \ell w$$

$$= 19(12)$$

$$= 228$$

**Area of Semicircle**

$$A = \frac{\pi r^2}{2}$$

$$\approx \frac{3.14 \cdot 6^2}{2}$$

$$= 56.52$$

> The semicircle has a radius of $\frac{12}{2} = 6$ feet.

⋮ So, the area is about $228 + 56.52 = 284.52$ square feet.

**EXAMPLE 3 Finding an Area**

**Find the area of the figure.**

The figure is made up of a triangle, a rectangle, and a parallelogram. Find the area of each figure.

**Area of Triangle**

$$A = \frac{1}{2}bh$$

$$= \frac{1}{2}(11.2)(4.5)$$

$$= 25.2$$

**Area of Rectangle**

$$A = \ell w$$

$$= 8(4.5)$$

$$= 36$$

**Area of Parallelogram**

$$A = bh$$

$$= 8(6.7)$$

$$= 53.6$$

⋮ So, the area is $25.2 + 36 + 53.6 = 114.8$ square centimeters.

**On Your Own**

**Now You're Ready**
Exercises 9 and 10

**Find the area of the figure.**

3.

4.

##  Vocabulary and Concept Check

1. **REASONING** Describe two different ways to find the area of the figure. Name the types of figures you used and the dimensions of each.

2. **REASONING** Draw a trapezoid. Explain how you can think of the trapezoid as a composite figure to find its area.

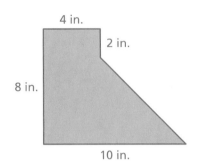

4 in.
2 in.
8 in.
10 in.

##  Practice and Problem Solving

**Find the area of the figure.**

① 3.

4.

5.

6.

7.

8.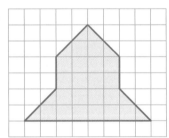

**Find the area of the figure.**

② ③ 9.

7 cm
4 cm    4 cm
10 cm    8 cm    10 cm
19 cm

10.

15 ft    15 ft
4 ft

11. **OPEN-ENDED** Trace your hand and your foot on grid paper. Then estimate the area of each. Which one has the greater area?

**Find the area of the figure.**

**12.**

13 m
6 m
8 m
4 m   4 m

**13.**

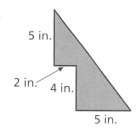

5 in.
2 in.   4 in.
5 in.

**14.**

6 ft
6 ft

**15. STRUCTURE** The figure is made up of a square and a rectangle. Find the area of the shaded region.

7 m
16 m
3 m

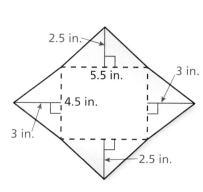

20 ft   20 ft

**16. FOUNTAIN** The fountain is made up of two semicircles and a quarter circle. Find the perimeter and the area of the fountain.

**17.** Critical Thinking   You are deciding on two different designs for envelopes.

2.5 in.
5.5 in.
3 in.
4.5 in.
3 in.
2.5 in.

5.5 in.
2 in.
0.75 in.   0.75 in.
4 in.   4.5 in.   4 in.
3.5 in.

**a.** Which design has the greater area?

**b.** You make 500 envelopes using the design with the greater area. Using the same amount of paper, how many more envelopes can you make with the other design?

**Fair Game Review** What you learned in previous grades & lessons

**Write the phrase as an expression.** *(Skills Review Handbook)*

**18.** 12 less than a number $x$

**19.** a number $y$ divided by 6

**20.** a number $b$ increased by 3

**21.** the product of 7 and a number $w$

**22. MULTIPLE CHOICE** What number is 0.02% of 50? *(Section 6.4)*

Ⓐ 0.01   Ⓑ 0.1   Ⓒ 1   Ⓓ 100

Check It Out
Progress Check
BigIdeasMath ✓.com

**Find the area of the figure.**    *(Section 8.4)*

**1.**

**2.**

**3.**

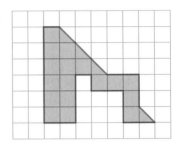

**Find the area of the circle. Use 3.14 or $\frac{22}{7}$ for $\pi$.**    *(Section 8.3)*

**4.**

12 in.

**5.**

6 cm

**6.**

$3\frac{1}{2}$ in.

**Find the area of the figure.**    *(Section 8.4)*

**7.**

12 ft, 9 ft, 12 ft

**8.**

4 m, 2 m, 6 m

**9.**

10 cm, 10 cm, 15 cm, 50 cm, 40 cm, 40 cm

**10. POT HOLDER** A knitted pot holder is shaped like a circle. Its radius is 3.5 inches. What is its area?    *(Section 8.3)*

**11. CARD** The heart-shaped card is made up of a square and two semicircles. What is the area of the card?    *(Section 8.4)*

8 cm    8 cm

14 ft

**12. DESK** A desktop is shaped like a semicircle with a diameter of 28 inches. What is the area of the desktop?    *(Section 8.3)*

**13. RUG** The circular rug is placed on a square floor. The rug touches all four walls. How much of the floor space is *not* covered by the rug?    *(Section 8.4)*

**Check It Out**
Vocabulary Help
BigIdeasMath com

## Review Key Vocabulary

circle, *p. 318*
center, *p. 318*
radius, *p. 318*
diameter, *p. 318*

circumference, *p. 319*
pi, *p. 319*
semicircle, *p. 320*
composite figure, *p. 326*

## Review Examples and Exercises

### 8.1 Circles and Circumference (pp. 316–323)

**Find the circumference of the circle. Use 3.14 for $\pi$.**

The radius is 4 millimeters.

$$C = 2\pi r \qquad \text{Write formula for circumference.}$$

$$\approx 2 \cdot 3.14 \cdot 4 \qquad \text{Substitute 3.14 for } \pi \text{ and 4 for } r.$$

$$= 25.12 \qquad \text{Multiply.}$$

∴ The circumference is about 25.12 millimeters.

### Exercises

**Find the radius of the circle with the given diameter.**

**1.** 8 inches

**2.** 60 millimeters

**3.** 100 meters

**4.** 3 yards

**Find the diameter of the circle with the given radius.**

**5.** 20 feet

**6.** 5 meters

**7.** 1 inch

**8.** 25 millimeters

**Find the circumference of the circle. Use 3.14 or $\dfrac{22}{7}$ for $\pi$.**

**9.**
3 ft

**10.**
21 cm

**11.**
42 in.

**8.2** **Perimeters of Composite Figures** *(pp. 324–329)*

**The figure is made up of a semicircle and a square. Find the perimeter.**

The distance around the square part is 6 + 6 + 6 = 18 meters. The distance around the semicircle is one-half the circumference of a circle with *d* = 6 meters.

$$\frac{C}{2} = \frac{\pi d}{2}$$      Divide the circumference by 2.

$$\approx \frac{3.14 \cdot 6}{2}$$      Substitute 3.14 for $\pi$ and 6 for *d*.

$$= 9.42$$      Simplify.

So, the perimeter is about 18 + 9.42 = 27.42 meters.

## Exercises

**Find the perimeter of the figure.**

**12.**

**13.**

**14.**

**15.**

**16.**

**17.**

**8.3** **Areas of Circles** *(pp. 332–337)*

**Find the area of the circle. Use 3.14 for $\pi$.**

$$A = \pi r^2$$      Write formula for area.

$$\approx 3.14 \cdot 20^2$$      Substitute 3.14 for $\pi$ and 20 for *r*.

$$= 1256$$      Multiply.

The area is about 1256 square yards.

## Exercises

**Find the area of the circle. Use 3.14 or $\frac{22}{7}$ for $\pi$.**

**18.**

4 in.

**19.**

11 cm

**20.**

42 mm

### 8.4 Areas of Composite Figures *(pp. 338–343)*

**Find the area of the figure.**

13 mi

10 mi

26 mi    24 mi

The figure is made up of a rectangle, a triangle and a semicircle.
Find the area of each figure.

**Area of Rectangle**

$A = \ell w$

$= 26(10)$

$= 260$

**Area of Triangle**

$A = \frac{1}{2}bh$

$= \frac{1}{2}(10)(24)$

$= 120$

**Area of Semicircle**

$A = \frac{\pi r^2}{2}$

$\approx \frac{3.14 \cdot 13^2}{2}$

$= 265.33$

So, the area is about $260 + 120 + 265.33 = 645.33$ square miles.

## Exercises

**Find the area of the figure.**

**21.**

4 in.

10 in.

**22.**

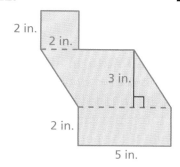

2 in.

2 in.

3 in.

2 in.

5 in.

**23.**

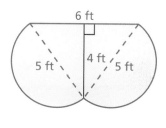

6 ft

5 ft   4 ft   5 ft

**Check It Out**
Test Practice
BigIdeasMath ✓com

**Find the radius of the circle with the given diameter.**

**1.** 10 inches

**2.** 5 yards

**Find the diameter of the circle with the given radius.**

**3.** 34 feet

**4.** 19 meters

**Find the circumference and the area of the circle. Use 3.14 or $\frac{22}{7}$ for $\pi$.**

**5.**
4 ft

**6.**
1 m

**7.**
70 in.

**8.** Estimate the perimeter of the figure. Then find the area.

**Find the perimeter and the area of the figure. Use 3.14 or $\frac{22}{7}$ for $\pi$.**

**9.**
5 m   5 m
3 m
4 m
8 m

**10.**
6 in.   2 in.
10 in.   10 in.
6 in.
12 in.   8 in.

**11.**
7 m   14 m   7 m
14 m

**12. MUSEUM** A museum plans to rope off the perimeter of the L-shaped exhibit. How much rope does it need?

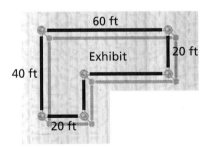
60 ft
Exhibit
20 ft
40 ft
20 ft

**13. ANIMAL PEN** You unfold chicken wire to make a circular pen with a diameter of 2.9 meters. How many meters of chicken wire do you need?

**14. YIN AND YANG** In the Chinese symbol for yin and yang, the dashed curve shows two semicircles formed by the curve separating the yin (dark) and the yang (light). Is the circumference of the entire yin and yang symbol *less than*, *greater than*, or *equal to* the perimeter of the yin?

1. To make 6 servings of soup, you need 5 cups of chicken broth. You want to know how many servings you can make with 2 quarts of chicken broth. Which proportion should you use? *(7.RP.2c)*

   A. $\dfrac{6}{5} = \dfrac{2}{x}$

   B. $\dfrac{6}{5} = \dfrac{x}{2}$

   C. $\dfrac{6}{5} = \dfrac{x}{8}$

   D. $\dfrac{5}{6} = \dfrac{x}{8}$

**Test-Taking Strategy**

**Answer Easy Questions First**

What is the radius of a cat food can that has a diameter of 4 inches?
Ⓐ 1 in.  Ⓑ 2 in.  Ⓒ 3 in.  Ⓓ 4 in.

I love easy questions!

"Scan the test and answer the easy questions first. You know that the radius is half the diameter."

2. What is the value of $x$? *(7.G.5)*

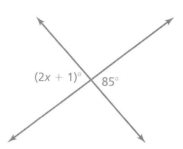

   $(2x + 1)°$   $85°$

3. Your mathematics teacher described an equation in words. Her description is in the box below.

   > "5 less than the product of 7 and an unknown number is equal to 42."

   Which equation matches your mathematics teacher's description? *(7.EE.4a)*

   F. $(5 - 7)n = 42$

   G. $(7 - 5)n = 42$

   H. $5 - 7n = 42$

   I. $7n - 5 = 42$

4. What is the area of the circle below? $\left(\text{Use } \dfrac{22}{7} \text{ for } \pi.\right)$ *(7.G.4)*

   84 cm

   A. $132 \text{ cm}^2$

   B. $264 \text{ cm}^2$

   C. $5544 \text{ cm}^2$

   D. $22{,}176 \text{ cm}^2$

**5.** John was finding the area of the figure below.

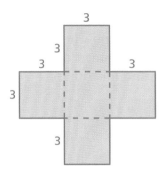

John's work is in the box below.

> area of horizontal rectangle
>
> A = 3 × (3 + 3 + 3)
>
>   = 3 × 9
>
>   = 27 square units
>
> area of vertical rectangle
>
> A = (3 + 3 + 3) × 3
>
>   = 9 × 3
>
>   = 27 square units
>
> total area of figure
>
> A = 27 + 27
>
>   = 54 square units

What should John do to correct the error that he made? *(7.G.6)*

  **F.** Add the area of the center square to the 54 square units.

  **G.** Find the area of one square and multiply this number by 4.

  **H.** Subtract the area of the center square from the 54 square units.

  **I.** Subtract 54 from the area of a large square that is 9 units on each side.

**6.** Which value of $x$ makes the equation below true? *(7.EE.4a)*

$$5x - 3 = 11$$

  **A.** 1.6                                    **C.** 40

  **B.** 2.8                                    **D.** 70

**7.** What is the perimeter of the figure below? (Use 3.14 for $\pi$.)   *(7.G.4)*

6

7

**8.** Which inequality has 5 in its solution set?   *(7.EE.4b)*

**F.** $5 - 2x \geq 3$

**H.** $8 - 3x > -7$

**G.** $3x - 4 \geq 8$

**I.** $4 - 2x < -6$

**9.** Four jewelry stores are selling an identical pair of earrings.

- Store A:  original price of $75; 20% off during sale
- Store B:  original price of $100; 35% off during sale
- Store C:  original price of $70; 10% off during sale
- Store D:  original price of $95; 30% off during sale

Which store has the least sale price for the pair of earrings?   *(7.RP.3)*

**A.** Store A

**C.** Store C

**B.** Store B

**D.** Store D

**10.** A lawn sprinkler sprays water onto part of a circular region, as shown below.   *(7.G.4)*

⊢ 20 ft ⊣

*Part A*   What is the area, in square feet, of the region that the sprinkler sprays with water? Show your work and explain your reasoning. (Use 3.14 for $\pi$.)

*Part B*   What is the perimeter, in feet, of the region that the sprinkler sprays with water? Show your work and explain your reasoning. (Use 3.14 for $\pi$.)

# 9 Surface Area and Volume

"I was thinking that I want the Pagodal roof instead of the Swiss chalet roof for my new doghouse."

"Sometimes you are a true genius."

"Because PAGODAL rearranges to spell 'A DOG PAL.'"

"Take a deep breath and hold it."

"Can I breathe now?"

"Now, do you feel like your surface area or your volume is increasing more?"

# What You Learned Before

"Descartes, how would you like it if I could double the height of your cat food can?"

## Finding Areas of Squares and Rectangles (4.MD.3, 5.NF.4b)

**Example 1** Find the area of the rectangle.

3 mm

7 mm

$\text{Area} = \ell w$     Write formula for area.

$= 7(3)$     Substitute 7 for $\ell$ and 3 for $w$.

$= 21$     Multiply.

∴ The area of the rectangle is 21 square millimeters.

### Try It Yourself
**Find the area of the square or rectangle.**

1.

9 m

11 m

2.
4.2 ft

8.5 ft

3.
$\frac{2}{3}$ in.

$\frac{2}{3}$ in.

## Finding Areas of Triangles (6.G.1)

**Example 2** Find the area of the triangle.

$A = \dfrac{1}{2}bh$     Write formula.

$= \dfrac{1}{2}(6)(7)$     Substitute 6 for $b$ and 7 for $h$.

$= \dfrac{1}{2}(42)$     Multiply 6 and 7.

$= 21$     Multiply $\dfrac{1}{2}$ and 42.

7 in.

6 in.

∴ The area of the triangle is 21 square inches.

### Try It Yourself
**Find the area of the triangle.**

4.

6 ft

13 ft

5.

14 m

20 m

6.

30 cm

15 cm

## Essential Question How can you find the surface area of a prism?

**ACTIVITY: Surface Area of a Rectangular Prism**

**Work with a partner. Copy the net for a rectangular prism. Label each side as $h$, $w$, or $\ell$. Then use your drawing to write a formula for the surface area of a rectangular prism.**

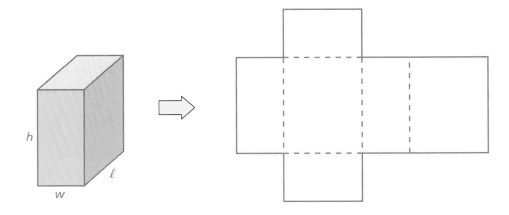

**ACTIVITY: Surface Area of a Triangular Prism**

**Work with a partner.**

a. Find the surface area of the solid shown by the net. Copy the net, cut it out, and fold it to form a solid. Identify the solid.

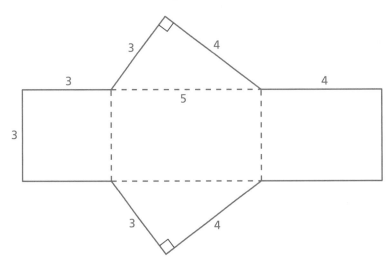

**COMMON CORE**

**Geometry**

In this lesson, you will
- use two-dimensional nets to represent three-dimensional solids.
- find surface areas of rectangular and triangular prisms.
- solve real-life problems.

Learning Standard
7.G.6

b. Which of the surfaces of the solid are bases? Why?

## ACTIVITY: Forming Rectangular Prisms

**Math Practice** 3

**Construct Arguments**

What method did you use to find the surface area of the rectangular prism? Explain.

**Work with a partner.**

- Use 24 one-inch cubes to form a rectangular prism that has the given dimensions.

- Draw each prism.

- Find the surface area of each prism.

**a.** $4 \times 3 \times 2$      *Drawing*      *Surface Area*

    in.$^2$

**b.** $1 \times 1 \times 24$      **c.** $1 \times 2 \times 12$      **d.** $1 \times 3 \times 8$

**e.** $1 \times 4 \times 6$      **f.** $2 \times 2 \times 6$      **g.** $2 \times 4 \times 3$

## What Is Your Answer?

4. Use your formula from Activity 1 to verify your results in Activity 3.

5. **IN YOUR OWN WORDS** How can you find the surface area of a prism?

6. **REASONING** When comparing ice blocks with the same volume, the ice with the greater surface area will melt faster. Which will melt faster, the bigger block or the three smaller blocks? Explain your reasoning.

3 ft

1 ft   1 ft

1 ft

1 ft

1 ft

1 ft

**Practice**    Use what you learned about the surface areas of rectangular prisms to complete Exercises 4–6 on page 359.

**Key Vocabulary**

lateral surface area,
p. 358

 **Key Idea**

**Surface Area of a Rectangular Prism**

**Words**  The surface area $S$ of a rectangular prism is the sum of the areas of the bases and the lateral faces.

**Algebra**  $S = 2\ell w + 2\ell h + 2wh$

Areas of bases

Areas of lateral faces

**EXAMPLE** **1**  **Finding the Surface Area of a Rectangular Prism**

**Find the surface area of the prism.**

Draw a net.

$S = 2\ell w + 2\ell h + 2wh$

$= 2(3)(5) + 2(3)(6) + 2(5)(6)$

$= 30 + 36 + 60$

$= 126$

⋰ The surface area is 126 square inches.

 **On Your Own**

**Find the surface area of the prism.**

**Now You're Ready**
Exercises 7–9

1.

2.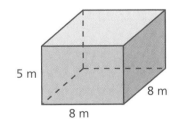

🔊 Multi-Language Glossary at BigIdeasMath✓com

**Key Idea**

**Surface Area of a Prism**

The surface area $S$ of any prism is the sum of the areas of the bases and the lateral faces.

$$S = \text{areas of bases} + \text{areas of lateral faces}$$

**EXAMPLE** ② **Finding the Surface Area of a Triangular Prism**

**Find the surface area of the prism.**

Draw a net.

**Remember**

The area $A$ of a triangle with base $b$ and height $h$ is $A = \dfrac{1}{2}bh$.

***Area of a Base***

Red base: $\dfrac{1}{2} \cdot 3 \cdot 4 = 6$

***Areas of Lateral Faces***

Green lateral face: $3 \cdot 6 = 18$

Purple lateral face: $5 \cdot 6 = 30$

Blue lateral face: $4 \cdot 6 = 24$

Add the areas of the bases and the lateral faces.

$S = \text{areas of bases} + \text{areas of lateral faces}$

$= \underbrace{6 + 6} + 18 + 30 + 24$

> There are two identical bases. Count the area twice.

$= 84$

∴ The surface area is 84 square meters.

● **On Your Own**

*Now You're Ready*
Exercises 10–12

**Find the surface area of the prism.**

**3.**

**4.**

**Remember**

A cube has 6 congruent square faces.

When all the edges of a rectangular prism have the same length *s*, the rectangular prism is a cube. The formula for the surface area of a cube is

$$S = 6s^2.$$ Formula for surface area of a cube

**EXAMPLE ③ Finding the Surface Area of a Cube**

**Find the surface area of the cube.**

| | |
|---|---|
| $S = 6s^2$ | Write formula for surface area of a cube. |
| $= 6(12)^2$ | Substitute 12 for *s*. |
| $= 864$ | Simplify. |

∴ The surface area of the cube is 864 square meters.

The **lateral surface area** of a prism is the sum of the areas of the lateral faces.

**EXAMPLE ④ Real-Life Application**

**The outsides of purple traps are coated with glue to catch emerald ash borers. You make your own trap in the shape of a rectangular prism with an open top and bottom. What is the surface area that you need to coat with glue?**

20 in.

12 in.   10 in.

Find the lateral surface area.

> Do not include the areas of the bases in the formula.

| | |
|---|---|
| $S = 2\ell h + 2wh$ | |
| $= 2(12)(20) + 2(10)(20)$ | Substitute. |
| $= 480 + 400$ | Multiply. |
| $= 880$ | Add. |

∴ So, you need to coat 880 square inches with glue.

**On Your Own**

**Now You're Ready**
Exercises 13–15

**5.** Which prism has the greater surface area?

**6. WHAT IF?** In Example 4, both the length and the width of your trap are 12 inches. What is the surface area that you need to coat with glue?

9 cm

9 cm

9 cm

7 cm

15 cm

5 cm

## Vocabulary and Concept Check

1. **VOCABULARY** Describe two ways to find the surface area of a rectangular prism.

2. **WRITING** Compare and contrast a rectangular prism to a cube.

3. **DIFFERENT WORDS, SAME QUESTION** Which is different? Find "both" answers.

   Find the surface area of the prism.

   Find the area of the bases of the prism.

   Find the area of the net of the prism.

   Find the sum of the areas of the bases and the lateral faces of the prism.

7 in.
4 in.
3 in.

## Practice and Problem Solving

**Use one-inch cubes to form a rectangular prism that has the given dimensions. Then find the surface area of the prism.**

4. $1 \times 2 \times 3$

5. $3 \times 4 \times 1$

6. $2 \times 3 \times 2$

**Find the surface area of the prism.**

7.

3 m
16 m
6 m

8.

7 mm
4 mm
5 mm

9.

3 yd
5 yd
$1\frac{1}{5}$ yd

10.

17 ft
8 ft
20 ft
15 ft

11.

5 m
5 m
4 m
7 m
6 m

12.

13.5 in.
9 in.
9 in.
10 in.

13.

7 yd
7 yd
7 yd

14.

0.5 cm
0.5 cm
0.5 cm

15.
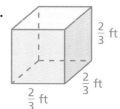
$\frac{2}{3}$ ft
$\frac{2}{3}$ ft
$\frac{2}{3}$ ft

16. **ERROR ANALYSIS** Describe and correct the error in finding the surface area of the prism.

4 in.
5 in.
3 in.

$S = 2(5)(3) + 2(3)(4) + 2(5)(3)$
$= 30 + 24 + 30$
$= 84$ in.$^2$

17. **GAME** Find the surface area of the tin game case.

10 in.    10 in.
8.7 in.
3 in.    10 in.

18. **WRAPPING PAPER** A cube-shaped gift is 11 centimeters long. What is the least amount of wrapping paper you need to wrap the gift?

3 in.
9 in.
13 in.

19. **FROSTING** One can of frosting covers about 280 square inches. Is one can of frosting enough to frost the cake? Explain.

**Find the surface area of the prism.**

20.

12 in.    4 in.
3 in.
5 in.    5 in.
6 in.

21.

2.5 m
2 m
4 m
4 m

22. **OPEN-ENDED** Draw and label a rectangular prism that has a surface area of 158 square yards.

23. **LABEL** A label that wraps around a box of golf balls covers 75% of its lateral surface area. What is the value of $x$?

3 in.
SUPER
Golf Balls
SUPER
Golf Balls
because
YOU are a
super golfer
2 in.
2 in.
$x$ in.

$h$
10 cm
10 cm

24. **BREAD** Fifty percent of the surface area of the bread is crust. What is the height $h$?

**Compare the dimensions of the prisms. How many times greater is the surface area of the red prism than the surface area of the blue prism?**

25.

3 m
4 m
2 m

9 m
6 m
12 m

26.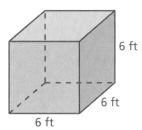

4 ft
4 ft
4 ft

6 ft
6 ft
6 ft

27. **STRUCTURE** You are painting the prize pedestals shown (including the bottoms). You need 0.5 pint of paint to paint the red pedestal.

a. The side lengths of the green pedestal are one-half the side lengths of the red pedestal. How much paint do you need to paint the green pedestal?

b. The side lengths of the blue pedestal are triple the side lengths of the green pedestal. How much paint do you need to paint the blue pedestal?

24 in.

c. Compare the ratio of paint amounts to the ratio of side lengths for the green and red pedestals. Repeat for the green and blue pedestals. What do you notice?

16 in.

16 in.

28. **Number Sense** A keychain-sized puzzle cube is made up of small cubes. Each small cube has a surface area of 1.5 square inches.

a. What is the side length of each small cube?

b. What is the surface area of the entire puzzle cube?

# Fair Game Review  *What you learned in previous grades & lessons*

**Find the area of the triangle**  *(Skills Review Handbook)*

29.

16 ft
20 ft

30.

9 m
12 m

31.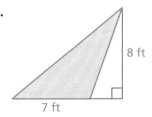

8 ft
7 ft

32. **MULTIPLE CHOICE** What is the circumference of the basketball? Use 3.14 for $\pi$.  *(Section 8.1)*

9 in.

　Ⓐ　14.13 in.　　Ⓑ　28.26 in.　　Ⓒ　56.52 in.　　Ⓓ　254.34 in.

## Essential Question How can you find the surface area of a pyramid?

Even though many well-known pyramids have square bases, the base of a pyramid can be any polygon.

**Triangular Base**

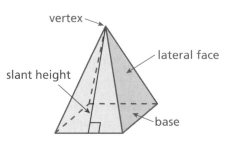

vertex

lateral face

slant height

base

**Square Base**

**Hexagonal Base**

### 1 ACTIVITY: Making a Scale Model

**Work with a partner. Each pyramid has a square base.**

- **Draw a net for a scale model of one of the pyramids. Describe your scale.**
- **Cut out the net and fold it to form a pyramid.**
- **Find the lateral surface area of the real-life pyramid.**

**a.** Cheops Pyramid in Egypt

Side = 230 m, Slant height ≈ 186 m

**c.** Louvre Pyramid in Paris

Side = 35 m, Slant height ≈ 28 m

**b.** Muttart Conservatory in Edmonton

Side = 26 m, Slant height ≈ 27 m

**d.** Pyramid of Caius Cestius in Rome

Side = 22 m, Slant height ≈ 29 m

COMMON
CORE

**Geometry**
In this lesson, you will
- find surface areas of regular pyramids.
- solve real-life problems.
Learning Standard
7.G.6

## 2 ACTIVITY: Estimation

**Work with a partner. There are many different types of gemstone cuts. Here is one called a brilliant cut.**

*Top View*    *Side View*    *Bottom View*

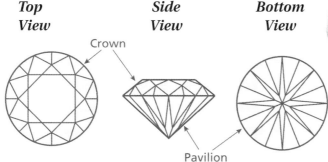

Crown

Pavilion

**The size and shape of the pavilion can be approximated by an octagonal pyramid.**

**a.** What does *octagonal* mean?

**b.** Draw a net for the pyramid.

**c.** Find the lateral surface area of the pyramid.

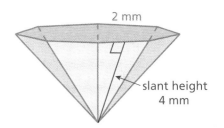

2 mm

slant height 4 mm

## 3 ACTIVITY: Comparing Surface Areas

**Work with a partner. Both pyramids have the same side lengths of the base and the same slant heights.**

**a. REASONING** Without calculating, which pyramid has the greater surface area? Explain.

**b.** Verify your answer to part (a) by finding the surface area of each pyramid.

14 in.

8 in.

14 in.

8 in.

6.9 in.

## What Is Your Answer?

**4. IN YOUR OWN WORDS** How can you find the surface area of a pyramid? Draw a diagram with your explanation.

**Practice**

Use what you learned about the surface area of a pyramid to complete Exercises 4–6 on page 366.

**Key Vocabulary** ◄))
regular pyramid,
    p. 364
slant height, p. 364

A **regular pyramid** is a pyramid whose base is a regular polygon. The lateral faces are triangles. The height of each triangle is the **slant height** of the pyramid.

 **Key Idea**

**Remember**

In a regular polygon, all the sides are congruent and all the angles are congruent.

**Surface Area of a Pyramid**

The surface area *S* of a pyramid is the sum of the areas of the base and the lateral faces.

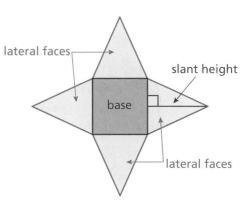

*S* = area of base + areas of lateral faces

**EXAMPLE** ① **Finding the Surface Area of a Square Pyramid**

**Find the surface area of the regular pyramid.**

Draw a net.

| *Area of Base* | *Area of a Lateral Face* |
|---|---|
| $5 \cdot 5 = 25$ | $\dfrac{1}{2} \cdot 5 \cdot 8 = 20$ |

Find the sum of the areas of the base and the lateral faces.

$$S = \text{area of base} + \text{areas of lateral faces}$$
$$= 25 + \underbrace{20 + 20 + 20 + 20}$$
$$= 105$$

There are 4 identical lateral faces. Count the area 4 times.

∴∴ The surface area is 105 square inches.

**On Your Own**

1. What is the surface area of a square pyramid with a base side length of 9 centimeters and a slant height of 7 centimeters?

EXAMPLE 2 **Finding the Surface Area of a Triangular Pyramid**

**Find the surface area of the regular pyramid.**

Draw a net.

*Area of Base*

$$\frac{1}{2} \cdot 10 \cdot 8.7 = 43.5$$

*Area of a Lateral Face*

$$\frac{1}{2} \cdot 10 \cdot 14 = 70$$

Find the sum of the areas of the base and the lateral faces.

$$S = \text{area of base} + \text{areas of lateral faces}$$

$$= 43.5 + \underbrace{70 + 70 + 70}$$

$$= 253.5$$

> There are 3 identical lateral faces. Count the area 3 times.

∴ The surface area is 253.5 square meters.

EXAMPLE 3 **Real-Life Application**

**A roof is shaped like a square pyramid. One bundle of shingles covers 25 square feet. How many bundles should you buy to cover the roof?**

The base of the roof does not need shingles. So, find the sum of the areas of the lateral faces of the pyramid.

*Area of a Lateral Face*

$$\frac{1}{2} \cdot 18 \cdot 15 = 135$$

There are four identical lateral faces. So, the lateral surface area is

$$135 + 135 + 135 + 135 = 540.$$

Because one bundle of shingles covers 25 square feet, it will take $540 \div 25 = 21.6$ bundles to cover the roof.

∴ So, you should buy 22 bundles of shingles.

**On Your Own**

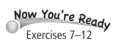
Now You're Ready
Exercises 7–12

**2.** What is the surface area of the regular pyramid at the right?

**3.** **WHAT IF?** In Example 3, one bundle of shingles covers 32 square feet. How many bundles should you buy to cover the roof?

## Vocabulary and Concept Check

1. **VOCABULARY** Can a pyramid have rectangles as lateral faces? Explain.

2. **CRITICAL THINKING** Why is it helpful to know the slant height of a pyramid to find its surface area?

3. **WHICH ONE DOESN'T BELONG?** Which description of the solid does *not* belong with the other three? Explain your answer.

| square pyramid | regular pyramid |
|---|---|
| rectangular pyramid | triangular pyramid |

5 m
5 m

## Practice and Problem Solving

**Use the net to find the surface area of the regular pyramid.**

4.

3 in.

4 in.

5.

9 mm

10 mm

Area of base is 43.3 mm².

6.

6 m

6 m

Area of base is 61.9 m².

**In Exercises 7–11, find the surface area of the regular pyramid.**

① ② 7.

9 ft

6 ft

8.

6 cm

4 cm

9.

10 yd

9 yd

7.8 yd

10.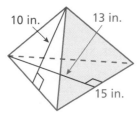

10 in. 13 in.

15 in.

11.

20 mm

16 mm

Area of base is 440.4 mm².

③ 12. **LAMPSHADE** The base of the lampshade is a regular hexagon with a side length of 8 inches. Estimate the amount of glass needed to make the lampshade.

13. **GEOMETRY** The surface area of a square pyramid is 85 square meters. The base length is 5 meters. What is the slant height?

10 in.

**Find the surface area of the composite solid.**

**14.**

6 ft
4 ft
5 ft
5 ft

**15.**

4 cm
10 cm
10 cm
6 cm
8.7 cm
10 cm

**16.**

5 ft
7 ft
4 ft
5 ft
12 ft

**17. PROBLEM SOLVING** You are making an umbrella that is shaped like a regular octagonal pyramid.

   **a.** Estimate the amount of fabric that you need to make the umbrella.

   **b.** The fabric comes in rolls that are 72 inches wide. You don't want to cut the fabric "on the bias." Find out what this means. Then draw a diagram of how you can cut the fabric most efficiently.

   **c.** How much fabric is wasted?

5 ft
4 ft

**18. REASONING** The *height* of a pyramid is the perpendicular distance between the base and the top of the pyramid. Which is greater, the height of a pyramid or the slant height? Explain your reasoning.

pyramid height

**19. TETRAHEDRON** A tetrahedron is a triangular pyramid whose four faces are identical equilateral triangles. The total lateral surface area is 93 square centimeters. Find the surface area of the tetrahedron.

**20.**  **Reasoning** Is the total area of the lateral faces of a pyramid *greater than*, *less than*, or *equal to* the area of the base? Explain.

 **Fair Game Review** What you learned in previous grades & lessons

**Find the area and the circumference of the circle. Use 3.14 for π.**
*(Section 8.1 and Section 8.3)*

**21.**
12

**22.**
8

**23.**
27

**24. MULTIPLE CHOICE** The distance between bases on a youth baseball field is proportional to the distance between bases on a professional baseball field. The ratio of the youth distance to the professional distance is 2 : 3. Bases on a youth baseball field are 60 feet apart. What is the distance between bases on a professional baseball field? *(Section 5.4)*

Ⓐ 40 ft          Ⓑ 90 ft          Ⓒ 120 ft          Ⓓ 180 ft

## Essential Question How can you find the surface area of a cylinder?

A *cylinder* is a solid that has two parallel, identical circular bases.

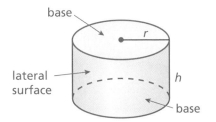

base
lateral surface
base
r
h

### 1 ACTIVITY: Finding Area

**Work with a partner. Use a cardboard cylinder.**

- **Talk about how you can find the area of the outside of the roll.**

- **Estimate the area using the methods you discussed.**

- **Use the roll and the scissors to find the actual area of the cardboard.**

- **Compare the actual area to your estimates.**

### 2 ACTIVITY: Finding Surface Area

**Work with a partner.**

**COMMON CORE**

**Geometry**
In this lesson, you will
- find surface areas of cylinders.
Applying Standard
7.G.4

- **Make a net for the can. Name the shapes in the net.**

- **Find the surface area of the can.**

- **How are the dimensions of the rectangle related to the dimensions of the can?**

Math
Practice 7

View as
Components

How can you use
the results of
Activity 2 to help
you identify the
components of the
surface area?

Work with a partner. From memory, estimate the dimensions of the real-life item in inches. Then use the dimensions to estimate the surface area of the item in square inches.

a.

b.

c.

d.

## What Is Your Answer?

4. **IN YOUR OWN WORDS** How can you find the surface area of a cylinder? Give an example with your description. Include a drawing of the cylinder.

5. To eight decimal places, $\pi \approx 3.14159265$. Which of the following is closest to $\pi$?

   a. 3.14

   b. $\dfrac{22}{7}$

   c. $\dfrac{355}{113}$

"To approximate $\pi \approx 3.141593$, I simply remember 1, 1, 3, 3, 5, 5."

"Then I compute $\frac{355}{113} \approx 3.141593$."

Practice ▶ Use what you learned about the surface area of a cylinder to complete Exercises 3–5 on page 372.

### Key Idea

**Surface Area of a Cylinder**

**Words** The surface area $S$ of a cylinder is the sum of the areas of the bases and the lateral surface.

**Remember**

Pi can be approximated as 3.14 or $\frac{22}{7}$.

**Algebra** $S = 2\pi r^2 + 2\pi rh$

Areas of bases

Area of lateral surface

**EXAMPLE** **1** **Finding the Surface Area of a Cylinder**

Find the surface area of the cylinder. Round your answer to the nearest tenth.

Draw a net.

$$S = 2\pi r^2 + 2\pi rh$$
$$= 2\pi(4)^2 + 2\pi(4)(3)$$
$$= 32\pi + 24\pi$$
$$= 56\pi$$
$$\approx 175.8$$

The surface area is about 175.8 square millimeters.

### On Your Own

**Now You're Ready**
Exercises 6–8

Find the surface area of the cylinder. Round your answer to the nearest tenth.

1.

2.

**EXAMPLE** 2 **Finding Surface Area**

**How much paper is used for the label on the can of peas?**

1 in.

2 in.

Find the lateral surface area of the cylinder.

$S = 2\pi rh$ ← Do not include the areas of the bases in the formula.

$\quad = 2\pi(1)(2)$      Substitute.

$\quad = 4\pi \approx 12.56$      Multiply.

∴ About 12.56 square inches of paper is used for the label.

**EXAMPLE** 3 **Real-Life Application**

2 in.

5.5 in.

**You earn $0.01 for recycling the can in Example 2. How much can you expect to earn for recycling the tomato can? Assume that the recycle value is proportional to the surface area.**

Find the surface area of each can.

| *Tomatoes* | *Peas* |
|---|---|
| $S = 2\pi r^2 + 2\pi rh$ | $S = 2\pi r^2 + 2\pi rh$ |
| $\quad = 2\pi(2)^2 + 2\pi(2)(5.5)$ | $\quad = 2\pi(1)^2 + 2\pi(1)(2)$ |
| $\quad = 8\pi + 22\pi$ | $\quad = 2\pi + 4\pi$ |
| $\quad = 30\pi$ | $\quad = 6\pi$ |

Use a proportion to find the recycle value $x$ of the tomato can.

$$\frac{30\pi \text{ in.}^2}{x} = \frac{6\pi \text{ in.}^2}{\$0.01}$$    ← surface area    ← recycle value

$30\pi \cdot 0.01 = x \cdot 6\pi$      Cross Products Property

$\quad\quad 5 \cdot 0.01 = x$      Divide each side by $6\pi$.

$\quad\quad\quad 0.05 = x$      Simplify.

∴ You can expect to earn $0.05 for recycling the tomato can.

**On Your Own**

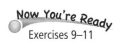

Now You're Ready
Exercises 9–11

**3. WHAT IF?** In Example 3, the height of the can of peas is doubled.

   **a.** Does the amount of paper used in the label double?

   **b.** Does the recycle value double? Explain.

## Vocabulary and Concept Check

1. **CRITICAL THINKING** Which part of the formula $S = 2\pi r^2 + 2\pi rh$ represents the lateral surface area of a cylinder?

2. **CRITICAL THINKING** You are given the height and the circumference of the base of a cylinder. Describe how to find the surface area of the entire cylinder.

 ## Practice and Problem Solving

**Make a net for the cylinder. Then find the surface area of the cylinder. Round your answer to the nearest tenth.**

3.

3 ft
2 ft

4.
4 m
1 m

5.
7 ft
5 ft

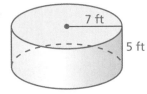

**Find the surface area of the cylinder. Round your answer to the nearest tenth.**

① 6.
5 mm
2 mm

7.
6 ft
7 ft

8.
12 cm
6 cm

**Find the lateral surface area of the cylinder. Round your answer to the nearest tenth.**

② 9.
10 ft
6 ft

10.
9 in.
4 in.

11.
14 m
2 m

12. **ERROR ANALYSIS** Describe and correct the error in finding the surface area of the cylinder.

5 yd
10.6 yd

$$\times \quad \begin{aligned} S &= \pi r^2 + 2\pi rh \\ &= \pi(5)^2 + 2\pi(5)(10.6) \\ &= 25\pi + 106\pi \\ &= 131\pi \approx 411.3 \text{ yd}^2 \end{aligned}$$

13. **TANKER** The truck's tank is a stainless steel cylinder. Find the surface area of the tank.

50 ft
radius = 4 ft

14. **OTTOMAN** What percent of the surface area of the ottoman is green (not including the bottom)?

16 in.
6 in.
8 in.

15. **REASONING** You make two cylinders using 8.5-by-11-inch pieces of paper. One has a height of 8.5 inches, and the other has a height of 11 inches. Without calculating, compare the surface areas of the cylinders.

10 cm
3.5 cm
24.5 cm
5.5 cm

16. **INSTRUMENT** A *ganza* is a percussion instrument used in samba music.

   a. Find the surface area of each of the two labeled ganzas.

   b. The weight of the smaller ganza is 1.1 pounds. Assume that the surface area is proportional to the weight. What is the weight of the larger ganza?

17. **BRIE CHEESE** The cut wedge represents one-eighth of the cheese.

   a. Find the surface area of the cheese before it is cut.

   b. Find the surface area of the remaining cheese after the wedge is removed. Did the surface area increase, decrease, or remain the same?

3 in.
1 in.

18. **Repeated Reasoning** A cylinder has radius $r$ and height $h$.

   a. How many times greater is the surface area of a cylinder when both dimensions are multiplied by a factor of 2? 3? 5? 10?

   b. Describe the pattern in part (a). How many times greater is the surface area of a cylinder when both dimensions are multiplied by a factor of 20?

r
h

**Fair Game Review** What you learned in previous grades & lessons

**Find the area.** *(Skills Review Handbook)*

19.
2 ft
5 ft

20.
4 cm
8 cm

21.
7 in.
5 in.
12 in.

22. **MULTIPLE CHOICE** 40% of what number is 80? *(Section 6.4)*

   Ⓐ 32          Ⓑ 48          Ⓒ 200          Ⓓ 320

# 9 Study Help

You can use an **information frame** to help you organize and remember concepts. Here is an example of an information frame for surface areas of rectangular prisms.

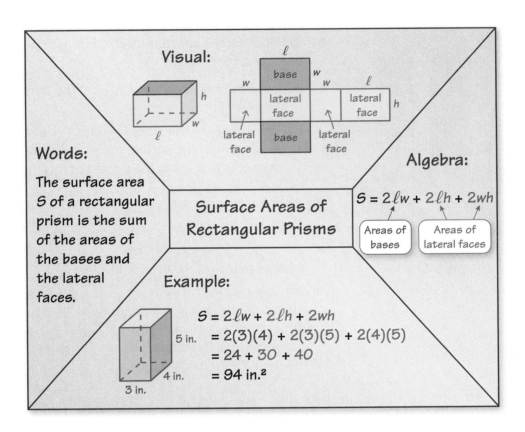

## On Your Own

**Make information frames to help you study the topics.**

1. surface areas of prisms

2. surface areas of pyramids

3. surface areas of cylinders

**After you complete this chapter, make information frames for the following topics.**

4. volumes of prisms

5. volumes of pyramids

"I'm having trouble thinking of a good title for my information frame."

**Check It Out**
Progress Check
BigIdeasMath  com

**Find the surface area of the prism.** *(Section 9.1)*

**1.**

3 cm    4 cm
10 cm
5 cm

**2.**

4 mm
2 mm
7 mm

**Find the surface area of the regular pyramid.** *(Section 9.2)*

**3.**

12 m
Area of base is 65.0 m².
5 m

**4.**

6 cm
2 cm

**Find the surface area of the cylinder. Round your answer to the nearest tenth.** *(Section 9.3)*

**5.**

10 ft
3 ft

**6.**

5 m
6 m

**Find the lateral surface area of the cylinder. Round your answer to the nearest tenth.** *(Section 9.3)*

**7.**

9 cm
7 cm

**8.**

12.2 mm
8 mm

**9. SKYLIGHT** You are making a skylight that has 12 triangular pieces of glass and a slant height of 3 feet. Each triangular piece has a base of 1 foot. *(Section 9.2)*

   **a.** How much glass will you need to make the skylight?

   **b.** Can you cut the 12 glass triangles from a sheet of glass that is 4 feet by 8 feet? If so, draw a diagram showing how this can be done.

**10. MAILING TUBE** What is the least amount of material needed to make the mailing tube? *(Section 9.3)*

3 ft
3 in.

**11. WOODEN CHEST** All the faces of the wooden chest will be painted except for the bottom. Find the area to be painted, in *square inches*. *(Section 9.1)*

4 ft
4 ft    4 ft

## Essential Question   How can you find the volume of a prism?

**1   ACTIVITY: Pearls in a Treasure Chest**

Work with a partner. A treasure chest is filled with valuable pearls. Each pearl is about 1 centimeter in diameter and is worth about $80.

Use the diagrams below to describe two ways that you can estimate the number of pearls in the treasure chest.

**a.**

1 cm

60 cm

60 cm

120 cm

**b.**

**c.**   Use the method in part (a) to estimate the value of the pearls in the chest.

**COMMON CORE**

**Geometry**
In this lesson, you will
• find volumes of prisms.
• solve real-life problems.
Learning Standard
7.G.6

**2   ACTIVITY: Finding a Formula for Volume**

Work with a partner. You know that the formula for the volume of a rectangular prism is $V = \ell wh$.

**a.**   Write a formula that gives the volume in terms of the area of the base $B$ and the height $h$.

**b.**   Use both formulas to find the volume of each prism. Do both formulas give you the same volume?

**ACTIVITY: Finding a Formula for Volume**

Work with a partner. Use the concept in Activity 2 to find a formula that gives
the volume of any prism.

Triangular Prism

Rectangular Prism

Pentagonal Prism

Triangular Prism

Hexagonal Prism

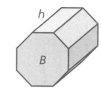

Octagonal Prism

**4** **ACTIVITY: Using a Formula**

**Work with a partner. A ream of paper has 500 sheets.**

**a.** Does a single sheet of paper have a volume? Why or why not?

**b.** If so, explain how you can find the volume of a single sheet of paper.

## What Is Your Answer?

5. **IN YOUR OWN WORDS** How can you find the volume of a prism?

6. **STRUCTURE** Draw a prism that has a trapezoid as its base. Use your formula
to find the volume of the prism.

Use what you learned about the volumes of prisms to complete
Exercises 4–6 on page 380.

The *volume* of a three-dimensional figure is a measure of the amount of space that it occupies. Volume is measured in cubic units.

 **Key Idea**

**Volume of a Prism**

**Words** The volume $V$ of a prism is the product of the area of the base and the height of the prism.

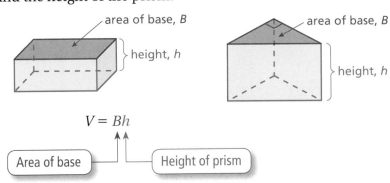

> **Remember**
>
> The volume $V$ of a cube with an edge length of $s$ is $V = s^3$.

**Algebra** $V = Bh$

Area of base ⟶ ⟵ Height of prism

---

**EXAMPLE** **1** **Finding the Volume of a Prism**

> **Study Tip**
>
> The area of the base of a rectangular prism is the product of the length $\ell$ and the width $w$.
>
> You can use $V = \ell wh$ to find the volume of a rectangular prism.

**Find the volume of the prism.**

$$V = Bh \qquad \text{Write formula for volume.}$$
$$= 6(8) \cdot 15 \qquad \text{Substitute.}$$
$$= 48 \cdot 15 \qquad \text{Simplify.}$$
$$= 720 \qquad \text{Multiply.}$$

∴ The volume is 720 cubic yards.

15 yd

8 yd

6 yd

---

**EXAMPLE** **2** **Finding the Volume of a Prism**

**Find the volume of the prism.**

$$V = Bh \qquad \text{Write formula for volume.}$$
$$= \frac{1}{2}(5.5)(2) \cdot 4 \qquad \text{Substitute.}$$
$$= 5.5 \cdot 4 \qquad \text{Simplify.}$$
$$= 22 \qquad \text{Multiply.}$$

∴ The volume is 22 cubic inches.

2 in.

4 in.

5.5 in.

## On Your Own

**Now You're Ready**
Exercises 4–12

**Find the volume of the prism.**

1.

4 ft
4 ft
4 ft

2.

5 m
12 m
9 m

---

**EXAMPLE 3 Real-Life Application**

A movie theater designs two bags to hold 96 cubic inches of popcorn. (a) Find the height of each bag. (b) Which bag should the theater choose to reduce the amount of paper needed? Explain.

Bag A

Bag B

3 in.
4 in.

4 in.
4 in.

**a.** Find the height of each bag.

| **Bag A** | **Bag B** |
|---|---|
| $V = Bh$ | $V = Bh$ |
| $96 = 4(3)(h)$ | $96 = 4(4)(h)$ |
| $96 = 12h$ | $96 = 16h$ |
| $8 = h$ | $6 = h$ |

∴ The height is 8 inches.   ∴ The height is 6 inches.

**b.** To determine the amount of paper needed, find the surface area of each bag. Do not include the top base.

| **Bag A** | **Bag B** |
|---|---|
| $S = \ell w + 2\ell h + 2wh$ | $S = \ell w + 2\ell h + 2wh$ |
| $= 4(3) + 2(4)(8) + 2(3)(8)$ | $= 4(4) + 2(4)(6) + 2(4)(6)$ |
| $= 12 + 64 + 48$ | $= 16 + 48 + 48$ |
| $= 124 \text{ in.}^2$ | $= 112 \text{ in.}^2$ |

∴ The surface area of Bag B is less than the surface area of Bag A. So, the theater should choose Bag B.

## On Your Own

Bag C

3. You design Bag C that has a volume of 96 cubic inches. Should the theater in Example 3 choose your bag? Explain.

4 in.
4.8 in.

h

### Vocabulary and Concept Check

1. **VOCABULARY** What types of units are used to describe volume?

2. **VOCABULARY** Explain how to find the volume of a prism.

3. **CRITICAL THINKING** How are volume and surface area different?

### Practice and Problem Solving

**Find the volume of the prism.**

**4.**
9 in.
9 in.
9 in.

**5.**
8 cm
12 cm
6 cm

**6.**
$8\frac{1}{2}$ m
7 m
4 m

**7.**
6 yd
$4\frac{1}{5}$ yd
$8\frac{1}{3}$ yd

**8.**
6 ft
9 ft
4.5 ft

**9.**
8 mm
10 mm
10.5 mm

**10.**
4.8 m
10 m
7.2 m

**11.**
15 mm
$B = 43$ mm$^2$

**12.**
20 ft
$B = 166$ ft$^2$

13. **ERROR ANALYSIS** Describe and correct the error in finding the volume of the triangular prism.

7 cm
10 cm
5 cm

$$V = Bh$$
$$= 10(5)(7)$$
$$= 50 \cdot 7$$
$$= 350 \text{ cm}^3$$

**School Locker**

60 in.
12 in.
10 in.

**Gym Locker**
48 in.
12 in.
15 in.

14. **LOCKER** Each locker is shaped like a rectangular prism. Which has more storage space? Explain.

15. **CEREAL BOX** A cereal box is 9 inches by 2.5 inches by 10 inches. What is the volume of the box?

**Find the volume of the prism.**

**16.**

12 in.

12 in.    10 in.

**17.**

24 ft

30 ft

20 ft

**18. LOGIC** Two prisms have the same volume. Do they *always*, *sometimes*, or *never* have the same surface area? Explain.

**19. CUBIC UNITS** How many cubic inches are in a cubic foot? Use a sketch to explain your reasoning.

**20. CAPACITY** As a gift, you fill the calendar with packets of chocolate candy. Each packet has a volume of 2 cubic inches. Find the maximum number of packets you can fit inside the calendar.

6 in.

8 in.    4 in.

**21. PRECISION** Two liters of water are poured into an empty vase shaped like an octagonal prism. The base area is 100 square centimeters. What is the height of the water? $(1 \text{ L} = 1000 \text{ cm}^3)$

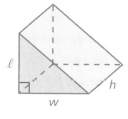

**22. GAS TANK** The gas tank is 20% full. Use the current price of regular gasoline in your community to find the cost to fill the tank. $(1 \text{ gal} = 231 \text{ in.}^3)$

**23. OPEN-ENDED** You visit an aquarium. One of the tanks at the aquarium holds 450 gallons of water. Draw a diagram to show one possible set of dimensions of the tank. $(1 \text{ gal} = 231 \text{ in.}^3)$

11 in.

1.25 ft    1.75 ft

**24.**  **Critical Thinking** How many times greater is the volume of a triangular prism when one of its dimensions is doubled? when all three dimensions are doubled?

$\ell$

$w$    $h$

## Fair Game Review  What you learned in previous grades & lessons

**Find the selling price.**  *(Section 6.6)*

**25.** Cost to store:  $75
Markup:  20%

**26.** Cost to store:  $90
Markup:  60%

**27.** Cost to store:  $130
Markup:  85%

**28. MULTIPLE CHOICE** What is the approximate surface area of a cylinder with a radius of 3 inches and a height of 10 inches?  *(Section 9.3)*

Ⓐ  30 in.$^2$

Ⓑ  87 in.$^2$

Ⓒ  217 in.$^2$

Ⓓ  245 in.$^2$

# 9.5 Volumes of Pyramids

**Essential Question** How can you find the volume of a pyramid?

### 1 ACTIVITY: Finding a Formula Experimentally

**Work with a partner.**

- Draw the two nets on cardboard and cut them out.

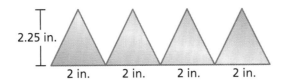

2.25 in.

2 in.    2 in.    2 in.    2 in.

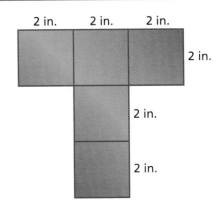

2 in.    2 in.    2 in.

2 in.

2 in.

2 in.

- Fold and tape the nets to form an open square box and an open pyramid.

- Both figures should have the same size square base and the same height.

- Fill the pyramid with pebbles. Then pour the pebbles into the box. Repeat this until the box is full. How many pyramids does it take to fill the box?

- Use your result to find a formula for the volume of a pyramid.

### 2 ACTIVITY: Comparing Volumes

**Work with a partner.** You are an archaeologist studying two ancient pyramids. What factors would affect how long it took to build each pyramid? Given similar conditions, which pyramid took longer to build? Explain your reasoning.

**COMMON CORE**

**Geometry**
In this lesson, you will
- find volumes of pyramids.
- solve real-life problems.
Learning Standard
7.G.6

The Sun Pyramid in Mexico
Height: about 246 ft
Base: about 738 ft by 738 ft

Cheops Pyramid in Egypt
Height: about 480 ft
Base: about 755 ft by 755 ft

**3** **ACTIVITY: Finding and Using a Pattern**

**Math Practice** **7**

**Look for Patterns**

As the height and the base lengths increase, how does this pattern affect the volume? Explain.

Work with a partner.

- Find the volumes of the pyramids.
- Organize your results in a table.
- Describe the pattern.
- Use your pattern to find the volume of a pyramid with a base length and a height of 20.

    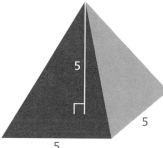

**4** **ACTIVITY: Breaking a Prism into Pyramids**

Work with a partner. The rectangular prism can be cut to form three pyramids. Show that the sum of the volumes of the three pyramids is equal to the volume of the prism.

a.

b.

c.

## What Is Your Answer?

5. **IN YOUR OWN WORDS** How can you find the volume of a pyramid?

6. **STRUCTURE** Write a general formula for the volume of a pyramid.

Use what you learned about the volumes of pyramids to complete Exercises 4–6 on page 386.

### Key Idea

**Volume of a Pyramid**

**Study Tip**

The *height* of a pyramid is the perpendicular distance from the base to the vertex.

**Words** The volume $V$ of a pyramid is one-third the product of the area of the base and the height of the pyramid.

height, $h$

area of base, $B$

Area of base

**Algebra** $V = \dfrac{1}{3}Bh$

Height of pyramid

**EXAMPLE 1** **Finding the Volume of a Pyramid**

**Find the volume of the pyramid.**

$$V = \frac{1}{3}Bh \qquad \text{Write formula for volume.}$$

$$= \frac{1}{3}(48)(9) \qquad \text{Substitute.}$$

$$= 144 \qquad \text{Multiply.}$$

9 mm

$B = 48$ mm²

∴ The volume is 144 cubic millimeters.

**EXAMPLE 2** **Finding the Volume of a Pyramid**

**Study Tip**

The area of the base of a rectangular pyramid is the product of the length $\ell$ and the width $w$.

You can use $V = \dfrac{1}{3}\ell wh$ to find the volume of a rectangular pyramid.

**Find the volume of the pyramid.**

**a.**

7 ft

4 ft

3 ft

$$V = \frac{1}{3}Bh$$

$$= \frac{1}{3}(4)(3)(7)$$

$$= 28$$

∴ The volume is 28 cubic feet.

**b.**

10 m

17.5 m

6 m

$$V = \frac{1}{3}Bh$$

$$= \frac{1}{3}\left(\frac{1}{2}\right)(17.5)(6)(10)$$

$$= 175$$

∴ The volume is 175 cubic meters.

## On Your Own

**Now You're Ready**
Exercises 4–11

**Find the volume of the pyramid.**

**1.**

6 ft

$B = 21 \text{ ft}^2$

**2.**

7 in.

8 in.

10 in.

**3.**

11 cm

7 cm

18 cm

---

**EXAMPLE 3  Real-Life Application**

**a. The volume of sunscreen in Bottle B is about how many times the volume in Bottle A?**

**b. Which is the better buy?**

**Bottle A**
**$9.96**

6 in.

2 in.

1 in.

**Bottle B**
**$14.40**

4 in.

3 in.

1.5 in.

**a.** Use the formula for the volume of a pyramid to estimate the amount of sunscreen in each bottle.

| **Bottle A** | **Bottle B** |
|---|---|
| $V = \dfrac{1}{3}Bh$ | $V = \dfrac{1}{3}Bh$ |
| $= \dfrac{1}{3}(2)(1)(6)$ | $= \dfrac{1}{3}(3)(1.5)(4)$ |
| $= 4 \text{ in.}^3$ | $= 6 \text{ in.}^3$ |

⋮⋰ So, the volume of sunscreen in Bottle B is about $\dfrac{6}{4} = 1.5$ times the volume in Bottle A.

**b.** Find the unit cost for each bottle.

| **Bottle A** | **Bottle B** |
|---|---|
| $\dfrac{\text{cost}}{\text{volume}} = \dfrac{\$9.96}{4 \text{ in.}^3}$ | $\dfrac{\text{cost}}{\text{volume}} = \dfrac{\$14.40}{6 \text{ in.}^3}$ |
| $= \dfrac{\$2.49}{1 \text{ in.}^3}$ | $= \dfrac{\$2.40}{1 \text{ in.}^3}$ |

⋮⋰ The unit cost of Bottle B is less than the unit cost of Bottle A. So, Bottle B is the better buy.

## On Your Own

**Now You're Ready**
Exercise 16

**4.** Bottle C is on sale for $13.20. Is Bottle C a better buy than Bottle B in Example 3? Explain.

**Bottle C**

3 in.

2 in.

3 in.

### Vocabulary and Concept Check

1. **WRITING** How is the formula for the volume of a pyramid different from the formula for the volume of a prism?

2. **OPEN-ENDED** Describe a real-life situation that involves finding the volume of a pyramid.

3. **REASONING** A triangular pyramid and a triangular prism have the same base and height. The volume of the prism is how many times the volume of the pyramid?

### Practice and Problem Solving

**Find the volume of the pyramid.**

 4.

5.

6.

7.

8.

9.

10.

11.

12. **PARACHUTE** In 1483, Leonardo da Vinci designed a parachute. It is believed that this was the first parachute ever designed. In a notebook, he wrote, "If a man is provided with a length of gummed linen cloth with a length of 12 yards on each side and 12 yards high, he can jump from any great height whatsoever without injury." Find the volume of air inside Leonardo's parachute.

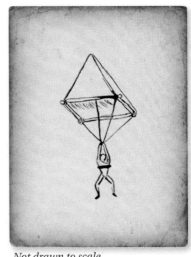

*Not drawn to scale*

**Find the volume of the composite solid.**

**13.**

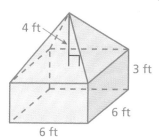

4 ft
3 ft
6 ft
6 ft

**14.**

8 m
4 m
6 m
6 m

**15.**

8 in. 7 in.
10 in.
6.9 in. 8 in.

6 in.
8 in.
$B = 30$ in.$^2$
$B = 24$ in.$^2$
Spire A
Spire B

**③ 16. SPIRE** Which sand-castle spire has a greater volume? How much more sand do you need to make the spire with the greater volume?

**17. PAPERWEIGHT** How much glass is needed to manufacture 1000 paperweights? Explain your reasoning.

4 in.
3 in.
3 in.
Paperweight

**18. PROBLEM SOLVING** Use the photo of the tepee.

    **a.** What is the shape of the base? How can you tell?

    **b.** The tepee's height is about 10 feet. Estimate the volume of the tepee.

**19. OPEN-ENDED** A pyramid has a volume of 40 cubic feet and a height of 6 feet. Find one possible set of dimensions of the rectangular base.

**20.** ❋Reasoning❋ Do the two solids have the same volume? Explain.

z
y
x

3z
y
x

**Fair Game Review** What you learned in previous grades & lessons

For the given angle measure, find the measure of a supplementary angle and the measure of a complementary angle, if possible. *(Section 7.2)*

**21.** $27°$             **22.** $82°$             **23.** $120°$

**24. MULTIPLE CHOICE** The circumference of a circle is 44 inches. Which estimate is closest to the area of the circle? *(Section 8.3)*

   Ⓐ 7 in.$^2$       Ⓑ 14 in.$^2$       Ⓒ 154 in.$^2$       Ⓓ 484 in.$^2$

**Check It Out**
Lesson Tutorials
BigIdeasMath ✓com

**Key Vocabulary** ◀))
cross section, *p. 388*

Consider a plane "slicing" through a solid. The intersection of the plane and the solid is a two-dimensional shape called a **cross section**. For example, the diagram shows that the intersection of the plane and the rectangular prism is a rectangle.

rectangular prism

intersection                    plane

## EXAMPLE 1 Describing the Intersection of a Plane and a Solid

**Describe the intersection of the plane and the solid.**

**COMMON CORE**

**Geometry**
In this extension, you will
• describe the intersections of planes and solids.
Learning Standard
7.G.3

a.

b.

c.
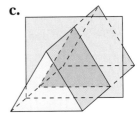

a.  The intersection is a triangle.

b.  The intersection is a rectangle.

c.  The intersection is a triangle.

## ● Practice

**Describe the intersection of the plane and the solid.**

1.

2.

3.

4.

5.

6.

7.  **REASONING** A plane that intersects a prism is parallel to the bases of the prism. Describe the intersection of the plane and the prism.

Example 1 shows how a plane intersects a polyhedron. Now consider the intersection of a plane and a solid having a curved surface, such as a cylinder or cone. As shown, a *cone* is a solid that has one circular base and one vertex.

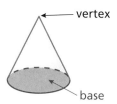
vertex
base

**EXAMPLE 2  Describing the Intersection of a Plane and a Solid**

**Math Practice 1**

**Analyze Givens**
What solid is shown? What are you trying to find? Explain.

Describe the intersection of the plane and the solid.

a.      b.

a. The intersection is a circle.

b. The intersection is a triangle.

## Practice

Describe the intersection of the plane and the solid.

8.      9.

10.      11.

Describe the shape that is formed by the cut made in the food shown.

12.      13.      14.

15. **REASONING** Explain how a plane can be parallel to the base of a cone and intersect the cone at exactly one point.

**Find the volume of the prism.** *(Section 9.4)*

**1.**

8 in.

3 in.

7 in.

**2.**

6 ft

8 ft

15 ft

**3.**

8 yd

12 yd

10 yd

**4.**

25 mm

$B = 197$ mm$^2$

**Find the volume of the solid. Round your answer to the nearest tenth if necessary.** *(Section 9.5)*

**5.**

12 ft

$B = 166$ ft$^2$

**6.**

3 m

2 m

5 m

**Describe the intersection of the plane and the solid.** *(Section 9.5)*

**7.**

**8.**

20 ft

40 ft   40 ft

**9. ROOF** A pyramid hip roof is a good choice for a house in a hurricane area. What is the volume of the roof to the nearest tenth? *(Section 9.5)*

**10. CUBIC UNITS** How many cubic feet are in a cubic yard? Use a sketch to explain your reasoning. *(Section 9.4)*

# 9 Chapter Review

Check It Out
Vocabulary Help
BigIdeasMath ✓com

## Review Key Vocabulary

lateral surface area, *p. 358*       slant height, *p. 364*
regular pyramid, *p. 364*           cross section, *p. 388*

## Review Examples and Exercises

### 9.1 Surface Areas of Prisms *(pp. 354–361)*

**Find the surface area of the prism.**

Draw a net.

$$S = 2\ell w + 2\ell h + 2wh$$
$$= 2(6)(4) + 2(6)(5) + 2(4)(5)$$
$$= 48 + 60 + 40$$
$$= 148$$

∴ The surface area is 148 square feet.

### Exercises

**Find the surface area of the prism.**

1.
   5 in.
   8 in.   3 in.

2.
   15 cm   17 cm
   8 cm   7 cm

3.
   3 m   4 m
   8 m
   5 m

### 9.2 Surface Areas of Pyramids *(pp. 362–367)*

**Find the surface area of the regular pyramid.**

Draw a net.

10 yd
5.2 yd
6 yd

10 yd
6 yd   5.2 yd

| *Area of Base* | *Area of a Lateral Face* |
|---|---|
| $\dfrac{1}{2} \cdot 6 \cdot 5.2 = 15.6$ | $\dfrac{1}{2} \cdot 6 \cdot 10 = 30$ |

Find the sum of the areas of the base and all three lateral faces.

$$S = 15.6 + \underbrace{30 + 30 + 30}$$
$$= 105.6$$

> There are 3 identical lateral faces. Count the area 3 times.

∴ The surface area is 105.6 square yards.

## Exercises

**Find the surface area of the regular pyramid.**

**4.**

3 in.

2 in.

**5.**

10 m

8 m    6.9 m

**6.**

9 cm

7 cm

Area of base
is 84.3 cm².

**9.3  Surface Areas of Cylinders**  *(pp. 368–373)*

**Find the surface area of the cylinder. Round your answer to the nearest tenth.**

Draw a net.

4 ft

5 ft

$S = 2\pi r^2 + 2\pi rh$

$= 2\pi(4)^2 + 2\pi(4)(5)$

$= 32\pi + 40\pi$

$= 72\pi \approx 226.1$

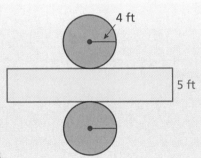

4 ft

5 ft

The surface area is about 226.1 square millimeters.

## Exercises

**Find the surface area of the cylinder. Round your answer to the nearest tenth.**

**7.**

3 yd

6 yd

**8.**

0.8 cm

*Lip Balm*
Protects your lips from the sun and wind
*Lip Balm*

6 cm

**9. ORANGES** Find the lateral surface area of the can of
mandarin oranges.

4 cm

*Sunnyview Farms*
*Mandarin*
ORANGE SEGMENTS

11 cm

NET WT. 16 OZ

**9.4  Volumes of Prisms**  *(pp. 376–381)*

**Find the volume of the prism.**

$V = Bh$          Write formula for volume.

$= \dfrac{1}{2}(7)(3) \cdot 5$     Substitute.

$= 52.5$          Multiply.

3 ft

7 ft     5 ft

The volume is 52.5 cubic feet.

## Exercises

**Find the volume of the prism.**

**10.**

6 in.
2 in.
8 in.

**11.**

7.5 m
8 m
4 m

**12.**

9 mm
15 mm
4.5 mm

### 9.5 Volumes of Pyramids *(pp. 382–389)*

**a. Find the volume of the pyramid.**

$V = \dfrac{1}{3}Bh$     Write formula for volume.

$= \dfrac{1}{3}(6)(5)(10)$     Substitute.

$= 100$     Multiply.

∴ The volume is 100 cubic yards.

10 yd
5 yd
6 yd

**b. Describe the intersection of the plane and the solid.**

**i.**

The intersection is a hexagon.

**ii.**

The intersection is a circle.

## Exercises

**Find the volume of the pyramid.**

**13.**

20 ft
17 ft   15 ft

**14.**

30 in.
$B = 210$ in.$^2$

**15.**

9 mm
8 mm
8 mm

**Describe the intersection of the plane and the solid.**

**16.**

**17.**

Check It Out
Test Practice
BigIdeasMath ✓com

**Find the surface area of the prism or regular pyramid.**

**1.**

3 ft
2 ft
5 ft

**2.**

2 in.
1 in.

**3.**

15 m
11 m
9.5 m

**Find the surface area of the cylinder. Round your answer to the nearest tenth.**

**4.**

2 cm
3 cm

**5.**

22 in.
12.5 in.

**Find the volume of the solid.**

**6.**

6 in.
9 in.
12 in.

**7.**

5.2 yd
2 yd
4 yd

**8.**

6 m
3 m
8 m

**9. SKATEBOARD RAMP** A quart of paint covers 80 square feet. How many quarts should you buy to paint the ramp with two coats? (Assume you will not paint the bottom of the ramp.)

15.2 ft
6 ft
19.5 ft
14 ft

GRAHAM CRACKERS
"with the taste of REAL honey in every bite"
$h = 9$ in.
$\ell = 6$ in.
$w = 2$ in.

**10. GRAHAM CRACKERS** A manufacturer wants to double the volume of the graham cracker box. The manufacturer will either double the height or double the width.

**a.** Which option uses less cardboard? Justify your answer.

**b.** What is the volume of the new graham cracker box?

**11. SOUP** The label on the can of soup covers about 354.2 square centimeters. What is the height of the can? Round your answer to the nearest whole number.

4.7 cm
TOMATO SOUP

1. A gift box and its dimensions are shown below.

What is the least amount of wrapping paper that you could have used to wrap the box? *(7.G.6)*

**A.** 20 in.²

**C.** 64 in.²

**B.** 56 in.²

**D.** 112 in.²

**Test-Taking Strategy**

**After Answering Easy Questions, Relax**

Find the surface area.
Ⓐ 10 ft  Ⓒ 10 ft²
Ⓑ 10 ft³  Ⓓ 2 ft³

2ft, 1ft, 1ft

Neat! Didn't even use a formula.

"After answering the easy questions, relax and try the harder ones. For this, you know area is measured in square units."

2. A student scored 600 the first time she took the mathematics portion of her college entrance exam. The next time she took the exam, she scored 660. Her second score represents what percent increase over her first score? *(7.RP.3)*

**F.** 9.1%

**H.** 39.6%

**G.** 10%

**I.** 60%

3. Raj was solving the proportion in the box below.

$$\frac{3}{8} = \frac{x-3}{24}$$

$$3 \cdot 24 = (x-3) \cdot 8$$

$$72 = x - 24$$

$$96 = x$$

What should Raj do to correct the error that he made? *(7.RP.2c)*

**A.** Set the product of the numerators equal to the product of the denominators.

**B.** Distribute 8 to get $8x - 24$.

**C.** Add 3 to each side to get $\frac{3}{8} + 3 = \frac{x}{24}$.

**D.** Divide both sides by 24 to get $\frac{3}{8} \div 24 = x - 3$.

**4.** A line contains the two points plotted in the coordinate plane below.

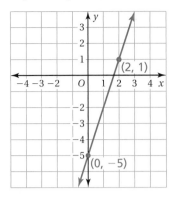

What is the slope of the line?  *(7.RP.2b)*

**F.** $\frac{1}{3}$

**H.** 3

**G.** 2

**I.** 6

**5.** James is getting ready for wrestling season. As part of his preparation, he plans to lose 5% of his body weight. James currently weighs 160 pounds. How much will he weigh, in pounds, after he loses 5% of his weight?  *(7.RP.3)*

**6.** How much material is needed to make the popcorn container?  *(7.G.4)*

4 in.

9.5 in.

**A.** $76\pi$ in.$^2$

**C.** $92\pi$ in.$^2$

**B.** $84\pi$ in.$^2$

**D.** $108\pi$ in.$^2$

**7.** To make 10 servings of soup you need 4 cups of broth. You want to know how many servings you can make with 8 pints of broth. Which proportion should you use?  *(7.RP.2c)*

**F.** $\frac{10}{4} = \frac{x}{8}$

**H.** $\frac{10}{4} = \frac{8}{x}$

**G.** $\frac{4}{10} = \frac{x}{16}$

**I.** $\frac{10}{4} = \frac{x}{16}$

**8.** A rectangular prism and its dimensions are shown below.

2 in.

3 in.

4 in.

What is the volume, in cubic inches, of a rectangular prism whose dimensions are three times greater?   *(7.G.6)*

**9.** What is the value of $x$?   *(7.G.5)*

$(2x + 4)°$

46°

**A.** 20

**C.** 44

**B.** 43

**D.** 65

**10.** Which of the following could be the angle measures of a triangle?   *(7.G.5)*

**F.** 60°, 50°, 20°

**H.** 30°, 60°, 90°

**G.** 40°, 80°, 90°

**I.** 0°, 90°, 90°

**11.** The table below shows the costs of buying matinee movie tickets.   *(7.RP.2b)*

Think
Solve
Explain

| Matinee Tickets, x | 2 | 3 | 4 | 5 |
|---|---|---|---|---|
| Cost, y | $9 | $13.50 | $18 | $22.50 |

*Part A*   Graph the data.

*Part B*   Find and interpret the slope of the line through the points.

*Part C*   How much does it cost to buy 8 matinee movie tickets?

# 10 Probability and Statistics

"If there are 7 cats in a sack and I draw one at random,..."

"... what is the probability that I will draw you?"

"I'm just about finished making my two number cubes."

"Now, here's how the game works. You toss the two cubes."

"If the sum is even, I win. If it's odd, you win."

# What You Learned Before

"Let's spin to decide what we will have for lunch."

M = Mouse
B = Dog Biscuit

Why do we always have to use this spinner?

## Writing Ratios (6.RP.1)

**Example 1**  There are 32 football players and 16 cheerleaders at your school. Write the ratio of cheerleaders to football players.

cheerleaders → $\dfrac{16}{32} = \dfrac{1}{2}$  Write in simplest form.

football players →

∴ So, the ratio of cheerleaders to football players is $\dfrac{1}{2}$.

**Example 2**

a. **Write the ratio of girls to boys in Classroom A.**

$$\frac{\text{Girls in Classroom A}}{\text{Boys in Classroom A}} = \frac{11}{14}$$

|  | Boys | Girls |
|---|---|---|
| **Classroom A** | 14 | 11 |
| **Classroom B** | 12 | 8 |

∴ So, the ratio of girls to boys in Classroom A is $\dfrac{11}{14}$.

b. **Write the ratio of boys in Classroom B to the total number of students in both classes.**

$$\frac{\text{Boys in Classroom B}}{\text{Total number of students}} = \frac{12}{14 + 11 + 12 + 8} = \frac{12}{45} = \frac{4}{15}$$  Write in simplest form.

∴ So, the ratio of boys in Classroom B to the total number of students is $\dfrac{4}{15}$.

## Try It Yourself

**Write the ratio in simplest form.**

1. baseballs to footballs

2. footballs to total pieces of equipment

3. sneakers to ballet slippers

4. sneakers to total number of shoes

5. green beads to blue beads

6. red beads : green beads

7. green beads : total number of beads

**Essential Question** In an experiment, how can you determine the number of possible results?

An *experiment* is an investigation or a procedure that has varying results. Flipping a coin, rolling a number cube, and spinning a spinner are all examples of experiments.

### 1 ACTIVITY: Conducting Experiments

**Work with a partner.**

a. You flip a dime.

There are ▮ possible results.

Out of 20 flips, you think you will flip heads ▮ times.

Flip a dime 20 times. Tally your results in a table. How close was your guess?

b. You spin the spinner shown.

There are ▮ possible results.

Out of 20 spins, you think you will spin orange ▮ times.

Spin the spinner 20 times. Tally your results in a table. How close was your guess?

c. You spin the spinner shown.

There are ▮ possible results.

Out of 20 spins, you think you will spin a 4 ▮ times.

Spin the spinner 20 times. Tally your results in a table. How close was your guess?

**COMMON CORE**

**Probability and Statistics**

In this lesson, you will

● identify and count the outcomes of experiments.

Preparing for Standard 7.SP.5

### 2 ACTIVITY: Comparing Different Results

**Work with a partner. Use the spinner in Activity 1(c).**

a. Do you have a better chance of spinning an even number or a multiple of 4? Explain your reasoning.

b. Do you have a better chance of spinning an even number or an odd number? Explain your reasoning.

**ACTIVITY: Rock Paper Scissors**

Rock

**Work with a partner.**

**a.** Play Rock Paper Scissors 30 times. Tally your results in the table.

**b.** How many possible results are there?

**c.** Of the possible results, in how many ways can Player A win? Player B win? the players tie?

**d.** Does one of the players have a better chance of winning than the other player? Explain your reasoning.

Paper

Scissors

**Math Practice** 1

**Interpret a Solution**

How do your results compare to the possible results? Explain.

Rock *breaks* scissors.
Paper *covers* rock.
Scissors *cut* paper.

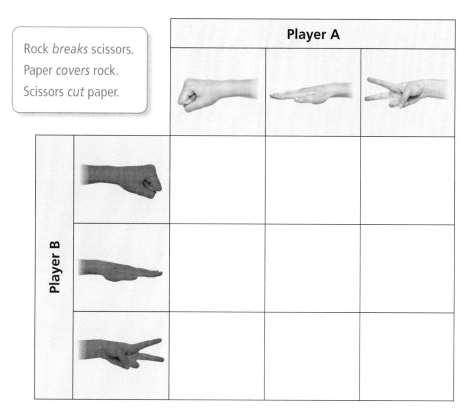

## What Is Your Answer?

**4. IN YOUR OWN WORDS** In an experiment, how can you determine the number of possible results?

 Use what you learned about experiments to complete Exercises 3 and 4 on page 404.

**Check It Out**
Lesson Tutorials
BigIdeasMath com

**Key Vocabulary** ◀))
experiment, *p. 402*
outcomes, *p. 402*
event, *p. 402*
favorable outcomes,
  *p. 402*

## Key Ideas

**Outcomes and Events**

An **experiment** is an investigation or a procedure that has varying results. The possible results of an experiment are called **outcomes**. A collection of one or more outcomes is an **event**. The outcomes of a specific event are called **favorable outcomes**.

For example, randomly selecting a marble from a group of marbles is an experiment. Each marble in the group is an outcome. Selecting a green marble from the group is an event.

**Reading**

When an experiment is performed *at random* or *randomly*, all of the possible outcomes are equally likely.

*Possible outcomes*

*Event:  Choosing a green marble*
*Number of favorable outcomes:  2*

**EXAMPLE 1** **Identifying Outcomes**

You roll the number cube.

**a.  What are the possible outcomes?**

  ⋮∙  The six possible outcomes are rolling a 1, 2, 3, 4, 5, and 6.

**b.  What are the favorable outcomes of rolling an even number?**

| even | *not* even |
|------|------------|
| 2, 4, 6 | 1, 3, 5 |

  ⋮∙  The favorable outcomes of the event are rolling a 2, 4, and 6.

**c.  What are the favorable outcomes of rolling a number greater than 5?**

| greater than 5 | *not* greater than 5 |
|----------------|----------------------|
| 6 | 1, 2, 3, 4, 5 |

  ⋮∙  The favorable outcome of the event is rolling a 6.

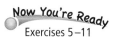

1.  You randomly choose a letter from a hat that contains the letters A through K.

    a.  What are the possible outcomes?

    b.  What are the favorable outcomes of choosing a vowel?

---

**EXAMPLE** **2** **Counting Outcomes**

**You spin the spinner.**

  a.  **How many possible outcomes are there?**

   The spinner has 6 sections. So, there are 6 possible outcomes.

  b.  **In how many ways can spinning red occur?**

   The spinner has 3 red sections. So, spinning red can occur in 3 ways.

  c.  **In how many ways can spinning *not* purple occur? What are the favorable outcomes of spinning *not* purple?**

   The spinner has 5 sections that are *not* purple. So, spinning *not* purple can occur in 5 ways.

   | purple | *not* purple |
   |--------|--------------|
   | purple | red, red, red, green, blue |

   The favorable outcomes of the event are red, red, red, green, and blue.

2.  You randomly choose a marble.

    a.  How many possible outcomes are there?

    b.  In how many ways can choosing blue occur?

    c.  In how many ways can choosing *not* yellow occur? What are the favorable outcomes of choosing *not* yellow?

**Vocabulary and Concept Check**

1. **VOCABULARY** Is rolling an even number on a number cube an *outcome* or an *event*? Explain.

2. **WRITING** Describe how an outcome and a favorable outcome are different.

 **Practice and Problem Solving**

**You spin the spinner shown.**

3. How many possible results are there?

4. Of the possible results, in how many ways can you spin an even number? an odd number?

5. **TILES** What are the possible outcomes of randomly choosing one of the tiles shown?

**You randomly choose one of the tiles shown above. Find the favorable outcomes of the event.**

6. Choosing a 6

7. Choosing an odd number

8. Choosing a number greater than 5

9. Choosing an odd number less than 5

10. Choosing a number less than 3

11. Choosing a number divisible by 3

**You randomly choose one marble from the bag. (a) Find the number of ways the event can occur. (b) Find the favorable outcomes of the event.**

12. Choosing blue

13. Choosing green

14. Choosing purple

15. Choosing yellow

16. Choosing *not* red

17. Choosing *not* blue

18. **ERROR ANALYSIS** Describe and correct the error in finding the number of ways that choosing *not* purple can occur.

| purple | *not* purple |
| --- | --- |
| purple | red, blue, green, yellow |

Choosing *not* purple can occur in 4 ways.

**19. COINS** You have 10 coins in your pocket. Five are Susan B. Anthony dollars, two are Kennedy half-dollars, and three are presidential dollars. You randomly choose a coin. In how many ways can choosing *not* a presidential dollar occur?

Susan B. Anthony dollar

Kennedy half-dollar

Presidential dollar

Spinner A

**Tell whether the statement is *true* or *false*. If it is false, change the italicized word to make the statement true.**

**20.** Spinning blue and spinning *green* have the same number of favorable outcomes on Spinner A.

**21.** Spinning blue has one *more* favorable outcome than spinning green on Spinner B.

**22.** There are *three* possible outcomes of spinning Spinner A.

**23.** Spinning *red* can occur in four ways on Spinner B.

**24.** Spinning not green can occur in *three* ways on Spinner B.

Spinner B

Dancer

Firefighter

Baker

Pirate

Bellhop

**25. MUSIC** A bargain bin contains classical and rock CDs. There are 60 CDs in the bin. Choosing a rock CD and *not* choosing a rock CD have the same number of favorable outcomes. How many rock CDs are in the bin?

**26. Precision** You randomly choose one of the cards and set it aside. Then you randomly choose a second card. Describe how the number of possible outcomes changes after the first card is chosen.

---

 **Fair Game Review** *What you learned in previous grades & lessons*

**Solve the proportion.** *(Section 5.4)*

**27.** $\dfrac{x}{10} = \dfrac{1}{5}$

**28.** $\dfrac{60}{n} = \dfrac{20}{7}$

**29.** $\dfrac{1}{3} = \dfrac{w}{36}$

**30.** $\dfrac{25}{17} = \dfrac{100}{b}$

**31. MULTIPLE CHOICE** What is the surface area of the rectangular prism? *(Section 9.1)*

ⓐ 162 in.$^2$

ⓑ 264 in.$^2$

ⓒ 324 in.$^2$

ⓓ 360 in.$^2$

5 in.

6 in.

12 in.

## 10.2 Probability

**Essential Question** How can you describe the likelihood of an event?

### 1 ACTIVITY: Black-and-White Spinner Game

**Work with a partner. You work for a game company. You need to create a game that uses the spinner below.**

a. Write rules for a game that uses the spinner. Then play it.

b. After playing the game, do you want to revise the rules? Explain.

**COMMON CORE**

**Probability and Statistics**

In this lesson, you will
- understand the concept of probability and the relationship between probability and likelihood.
- find probabilities of events.

Learning Standards
7.SP.5
7.SP.7a

c. **CHOOSE TOOLS** Using the center of the spinner as the vertex, measure the angle of each pie-shaped section. Is each section the same size? How do you think this affects the likelihood of spinning a given number?

d. Your friend is about to spin the spinner and wants to know how likely it is to spin a 3. How would you describe the likelihood of this event to your friend?

Work with a partner. For each spinner, do the following.

- Measure the angle of each pie-shaped section.
- Tell whether you are more likely to spin a particular number. Explain your reasoning.
- Tell whether your rules from Activity 1 make sense for these spinners. Explain your reasoning.

a.          b.

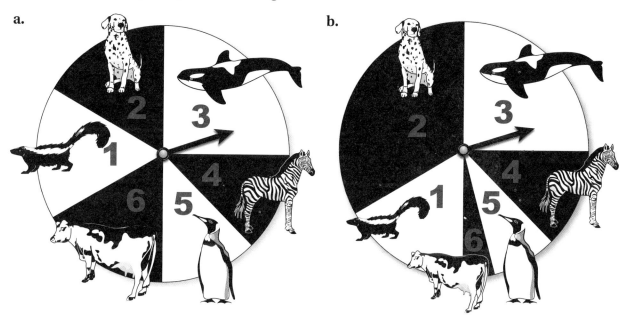

③ **ACTIVITY: Is This Game Fair?**

Work with a partner. Apply the following rules to each spinner in Activities 1 and 2. Is the game fair? Why or why not? If not, who has the better chance of winning?

- Take turns spinning the spinner.
- If you spin an odd number, Player 1 wins.
- If you spin an even number, Player 2 wins.

## What Is Your Answer?

4. **IN YOUR OWN WORDS** How can you describe the likelihood of an event?

5. Describe the likelihood of spinning an 8 in Activity 1.

6. Describe a career in which it is important to know the likelihood of an event.

**Practice** ➤ Use what you learned about the likelihood of an event to complete Exercises 4 and 5 on page 410.

**Key Vocabulary**
probability, *p. 408*

## Key Idea

**Probability**

The **probability** of an event is a number that measures the likelihood that the event will occur. Probabilities are between 0 and 1, including 0 and 1. The diagram relates likelihoods (above the diagram) and probabilities (below the diagram).

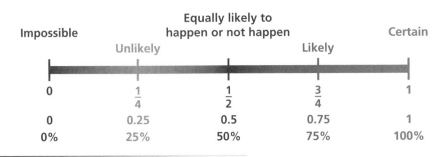

**Study Tip**

Probabilities can be written as fractions, decimals, or percents.

**EXAMPLE 1    Describing the Likelihood of an Event**

80% chance

**There is an 80% chance of thunderstorms tomorrow. Describe the likelihood of the event.**

The probability of thunderstorms tomorrow is 80%.

⋮⋰ Because 80% is close to 75%, it is *likely* that there will be thunderstorms tomorrow.

## On Your Own

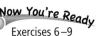
Now You're Ready
Exercises 6–9

**Describe the likelihood of the event given its probability.**

1.  The probability that you land a jump on a snowboard is $\frac{1}{2}$.

2.  There is a 100% chance that the temperature will be less than 120°F tomorrow.

## Key Idea

**Finding the Probability of an Event**

When all possible outcomes are equally likely, the probability of an event is the ratio of the number of favorable outcomes to the number of possible outcomes. The probability of an event is written as *P*(event).

$$P(\text{event}) = \frac{\text{number of favorable outcomes}}{\text{number of possible outcomes}}$$

**EXAMPLE** **2** **Finding a Probability**

You roll the number cube. What is the probability of rolling an odd number?

$$P(\text{event}) = \frac{\text{number of favorable outcomes}}{\text{number of possible outcomes}}$$

$$P(\text{odd}) = \frac{3}{6}$$ ← There are 3 odd numbers (1, 3, and 5).

← There is a total of 6 numbers.

$$= \frac{1}{2}$$ Simplify.

The probability of rolling an odd number is $\frac{1}{2}$, or 50%.

**EXAMPLE** **3** **Using a Probability**

The probability that you randomly draw a short straw from a group of 40 straws is $\frac{3}{20}$. How many are short straws?

Ⓐ 4      Ⓑ 6

Ⓒ 15      Ⓓ 34

$$P(\text{short}) = \frac{\text{number of short straws}}{\text{total number of straws}}$$

$$\frac{3}{20} = \frac{n}{40}$$ Substitute. Let $n$ be the number of short straws.

$$6 = n$$ Solve for $n$.

There are 6 short straws.

So, the correct answer is Ⓑ.

**On Your Own**

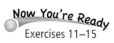

Exercises 11–15

**3.** In Example 2, what is the probability of rolling a number greater than 2?

**4.** In Example 2, what is the probability of rolling a 7?

**5.** The probability that you randomly draw a short straw from a group of 75 straws is $\frac{1}{15}$. How many are short straws?

## Vocabulary and Concept Check

1. **VOCABULARY** Explain how to find the probability of an event.

2. **REASONING** Can the probability of an event be 1.5? Explain.

3. **OPEN-ENDED** Give a real-life example of an event that is impossible. Give a real-life example of an event that is certain.

## Practice and Problem Solving

**You are playing a game using the spinners shown.**

4. You want to move down. On which spinner are you more likely to spin "Down"? Explain.

5. You want to move forward. Which spinner would you spin? Explain.

**Describe the likelihood of the event given its probability.**

① 6. Your soccer team wins $\frac{3}{4}$ of the time.

7. There is a 0% chance that you will grow 12 more feet.

8. The probability that the sun rises tomorrow is 1.

9. It rains on $\frac{1}{5}$ of the days in July.

10. **VIOLIN** You have a 50% chance of playing the correct note on a violin. Describe the likelihood of playing the correct note.

**You randomly choose one shirt from the shelves. Find the probability of the event.**

② 11. Choosing a red shirt

12. Choosing a green shirt

13. *Not* choosing a white shirt

14. *Not* choosing a black shirt

15. Choosing an orange shirt

16. **ERROR ANALYSIS** Describe and correct the error in finding the probability of *not* choosing a blue shirt from the shelves above.

$$P(not\ blue) = \frac{4}{10} = \frac{2}{5}$$

17. **CONTEST** The rules of a contest say that there is a 5% chance of winning a prize. Four hundred people enter the contest. Predict how many people will win a prize.

18. **RUBBER DUCKS** At a carnival, the probability that you choose a winning rubber duck from 25 ducks is 0.24.

   **a.** How many are *not* winning ducks?

   **b.** Describe the likelihood of *not* choosing a winning duck.

19. **DODECAHEDRON** A dodecahedron has twelve sides numbered 1 through 12. Find the probability and describe the likelihood of each event.

   **a.** Rolling a number less than 9

   **b.** Rolling a multiple of 3

   **c.** Rolling a number greater than 6

**A Punnett square is a grid used to show possible gene combinations for the offspring of two parents. In the Punnett square shown, a boy is represented by *XY*. A girl is represented by *XX*.**

20. Complete the Punnett square.

21. Explain why the probability of two parents having a boy or having a girl is equally likely.

22. **Critical Thinking** Two parents each have the gene combination *Cs*. The gene *C* is for curly hair. The gene *s* is for straight hair.

   **a.** Make a Punnett square for the two parents. When all outcomes are equally likely, what is the probability of a child having the gene combination *CC*?

   **b.** Any gene combination that includes a *C* results in curly hair. When all outcomes are equally likely, what is the probability of a child having curly hair?

 **Fair Game Review** What you learned in previous grades & lessons

**Solve the inequality. Graph the solution.** *(Section 4.2 and Section 4.3)*

**23.** $x + 5 < 9$     **24.** $b - 2 \geq -7$     **25.** $1 > -\dfrac{w}{3}$     **26.** $6 \leq -2g$

27. **MULTIPLE CHOICE** Find the value of *x*. *(Section 7.4)*

   (A) 85     (B) 90

   (C) 93     (D) 102

## Essential Question How can you use relative frequencies to find probabilities?

When you conduct an experiment, the **relative frequency** of an event is the fraction or percent of the time that the event occurs.

$$\text{relative frequency} = \frac{\text{number of times the event occurs}}{\text{total number of times you conduct the experiment}}$$

### 1 ACTIVITY: Finding Relative Frequencies

**Work with a partner.**

a. Flip a quarter 20 times and record your results. Then complete the table. Are the relative frequencies the same as the probability of flipping heads or tails? Explain.

| | Flipping Heads | Flipping Tails |
|---|---|---|
| **Relative Frequency** | | |

b. Compare your results with those of other students in your class. Are the relative frequencies the same? If not, why do you think they differ?

c. Combine all of the results in your class. Then complete the table again. Did the relative frequencies change? What do you notice? Explain.

d. Suppose everyone in your school conducts this experiment and you combine the results. How do you think the relative frequencies will change?

### 2 ACTIVITY: Using Relative Frequencies

**COMMON CORE**

**Probability and Statistics**
In this lesson, you will
• find relative frequencies.
• use experimental probabilities to make predictions.
• use theoretical probabilities to find quantities.
• compare experimental and theoretical probabilities.
Learning Standards
7.SP.5
7.SP.6
7.SP.7a
7.SP.7b

**Work with a partner. You have a bag of colored chips. You randomly select a chip from the bag and replace it. The table shows the number of times you select each color.**

| Red | Blue | Green | Yellow |
|---|---|---|---|
| 24 | 12 | 15 | 9 |

a. There are 20 chips in the bag. Can you use the table to find the exact number of each color in the bag? Explain.

b. You randomly select a chip from the bag and replace it. You do this 50 times, then 100 times, and you calculate the relative frequencies after each experiment. Which experiment do you think gives a better approximation of the exact number of each color in the bag? Explain.

**Work with a partner. You toss a thumbtack onto a table. There are two ways the thumbtack can land.**

Point up          On its side

**Math Practice 4**

**Analyze Relationships**

How can you use the results of your experiment to determine whether this is a uniform probability model?

**a.** Your friend says that because there are two outcomes, the probability of the thumbtack landing point up must be $\frac{1}{2}$. Do you think this conclusion is true? Explain.

**b.** Toss a thumbtack onto a table 50 times and record your results. In a *uniform probability model*, each outcome is equally likely to occur. Do you think this experiment represents a uniform probability model? Explain.

Use the relative frequencies to complete the following.

$$P(\text{point up}) = \boxed{\phantom{xx}} \qquad P(\text{on its side}) = \boxed{\phantom{xx}}$$

## What Is Your Answer?

4. **IN YOUR OWN WORDS** How can you use relative frequencies to find probabilities? Give an example.

5. Your friend rolls a number cube 500 times. How many times do you think your friend will roll an odd number? Explain your reasoning.

6. In Activity 2, your friend says, "There are no orange-colored chips in the bag." Do you think this conclusion is true? Explain.

7. Give an example of an experiment that represents a uniform probability model.

8. Tell whether you can use each spinner to represent a uniform probability model. Explain your reasoning.

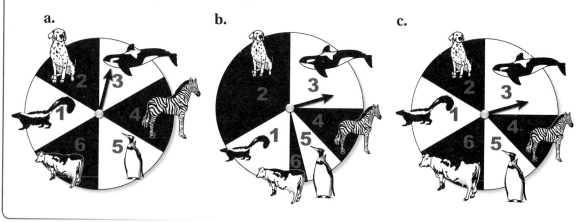

**a.**          **b.**          **c.**

**Practice**

Use what you learned about relative frequencies to complete Exercises 6 and 7 on page 417.

**Check It Out**
Lesson Tutorials
BigIdeasMath√com

 **Key Idea**

**Experimental Probability**

Probability that is based on repeated trials of an experiment is called **experimental probability**.

$$P(\text{event}) = \frac{\text{number of times the event occurs}}{\text{total number of trials}}$$

**EXAMPLE** **1** **Finding an Experimental Probability**

**Rolling a Number Cube**

**The bar graph shows the results of rolling a number cube 50 times. What is the experimental probability of rolling an odd number?**

The bar graph shows 10 ones, 8 threes, and 11 fives. So, an odd number was rolled $10 + 8 + 11 = 29$ times in a total of 50 rolls.

$$P(\text{event}) = \frac{\text{number of times the event occurs}}{\text{total number of trials}}$$

$$P(\text{odd}) = \frac{29}{50}$$

An odd number was rolled 29 times.

There was a total of 50 rolls.

∴ The experimental probability is $\frac{29}{50}$, 0.58, or 58%.

**EXAMPLE** **2** **Making a Prediction**

**It rains 2 out of the last 12 days in March. If this trend continues, how many rainy days would you expect in April?**

Find the experimental probability of a rainy day.

$$P(\text{event}) = \frac{\text{number of times the event occurs}}{\text{total number of trials}}$$

$$P(\text{rain}) = \frac{2}{12} = \frac{1}{6}$$

It rains 2 days.

There is a total of 12 days.

"April showers bring May flowers." Old Proverb, 1557

To make a prediction, multiply the probability of a rainy day by the number of days in April.

$$\frac{1}{6} \cdot 30 = 5$$

∴ So, you can predict that there will be 5 rainy days in April.

🔊 Multi-Language Glossary at BigIdeasMath√com

Exercises 8–14

 **On Your Own**

1. In Example 1, what is the experimental probability of rolling an even number?

2. At a clothing company, an inspector finds 5 defective pairs of jeans in a shipment of 200. If this trend continues, about how many pairs of jeans would you expect to be defective in a shipment of 5000?

---

 **Key Idea**

**Theoretical Probability**

When all possible outcomes are equally likely, the **theoretical probability** of an event is the ratio of the number of favorable outcomes to the number of possible outcomes.

$$P(\text{event}) = \frac{\text{number of favorable outcomes}}{\text{number of possible outcomes}}$$

**EXAMPLE 3  Finding a Theoretical Probability**

**You randomly choose one of the letters shown. What is the theoretical probability of choosing a vowel?**

$$P(\text{event}) = \frac{\text{number of favorable outcomes}}{\text{number of possible outcomes}}$$

$$P(\text{vowel}) = \frac{3}{7}$$

There are 3 vowels.

There is a total of 7 letters.

 The probability of choosing a vowel is $\frac{3}{7}$, or about 43%.

**EXAMPLE 4  Using a Theoretical Probability**

**The theoretical probability of winning a bobblehead when spinning a prize wheel is $\frac{1}{6}$. The wheel has 3 bobblehead sections. How many sections are on the wheel?**

$$P(\text{bobblehead}) = \frac{\text{number of bobblehead sections}}{\text{total number of sections}}$$

$$\frac{1}{6} = \frac{3}{s} \qquad \text{Substitute. Let } s \text{ be the total number of sections.}$$

$$s = 18 \qquad \text{Cross Products Property}$$

 So, there are 18 sections on the wheel.

## On Your Own

Now You're Ready
Exercises 15–23

3. In Example 3, what is the theoretical probability of choosing an X?

4. The theoretical probability of spinning an odd number on a spinner is 0.6. The spinner has 10 sections. How many sections have odd numbers?

5. The prize wheel in Example 4 was spun 540 times at a baseball game. About how many bobbleheads would you expect were won?

**EXAMPLE 5** **Comparing Experimental and Theoretical Probability**

**The bar graph shows the results of rolling a number cube 300 times.**

**a. What is the experimental probability of rolling an odd number?**

The bar graph shows 48 ones, 50 threes, and 49 fives. So, an odd number was rolled $48 + 50 + 49 = 147$ times in a total of 300 rolls.

$$P(\text{event}) = \frac{\text{number of times the event occurs}}{\text{total number of trials}}$$

$$P(\text{odd}) = \frac{147}{300}$$

An odd number was rolled 147 times.

There was a total of 300 rolls.

$$= \frac{49}{100}, \text{ or } 49\%$$

**b. How does the experimental probability compare with the theoretical probability of rolling an odd number?**

In Section 10.2, Example 2, you found that the theoretical probability of rolling an odd number is 50%. The experimental probability, 49%, is close to the theoretical probability.

**c. Compare the experimental probability in part (a) to the experimental probability in Example 1.**

As the number of trials increased from 50 to 300, the experimental probability decreased from 58% to 49%. So, it became closer to the theoretical probability of 50%.

## On Your Own

Now You're Ready
Exercises 25–27

6. Use the bar graph in Example 5 to find the experimental probability of rolling a number greater than 1. Compare the experimental probability to the theoretical probability of rolling a number greater than 1.

## Vocabulary and Concept Check

1. **VOCABULARY** Describe how to find the experimental probability of an event.

2. **REASONING** You flip a coin 10 times and find the experimental probability of flipping tails to be 0.7. Does this seem reasonable? Explain.

3. **VOCABULARY** An event has a theoretical probability of 0.5. What does this mean?

4. **OPEN-ENDED** Describe an event that has a theoretical probability of $\frac{1}{4}$.

5. **LOGIC** A pollster surveys randomly selected individuals about an upcoming election. Do you think the pollster will use experimental probability or theoretical probability to make predictions? Explain.

## Practice and Problem Solving

**Use the bar graph to find the relative frequency of the event.**

6. Spinning a 6

7. Spinning an even number

**Use the bar graph to find the experimental probability of the event.**

1️⃣ 8. Spinning a number less than 3

9. *Not* spinning a 1

10. Spinning a 1 or a 3

11. Spinning a 7

12. **EGGS** You check 20 cartons of eggs. Three of the cartons have at least one cracked egg. What is the experimental probability that a carton of eggs has at least one cracked egg?

2️⃣ 13. **BOARD GAME** There are 105 lettered tiles in a board game. You choose the tiles shown. How many of the 105 tiles would you expect to be vowels?

14. **CARDS** You have a package of 20 assorted thank-you cards. You pick the four cards shown. How many of the 20 cards would you expect to have flowers on them?

**Use the spinner to find the theoretical probability of the event.**

③ **15.** Spinning red

**16.** Spinning a 1

**17.** Spinning an odd number

**18.** Spinning a multiple of 2

**19.** Spinning a number less than 7

**20.** Spinning a 9

**21. LETTERS** Each letter of the alphabet is printed on an index card. What is the theoretical probability of randomly choosing any letter except Z?

④ **22. GAME SHOW** On a game show, a contestant randomly chooses a chip from a bag that contains numbers and strikes. The theoretical probability of choosing a strike is $\frac{3}{10}$. The bag contains 9 strikes. How many chips are in the bag?

**23. MUSIC** The theoretical probability that a pop song plays on your MP3 player is 0.45. There are 80 songs on your MP3 player. How many of the songs are pop songs?

**24. MODELING** There are 16 females and 20 males in a class.

    **a.** What is the theoretical probability that a randomly chosen student is female?

    **b.** One week later, there are 45 students in the class. The theoretical probability that a randomly chosen student is a female is the same as last week. How many males joined the class?

**The bar graph shows the results of spinning the spinner 200 times. Compare the theoretical and experimental probabilities of the event.**

⑤ **25.** Spinning a 4

**26.** Spinning a 3

**27.** Spinning a number greater than 4

**28.** Should you use *theoretical* or *experimental* probability to predict the number of times you will spin a 3 in 10,000 spins?

**29. NUMBER SENSE** The table at the right shows the results of flipping two coins 12 times each.

| HH | HT | TH | TT |
|----|----|----|----|
| 2  | 6  | 3  | 1  |

    **a.** What is the experimental probability of flipping two tails? Using this probability, how many times can you expect to flip two tails in 600 trials?

| HH | HT | TH | TT |
|----|----|----|----|
| 23 | 29 | 26 | 22 |

    **b.** The table at the left shows the results of flipping the same two coins 100 times each. What is the experimental probability of flipping two tails? Using this probability, how many times can you expect to flip two tails in 600 trials?

    **c.** Why is it important to use a large number of trials when using experimental probability to predict results?

**You roll a pair of number cubes 60 times. You record your results in the bar graph shown.**

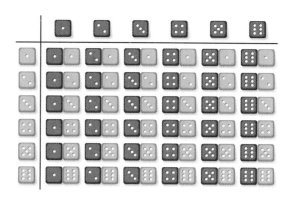

**30.** Use the bar graph to find the experimental probability of rolling each sum. Is each sum equally likely? Explain. If not, which is most likely?

**31.** Use the table to find the theoretical probability of rolling each sum. Is each sum equally likely? Explain. If not, which is most likely?

**32. PROBABILITIES** Compare the probabilities you found in Exercises 30 and 31.

**33. REASONING** Consider the results of Exercises 30 and 31.

   **a.** Which sum would you expect to be most likely after 500 trials? 1000 trials? 10,000 trials?

   **b.** Explain how experimental probability is related to theoretical probability as the number of trials increases.

**34.** **Project** When you toss a paper cup into the air, there are three ways for the cup to land: *open-end up*, *open-end down*, or *on its side*.

   **a.** Toss a paper cup 100 times and record your results. Do the outcomes for tossing the cup appear to be equally likely? Explain.

   **b.** What is the probability of the cup landing open-end up? open-end down? on its side?

   **c.** Use your results to predict the number of times the cup lands on its side in 1000 tosses.

   **d.** Suppose you tape a quarter to the bottom of the cup. Do you think the cup will be *more likely* or *less likely* to land open-end up? Justify your answer.

 **Fair Game Review** What you learned in previous grades & lessons

**Find the annual interest rate.** *(Section 6.7)*

**35.** $I = \$16$, $P = \$200$, $t = 2$ years

**36.** $I = \$26.25$, $P = \$500$, $t = 18$ months

**37. MULTIPLE CHOICE** The volume of a prism is 9 cubic yards. What is its volume in cubic feet? *(Section 9.4)*

   **(A)** $3 \text{ ft}^3$       **(B)** $27 \text{ ft}^3$       **(C)** $81 \text{ ft}^3$       **(D)** $243 \text{ ft}^3$

# 10.4 Compound Events

**Essential Question** How can you find the number of possible outcomes of one or more events?

### 1  ACTIVITY: Comparing Combination Locks

**Work with a partner. You are buying a combination lock. You have three choices.**

a. This lock has 3 wheels. Each wheel is numbered from 0 to 9.

   The least three-digit combination possible is ____ .

   The greatest three-digit combination possible is ____ .

   How many possible combinations are there?

b. Use the lock in part (a).

   There are ____ possible outcomes for the first wheel.

   There are ____ possible outcomes for the second wheel.

   There are ____ possible outcomes for the third wheel.

   How can you use multiplication to determine the number of possible combinations?

c. This lock is numbered from 0 to 39. Each combination uses three numbers in a right, left, right pattern. How many possible combinations are there?

**COMMON CORE**

**Probability and Statistics**

In this lesson, you will

- use tree diagrams, tables, or a formula to find the number of possible outcomes.
- find probabilities of compound events.

Learning Standards
7.SP.8a
7.SP.8b

d. This lock has 4 wheels.

   **Wheel 1:** 0–9
   **Wheel 2:** A–J
   **Wheel 3:** K–T
   **Wheel 4:** 0–9

   How many possible combinations are there?

e. For which lock is it most difficult to guess the combination? Why?

## 2

**ACTIVITY: Comparing Password Security**

**Work with a partner. Which password requirement is most secure? Explain your reasoning. Include the number of different passwords that are possible for each requirement.**

**a.** The password must have four digits.

**b.** The password must have five digits.

**c.** The password must have six letters.

**d.** The password must have eight digits or letters.

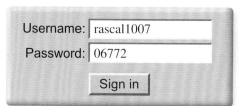

Username: funnydog
Password: 2335
Sign in

Username: rascal1007
Password: 06772
Sign in

Username: supergrowl
Password: AFYYWP
Sign in

Username: jupitermars
Password: 7TT3PX4W
Sign in

## What Is Your Answer?

**3. IN YOUR OWN WORDS** How can you find the number of possible outcomes of one or more events?

**4. SECURITY** A hacker uses a software program to guess the passwords in Activity 2. The program checks 600 passwords per minute. What is the greatest amount of time it will take the program to guess each of the four types of passwords?

**Practice** Use what you learned about the total number of possible outcomes of one or more events to complete Exercise 5 on page 425.

**Key Vocabulary**
sample space, *p. 422*
Fundamental
   Counting Principle,
   *p. 422*
compound event,
   *p. 424*

The set of all possible outcomes of one or more events is called the **sample space**.

You can use tables and tree diagrams to find the sample space of two or more events.

**EXAMPLE** **1** **Finding a Sample Space**

**Crust**

• Thin Crust
• Stuffed Crust

**Style**

• Hawaiian
• Mexican
• Pepperoni
• Veggie

**You randomly choose a crust and style of pizza. Find the sample space. How many different pizzas are possible?**

Use a tree diagram to find the sample space.

| Crust | Style | Outcome |
|---|---|---|
| | Hawaiian | Thin Crust Hawaiian |
| Thin | Mexican | Thin Crust Mexican |
| | Pepperoni | Thin Crust Pepperoni |
| | Veggie | Thin Crust Veggie |
| | Hawaiian | Stuffed Crust Hawaiian |
| Stuffed | Mexican | Stuffed Crust Mexican |
| | Pepperoni | Stuffed Crust Pepperoni |
| | Veggie | Stuffed Crust Veggie |

There are 8 different outcomes in the sample space. So, there are 8 different pizzas possible.

**On Your Own**

Now You're Ready
Exercises 6 and 7

1. **WHAT IF?** The pizza shop adds a deep dish crust. Find the sample space. How many pizzas are possible?

Another way to find the total number of possible outcomes is to use the **Fundamental Counting Principle**.

**Study Tip**

The Fundamental Counting Principle can be extended to more than two events.

 **Key Idea**

**Fundamental Counting Principle**

An event $M$ has $m$ possible outcomes. An event $N$ has $n$ possible outcomes. The total number of outcomes of event $M$ followed by event $N$ is $m \times n$.

Multi-Language Glossary at BigIdeasMath.com

**EXAMPLE** ② **Finding the Total Number of Possible Outcomes**

**Find the total number of possible outcomes of rolling a number cube and flipping a coin.**

**Method 1:** Use a table to find the sample space. Let H = heads and T = tails.

|  | **1** | **2** | **3** | **4** | **5** | **6** |
|---|---|---|---|---|---|---|
| (quarter heads) | 1H | 2H | 3H | 4H | 5H | 6H |
| (quarter tails) | 1T | 2T | 3T | 4T | 5T | 6T |

∴ There are 12 possible outcomes.

**Method 2:** Use the Fundamental Counting Principle. Identify the number of possible outcomes of each event.

*Event 1:* Rolling a number cube has 6 possible outcomes.

*Event 2:* Flipping a coin has 2 possible outcomes.

$$6 \times 2 = 12 \qquad \text{Fundamental Counting Principle}$$

∴ There are 12 possible outcomes.

**EXAMPLE** ③ **Finding the Total Number of Possible Outcomes**

**How many different outfits can you make from the T-shirts, jeans, and shoes in the closet?**

Use the Fundamental Counting Principle. Identify the number of possible outcomes for each event.

*Event 1:* Choosing a T-shirt has 7 possible outcomes.

*Event 2:* Choosing jeans has 4 possible outcomes.

*Event 3:* Choosing shoes has 3 possible outcomes.

$$7 \times 4 \times 3 = 84 \qquad \text{Fundamental Counting Principle}$$

∴ So, you can make 84 different outfits.

● **On Your Own**

*Now You're Ready*
Exercises 8–11

**2.** Find the total number of possible outcomes of spinning the spinner and choosing a number from 1 to 5.

**3.** How many different outfits can you make from 4 T-shirts, 5 pairs of jeans, and 5 pairs of shoes?

A **compound event** consists of two or more events. As with a single event, the probability of a compound event is the ratio of the number of favorable outcomes to the number of possible outcomes.

EXAMPLE 4 **Finding the Probability of a Compound Event**

**In Example 2, what is the probability of rolling a number greater than 4 and flipping tails?**

There are two favorable outcomes in the sample space for rolling a number greater than 4 and flipping tails: 5T and 6T.

$$P(\text{event}) = \frac{\text{number of favorable outcomes}}{\text{number of possible outcomes}}$$

$$P(\text{greater than 4 and tails}) = \frac{2}{12} \qquad \text{Substitute.}$$

$$= \frac{1}{6} \qquad \text{Simplify.}$$

∴ The probability is $\frac{1}{6}$, or $16\frac{2}{3}\%$.

EXAMPLE 5 **Finding the Probability of a Compound Event**

**You flip three nickels. What is the probability of flipping two heads and one tails?**

Use a tree diagram to find the sample space. Let H = heads and T = tails.

There are three favorable outcomes in the sample space for flipping two heads and one tails: HHT, HTH, and THH.

$$P(\text{event}) = \frac{\text{number of favorable outcomes}}{\text{number of possible outcomes}}$$

$$P(\text{2 heads and 1 tails}) = \frac{3}{8} \qquad \text{Substitute.}$$

∴ The probability is $\frac{3}{8}$, or 37.5%.

### On Your Own

Now You're Ready
Exercises 15–24

**4.** In Example 2, what is the probability of rolling at most 4 and flipping heads?

**5.** In Example 5, what is the probability of flipping at least two tails?

**6.** You roll two number cubes. What is the probability of rolling double threes?

**7.** In Example 1, what is the probability of choosing a stuffed crust Hawaiian pizza?

## Vocabulary and Concept Check

1. **VOCABULARY** What is the sample space of an event? How can you find the sample space of two or more events?

2. **WRITING** Explain how to use the Fundamental Counting Principle.

3. **WRITING** Describe two ways to find the total number of possible outcomes of spinning the spinner and rolling the number cube.

4. **OPEN-ENDED** Give a real-life example of a compound event.

## Practice and Problem Solving

5. **COMBINATIONS** The lock is numbered from 0 to 49. Each combination uses three numbers in a right, left, right pattern. Find the total number of possible combinations for the lock.

**Use a tree diagram to find the sample space and the total number of possible outcomes.**

① 6.

| Birthday Party | |
|---|---|
| Event | Miniature golf, Laser tag, Roller skating |
| Time | 1:00 P.M.–3:00 P.M., 6:00 P.M.–8:00 P.M. |

7.

| New School Mascot | |
|---|---|
| Type | Lion, Bear, Hawk, Dragon |
| Style | Realistic, Cartoon |

**Use the Fundamental Counting Principle to find the total number of possible outcomes.**

② 8.

| Beverage | |
|---|---|
| Size | Small, Medium, Large |
| Flavor | Milk, Orange juice, Apple juice, Iced tea, Lemonade, Water, Coffee |

9.

| MP3 Player | |
|---|---|
| Memory | 2 GB, 4 GB, 8 GB, 16 GB |
| Color | Silver, Green, Blue, Pink, Black |

③ 10.

| Clown | |
|---|---|
| Suit | Dots, Stripes, Checkers board |
| Wig | One color, Multicolor |
| Talent | Balloon animals, Juggling, Unicycle, Magic |

11.

| Meal | |
|---|---|
| Appetizer | Nachos, Soup, Spinach dip, Salad, Applesauce |
| Entrée | Chicken, Beef, Spaghetti, Fish |
| Dessert | Yogurt, Fruit, Ice cream |

12. **NOTE CARDS** A store sells three types of note cards. There are three sizes of each type. Show two ways to find the total number of note cards the store sells.

13. **ERROR ANALYSIS** A true-false quiz has five questions. Describe and correct the error in using the Fundamental Counting Principle to find the total number of ways that you can answer the quiz.

2 + 2 + 2 + 2 + 2 = 10
You can answer the quiz in 10 different ways.

14. **CHOOSE TOOLS** You randomly choose one of the marbles. Without replacing the first marble, you choose a second marble.

    **a.** Name two ways you can find the total number of possible outcomes.

    **b.** Find the total number of possible outcomes.

**You spin the spinner and flip a coin. Find the probability of the compound event.**

④ **15.** Spinning a 1 and flipping heads

  **16.** Spinning an even number and flipping heads

  **17.** Spinning a number less than 3 and flipping tails

  **18.** Spinning a 6 and flipping tails

  **19.** *Not* spinning a 5 and flipping heads

  **20.** Spinning a prime number and *not* flipping heads

**You spin the spinner, flip a coin, then spin the spinner again. Find the probability of the compound event.**

⑤ **21.** Spinning blue, flipping heads, then spinning a 1

  **22.** Spinning an odd number, flipping heads, then spinning yellow

  **23.** Spinning an even number, flipping tails, then spinning an odd number

  **24.** *Not* spinning red, flipping tails, then *not* spinning an even number

25. **TAKING A TEST** You randomly guess the answers to two questions on a multiple-choice test. Each question has three choices: A, B, and C.

    **a.** What is the probability that you guess the correct answers to both questions?

    **b.** Suppose you can eliminate one of the choices for each question. How does this change the probability that your guesses are correct?

26. **PASSWORD** You forget the last two digits of your password for a website.

    **a.** What is the probability that you randomly choose the correct digits?

    **b.** Suppose you remember that both digits are even. How does this change the probability that your choices are correct?

27. **COMBINATION LOCK** The combination lock has 3 wheels, each numbered from 0 to 9.

    **a.** What is the probability that someone randomly guesses the correct combination in one attempt?

    **b.** Explain how to find the probability that someone randomly guesses the correct combination in five attempts.

28. **TRAINS** Your model train has one engine and eight train cars. Find the total number of ways you can arrange the train. (The engine must be first.)

29. **REPEATED REASONING** You have been assigned a 9-digit identification number.

    **a.** Why should you use the Fundamental Counting Principle instead of a tree diagram to find the total number of possible identification numbers?

    **b.** How many identification numbers are possible?

    **c.** **RESEARCH** Use the Internet to find out why the possible number of Social Security numbers is not the same as your answer to part (b).

30.  **Problem Solving** From a group of 5 candidates, a committee of 3 people is selected. In how many different ways can the committee be selected?

---

**Fair Game Review** What you learned in previous grades & lessons

Name two pairs of adjacent angles and two pairs of vertical angles in the figure. *(Section 7.1)*

31.

32.

33. **MULTIPLE CHOICE** A drawing has a scale of 1 cm : 1 m. What is the scale factor of the drawing? *(Section 7.5)*

    (A) 1 : 1        (B) 1 : 100        (C) 10 : 1        (D) 100 : 1

**Essential Question** What is the difference between dependent and independent events?

 **ACTIVITY: Drawing Marbles from a Bag (With Replacement)**

**Work with a partner. You have three marbles in a bag. There are two green marbles and one purple marble. Randomly draw a marble from the bag. Then put the marble back in the bag and draw a second marble.**

**a.** Complete the tree diagram. Let G = green and P = purple. Find the probability that both marbles are green.

First draw:

Second draw:

GG

**b.** Does the probability of getting a green marble on the second draw *depend* on the color of the first marble? Explain.

 **ACTIVITY: Drawing Marbles from a Bag (Without Replacement)**

**Work with a partner. Using the same marbles from Activity 1, randomly draw two marbles from the bag.**

**a.** Complete the tree diagram. Let G = green and P = purple. Find the probability that both marbles are green.

First draw:

Second draw:

GP

Is this event more likely than the event in Activity 1? Explain.

**b.** Does the probability of getting a green marble on the second draw *depend* on the color of the first marble? Explain.

 **COMMON CORE**

**Probability and Statistics**

In this lesson, you will
- identify independent and dependent events.
- use formulas to find probabilities of independent and dependent events.

Applying Standards
7.SP.8a
7.SP.8b

## ③ ACTIVITY: Conducting an Experiment

**Work with a partner. Conduct two experiments.**

**a.** In the first experiment, randomly draw one marble from the bag. Put it back. Draw a second marble. Repeat this 36 times. Record each result. Make a bar graph of your results.

**b.** In the second experiment, randomly draw two marbles from the bag 36 times. Record each result. Make a bar graph of your results.

**Math Practice 3**

**Use Definitions**

In what other mathematical context have you seen the terms *independent* and *dependent*? How does knowing these definitions help you answer the questions in part (d)?

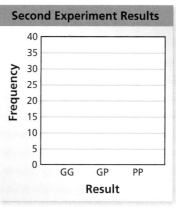

**c.** For each experiment, estimate the probability of drawing two green marbles.

**d.** Which experiment do you think represents *dependent events*? Which represents *independent events*? Explain your reasoning.

## What Is Your Answer?

**4.** **IN YOUR OWN WORDS** What is the difference between dependent and independent events? Describe a real-life example of each.

**In Questions 5–7, tell whether the events are *independent* or *dependent*. Explain your reasoning.**

**5.** You roll a 5 on a number cube and spin blue on a spinner.

**6.** Your teacher chooses one student to lead a group, and then chooses another student to lead another group.

**7.** You spin red on one spinner and green on another spinner.

**8.** In Activities 1 and 2, what is the probability of drawing a green marble on the first draw? on the second draw? How do you think you can use these two probabilities to find the probability of drawing two green marbles?

**Practice**

Use what you learned about independent and dependent events to complete Exercises 3 and 4 on page 433.

## 10.5 Lesson

**Key Vocabulary** 🔊
independent events,
  p. 430
dependent events,
  p. 431

Compound events may be *independent events* or *dependent events*. Events are **independent events** if the occurrence of one event *does not* affect the likelihood that the other event(s) will occur.

 **Key Idea**

**Probability of Independent Events**

**Words**     The probability of two or more independent events is the product of the probabilities of the events.

**Symbols**       $P(A \text{ and } B) = P(A) \cdot P(B)$

$$P(A \text{ and } B \text{ and } C) = P(A) \cdot P(B) \cdot P(C)$$

---

**EXAMPLE**   **1**   **Finding the Probability of Independent Events**

**You spin the spinner and flip the coin. What is the probability of spinning a prime number and flipping tails?**

The outcome of spinning the spinner does not affect the outcome of flipping the coin. So, the events are independent.

$P(\text{prime}) = \dfrac{3}{5}$    ← There are 3 prime numbers (2, 3, and 5).
            ← There is a total of 5 numbers.

$P(\text{tails}) = \dfrac{1}{2}$    ← There is 1 tails side.
            ← There is a total of 2 sides.

Use the formula for the probability of independent events.

$$P(A \text{ and } B) = P(A) \cdot P(B)$$

$$P(\text{prime and tails}) = P(\text{prime}) \cdot P(\text{tails})$$

$$= \frac{3}{5} \cdot \frac{1}{2} \qquad \text{Substitute.}$$

$$= \frac{3}{10} \qquad \text{Multiply.}$$

⋮ The probability of spinning a prime number and flipping tails is $\dfrac{3}{10}$, or 30%.

---

⬤ **On Your Own**

Now You're Ready
Exercises 5–8

**1.** What is the probability of spinning a multiple of 2 and flipping heads?

---

Events are **dependent events** if the occurrence of one event *does* affect the likelihood that the other event(s) will occur.

 **Key Idea**

**Probability of Dependent Events**

**Words** The probability of two dependent events *A* and *B* is the probability of *A* times the probability of *B* after *A* occurs.

**Symbols** $P(A \text{ and } B) = P(A) \cdot P(B \text{ after } A)$

**EXAMPLE** ② **Finding the Probability of Dependent Events**

**People are randomly chosen to be game show contestants from an audience of 100 people. You are with 5 of your relatives and 6 other friends. What is the probability that one of your relatives is chosen first, and then one of your friends is chosen second?**

Choosing an audience member changes the number of audience members left. So, the events are dependent.

$$P(\text{relative}) = \frac{5}{100} = \frac{1}{20}$$

> There are 5 relatives.

> There is a total of 100 audience members.

$$P(\text{friend}) = \frac{6}{99} = \frac{2}{33}$$

> There are 6 friends.

> There is a total of 99 audience members left.

Use the formula for the probability of dependent events.

$$P(A \text{ and } B) = P(A) \cdot P(B \text{ after } A)$$

$$P(\text{relative and friend}) = P(\text{relative}) \cdot P(\text{friend after relative})$$

$$= \frac{1}{20} \cdot \frac{2}{33} \qquad \text{Substitute.}$$

$$= \frac{1}{330} \qquad \text{Simplify.}$$

∴ The probability is $\frac{1}{330}$, or about 0.3%.

● **On Your Own**

Now You're Ready
Exercises 9–12

**2.** What is the probability that you, your relatives, and your friends are *not* chosen to be either of the first two contestants?

EXAMPLE 3 | **Finding the Probability of a Compound Event**

**A student randomly guesses the answer for each of the multiple-choice questions. What is the probability of answering all three questions correctly?**

**1.** In what year did the United States gain independence from Britain?
  **A.** 1492      **B.** 1776      **C.** 1788      **D.** 1795      **E.** 2000

**2.** Which amendment to the Constitution grants citizenship to all persons born in the United States and guarantees them equal protection under the law?
  **A.** 1st      **B.** 5th      **C.** 12th      **D.** 13th      **E.** 14th

**3.** In what year did the Boston Tea Party occur?
  **A.** 1607      **B.** 1773      **C.** 1776      **D.** 1780      **E.** 1812

Choosing the answer for one question does not affect the choice for the other questions. So, the events are independent.

**Method 1:** Use the formula for the probability of independent events.

$P(\#1 \text{ and } \#2 \text{ and } \#3 \text{ correct}) = P(\#1 \text{ correct}) \cdot P(\#2 \text{ correct}) \cdot$
$$P(\#3 \text{ correct})$$

$$= \frac{1}{5} \cdot \frac{1}{5} \cdot \frac{1}{5} \qquad \text{Substitute.}$$

$$= \frac{1}{125} \qquad \text{Multiply.}$$

∴ The probability of answering all three questions correctly is $\frac{1}{125}$, or 0.8%.

**Method 2:** Use the Fundamental Counting Principle.

There are 5 choices for each question, so there are $5 \cdot 5 \cdot 5 = 125$ possible outcomes. There is only 1 way to answer all three questions correctly.

$$P(\#1 \text{ and } \#2 \text{ and } \#3 \text{ correct}) = \frac{1}{125}$$

∴ The probability of answering all three questions correctly is $\frac{1}{125}$, or 0.8%.

● **On Your Own**

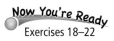
Now You're Ready
Exercises 18–22

**3.** The student can eliminate Choice A for all three questions. What is the probability of answering all three questions correctly? Compare this probability with the probability in Example 3. What do you notice?

## ✓ Vocabulary and Concept Check

1. **DIFFERENT WORDS, SAME QUESTION** You randomly choose one of the chips. Without replacing the first chip, you choose a second chip. Which question is different? Find "both" answers.

   > What is the probability of choosing a 1 and then a blue chip?

   > What is the probability of choosing a 1 and then an even number?

   > What is the probability of choosing a green chip and then a chip that is *not* red?

   > What is the probability of choosing a number less than 2 and then an even number?

2. **WRITING** How do you find the probability of two events *A* and *B* when *A* and *B* are independent? dependent?

##  Practice and Problem Solving

**Tell whether the events are *independent* or *dependent*. Explain.**

3. You roll a 4 on a number cube. Then you roll an even number on a different number cube.

4. You randomly draw a lane number for a 100-meter race. Then your friend randomly draws a lane number for the same race.

**You spin the spinner and flip a coin. Find the probability of the compound event.**

① 5. Spinning a 3 and flipping heads

6. Spinning an even number and flipping tails

7. Spinning a number greater than 1 and flipping tails

8. *Not* spinning a 2 and flipping heads

**You randomly choose one of the tiles. Without replacing the first tile, you choose a second tile. Find the probability of the compound event.**

② 9. Choosing a 5 and then a 6

10. Choosing an odd number and then a 20

11. Choosing a number less than 7 and then a multiple of 4

12. Choosing two even numbers

13. **ERROR ANALYSIS** Describe and correct the error in finding the probability.

You randomly choose one of the marbles. Without replacing the first marble, you choose a second marble. What is the probability of choosing red and then green?

$P(\text{red and green}) = \dfrac{1}{4} \cdot \dfrac{1}{4} = \dfrac{1}{16}$

First Draw    Second Draw

14. **LOGIC** A bag contains three marbles. Does the tree diagram show the outcomes for *independent* or *dependent* events? Explain.

15. **EARRINGS** A jewelry box contains two gold hoop earrings and two silver hoop earrings. You randomly choose two earrings. What is the probability that both are silver hoop earrings?

16. **HIKING** You are hiking to a ranger station. There is one correct path. You come to a fork and randomly take the path on the left. You come to another fork and randomly take the path on the right. What is the probability that you are still on the correct path?

17. **CARNIVAL** At a carnival game, you randomly throw two darts at the board and break two balloons. What is the probability that both of the balloons you break are purple?

**You spin the spinner, flip a coin, then spin the spinner again. Find the probability of the compound event.**

③ 18. Spinning a 4, flipping heads, then spinning a 7

19. Spinning an odd number, flipping heads, then spinning a 3

20. Spinning an even number, flipping tails, then spinning an odd number

21. *Not* spinning a 5, flipping heads, then spinning a 1

22. Spinning an odd number, *not* flipping heads, then *not* spinning a 6

**23. LANGUAGES** There are 16 students in your Spanish class. Your teacher randomly chooses one student at a time to take a verbal exam. What is the probability that you are *not* one of the first four students chosen?

**24. SHOES** Twenty percent of the shoes in a factory are black. One shoe is chosen and replaced. A second shoe is chosen and replaced. Then a third shoe is chosen. What is the probability that *none* of the shoes are black?

**25. PROBLEM SOLVING** Your teacher divides your class into two groups, and then randomly chooses a leader for each group. The probability that you are chosen to be a leader is $\frac{1}{12}$. The probability that both you and your best friend are chosen is $\frac{1}{132}$.

   **a.** Is your best friend in your group? Explain.

   **b.** What is the probability that your best friend is chosen as a group leader?

   **c.** How many students are in the class?

**26.** **Structure** After ruling out some of the answer choices, you randomly guess the answer for each of the story questions below.

> **1.** Who was the oldest?
>    **A.** Ned      **B.** Yvonne      **C.** Sun Li      **D.** Angel      **E.** Dusty
>
> **2.** What city was Stacey from?
>    **A.** Raleigh      **B.** New York      **C.** Roanoke      **D.** Dallas      **E.** San Diego

   **a.** How can the probability of getting both answers correct be 25%?

   **b.** How can the probability of getting both answers correct be $8\frac{1}{3}$%?

## Fair Game Review  *What you learned in previous grades & lessons*

**Draw a triangle with the given angle measures. Then classify the triangle.**  *(Section 7.3)*

**27.** 30°, 60°, 90°         **28.** 20°, 50°, 110°         **29.** 50°, 50°, 80°

**30. MULTIPLE CHOICE** Which set of numbers is in order from least to greatest?  *(Section 6.2)*

   Ⓐ $\frac{2}{3}$, 0.6, 67%                 Ⓑ 44.5%, $\frac{4}{9}$, $0.4\overline{6}$

   Ⓒ 0.269, 27%, $\frac{3}{11}$             Ⓓ $2\frac{1}{7}$, 214%, $2.\overline{14}$

**Check It Out**
Lesson Tutorials
BigIdeasMath⚓com

**Key Vocabulary** 🔊
simulation, *p. 436*

A **simulation** is an experiment that is designed to reproduce the conditions of a situation or process. Simulations allow you to study situations that are impractical to create in real life.

## EXAMPLE 1 Simulating Outcomes That Are Equally Likely

| | |
|---|---|
| HTH | HTT |
| HTT | HTH |
| HTT | TTT |
| (HHH) | HTT |
| HTT | TTT |
| HTT | HTH |
| HTH | (HHH) |
| HTT | HTT |
| TTT | HTH |
| HTH | HTT |

**A couple plans on having three children. The gender of each child is equally likely. (a) Design a simulation involving 20 trials that you can use to model the genders of the children. (b) Use your simulation to find the experimental probability that all three children are boys.**

**a.** Choose an experiment that has two equally likely outcomes for each event (gender), such as tossing three coins. Let heads (H) represent a boy and tails (T) represent a girl.

**b.** To find the experimental probability, you need repeated trials of the simulation. The table shows 20 trials.

$$P(\text{three boys}) = \frac{2}{20} = \frac{1}{10}$$

HHH occurred 2 times.

There is a total of 20 trials.

∴ The experimental probability is $\frac{1}{10}$, 0.1, or 10%.

## EXAMPLE 2 Simulating Outcomes That Are Not Equally Likely

**Study Tip**

In Example 2, the digits 1 through 6 represent 60% of the possible digits (0 through 9) in the tens place. Likewise, the digits 1 and 2 represent 20% of the possible digits in the ones place.

**There is a 60% chance of rain on Monday and a 20% chance of rain on Tuesday. Design and use a simulation involving 50 randomly generated numbers to find the experimental probability that it will rain on both days.**

Use the random number generator on a graphing calculator. Randomly generate 50 numbers from 0 to 99. The table below shows the results.

Let the digits 1 through 6 in the tens place represent rain on Monday. Let digits 1 and 2 in the ones place represent rain on Tuesday. Any number that meets these criteria represents rain on both days.

```
randInt(0,99,50)
{52 66 73 68 75…
```

| | | | | | | | | | |
|---|---|---|---|---|---|---|---|---|---|
| (52) | 66 | 73 | 68 | 75 | 28 | 35 | 47 | 48 | 2 |
| 16 | 68 | 49 | 3 | 77 | 35 | 92 | 78 | 6 | 6 |
| 58 | 18 | 89 | 39 | 24 | 80 | (32) | (41) | 77 | (21) |
| (32) | 40 | 96 | 59 | 86 | 1 | (12) | 0 | 94 | 73 |
| 40 | 71 | 28 | (61) | 1 | 24 | 37 | 25 | 3 | 25 |

$$P(\text{rain both days}) = \frac{7}{50}$$

7 numbers meet the criteria.

There is a total of 50 trials.

∴ The experimental probability is $\frac{7}{50}$, 0.14, or 14%.

🔊 Multi-Language Glossary at BigIdeasMath⚓com

EXAMPLE 3 **Using a Spreadsheet to Simulate Outcomes**

**Common Core**

**Probability and Statistics**

In this extension, you will
- use simulations to find experimental probabilities.

Learning Standard
7.SP.8c

**Each school year, there is a 50% chance that weather causes one or more days of school to be canceled. Design and use a simulation involving 50 randomly generated numbers to find the experimental probability that weather will cause school to be canceled in at least three of the next four school years.**

Use a random number table in a spreadsheet. Randomly generate 50 four-digit whole numbers. The spreadsheet below shows the results.

Let the digits 1 through 5 represent school years with a cancellation. The numbers in the spreadsheet that contain at least three digits from 1 through 5 represent four school years in which at least three of the years have a cancellation.

**Study Tip**

To create a four-digit random number table in a spreadsheet, follow these steps.
1. Highlight the group of cells to use for your table.
2. Format the cells to display four-digit whole numbers.
3. Enter the formula RAND()*10000 into each cell.

|   | A | B | C | D | E | F |
|---|------|------|------|------|------|---|
| 1 | 7584 | 3974 | 8614 | 2500 | 4629 | |
| 2 | 3762 | 3805 | 2725 | 7320 | 6487 | |
| 3 | 3024 | 1554 | 2708 | 1126 | 9395 | |
| 4 | 4547 | 6220 | 9497 | 7530 | 3036 | |
| 5 | 1719 | 0662 | 1814 | 6218 | 2766 | |
| 6 | 7938 | 9551 | 8552 | 4321 | 8043 | |
| 7 | 6951 | 0578 | 5560 | 0740 | 4479 | |
| 8 | 4714 | 4511 | 5115 | 6952 | 5609 | |
| 9 | 0797 | 3022 | 9067 | 2193 | 6553 | |
| 10 | 3300 | 5454 | 5351 | 6319 | 0387 | |
| 11 | | | | | | |

$$P\left(\begin{array}{c}\text{cancellation in at least three}\\\text{of the next four school years}\end{array}\right) = \frac{17}{50}$$

17 numbers contain at least three digits from 1 to 5.

There is a total of 50 trials.

∴ The experimental probability is $\frac{17}{50}$, 0.34, or 34%.

## Practice

1. **QUIZ** You randomly guess the answers to four true-false questions. (a) Design a simulation that you can use to model the answers. (b) Use your simulation to find the experimental probability that you answer all four questions correctly.

2. **BASEBALL** A baseball team wins 70% of its games. Assuming this trend continues, design and use a simulation to find the experimental probability that the team wins the next three games.

3. **WHAT IF?** In Example 3, there is a 40% chance that weather causes one or more days of school to be canceled each school year. Find the experimental probability that weather will cause school to be canceled in at least three of the next four school years.

4. **REASONING** In Examples 1–3 and Exercises 1–3, try to find the theoretical probability of the event. What do you think happens to the experimental probability when you increase the number of trials in the simulation?

# 10 Study Help

**Check It Out**
Graphic Organizer
BigIdeasMath ✓com

You can use a **notetaking organizer** to write notes, vocabulary, and questions about a topic. Here is an example of a notetaking organizer for probability.

Write important vocabulary or formulas in this space.

If *P*(event) = 0, the event is *impossible*.

If *P*(event) = 0.25, the event is *unlikely*.

If *P*(event) = 0.5, the event is *equally likely to happen or not happen*.

If *P*(event) = 0.75, the event is *likely*.

If *P*(event) = 1, the event is *certain*.

## Probability

A number that measures the likelihood that an event will occur

Can be written as a fraction, decimal, or percent

Always between 0 and 1, inclusive

Write your notes about the topic in this space.

Write your questions about the topic in this space.

How do you find the probability of two or more events?

## On Your Own

**Make notetaking organizers to help you study these topics.**

1. experimental probability
2. theoretical probability
3. Fundamental Counting Principle
4. independent events
5. dependent events

**After you complete this chapter, make notetaking organizers for the following topics.**

6. sample
7. population

"I am using a notetaking organizer to plan my autobiography."

# 10.1–10.5 Quiz

**You randomly choose one butterfly. Find the number of ways the event can occur.** *(Section 10.1)*

1. Choosing red

2. Choosing brown

3. Choosing *not* blue

6 Green
3 White
4 Red
2 Blue
5 Yellow

**You randomly choose one paper clip from the jar. Find the probability of the event.** *(Section 10.2)*

4. Choosing a green paper clip

5. Choosing a yellow paper clip

6. *Not* choosing a yellow paper clip

7. Choosing a purple paper clip

**Use the bar graph to find the experimental probability of the event.** *(Section 10.3)*

8. Rolling a 4

9. Rolling a multiple of 3

10. Rolling a 2 or a 3

11. Rolling a number less than 7

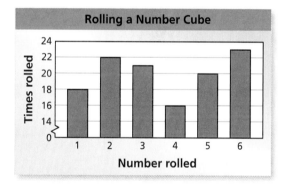

Rolling a Number Cube

**Use the Fundamental Counting Principle to find the total number of possible outcomes.** *(Section 10.4)*

12.

| Calculator | |
|------|------|
| **Type** | Basic display, Scientific, Graphing, Financial |
| **Color** | Black, White, Silver |

13.

| Vacation | |
|------|------|
| **Destination** | Florida, Italy, Mexico, England |
| **Length** | 1 week, 2 weeks |

14. **BLACK PENS** You randomly choose one of the pens shown. What is the theoretical probability of choosing a black pen? *(Section 10.3)*

15. **BLUE PENS** You randomly choose one of the five pens shown. Your friend randomly chooses one of the remaining pens. What is the probability that you and your friend both choose a blue pen? *(Section 10.5)*

# 10.6 Samples and Populations

## Essential Question How can you determine whether a sample accurately represents a population?

A **population** is an entire group of people or objects. A **sample** is a part of the population. You can use a sample to make an *inference*, or conclusion, about a population.

Identify a population. → Select a sample. → Interpret the data in the sample. → Make an inference about the population.

Population → Sample → Interpretation → Inference

### 1 ACTIVITY: Identifying Populations and Samples

**Work with a partner. Identify the population and the sample.**

a.

The students in a school    The students in a math class

b.

The grizzly bears with GPS collars in a park    The grizzly bears in a park

c.

150 quarters    All quarters in circulation

d.

All books in a library    10 fiction books in a library

**COMMON CORE**

**Probability and Statistics**

In this lesson, you will
- determine when samples are representative of populations.
- use data from random samples to make predictions about populations.

Learning Standards
7.SP.1
7.SP.2

### 2 ACTIVITY: Identifying Random Samples

**Work with a partner. When a sample is selected at random, each member of the population is equally likely to be selected. You want to know the favorite extracurricular activity of students at your school. Determine whether each method will result in a random sample. Explain your reasoning.**

a. You ask members of the school band.

b. You publish a survey in the school newspaper.

c. You ask every eighth student who enters the school in the morning.

d. You ask students in your class.

There are many different ways to select a sample from a population. To make valid inferences about a population, you must choose a random sample very carefully so that it accurately represents the population.

## 3  ACTIVITY: Identifying Representative Samples

**Work with a partner. A new power plant is being built outside a town. In each situation below, residents of the town are asked how they feel about the new power plant. Determine whether each conclusion is valid. Explain your reasoning.**

**a.** A local radio show takes calls from 500 residents. The table shows the results. The radio station concludes that most of the residents of the town oppose the new power plant.

| New Power Plant | |
| --- | --- |
| For | 70 |
| Against | 425 |
| Don't know | 5 |

New Power Plant

**Math Practice 2**

**Understand Quantities**

Can the size of a sample affect the validity of a conclusion about a population?

**b.** A news reporter randomly surveys 2 residents outside a supermarket. The graph shows the results. The reporter concludes that the residents of the town are evenly divided on the new power plant.

**c.** You randomly survey 250 residents at a shopping mall. The table shows the results. You conclude that there are about twice as many residents of the town against the new power plant than for the new power plant.

| New Power Plant | |
| --- | --- |
| For | 32% |
| Against | 62% |
| Don't know | 6% |

## What Is Your Answer?

**4. IN YOUR OWN WORDS** How can you determine whether a sample accurately represents a population?

**5. RESEARCH** Choose a topic that you would like to ask people's opinions about, and then write a survey question. How would you choose people to survey so that your sample is random? How many people would you survey? Conduct your survey and display your results. Would you change any part of your survey to make it more accurate? Explain.

**6.** Does increasing the size of a sample necessarily make the sample representative of a population? Give an example to support your explanation.

**Practice**

Use what you learned about populations and samples to complete Exercises 3 and 4 on page 444.

**Key Vocabulary** 🔊
population, *p. 440*
sample, *p. 440*
unbiased sample,
   *p. 442*
biased sample, *p. 442*

An **unbiased sample** is representative of a population. It is selected at random and is large enough to provide accurate data.

A **biased sample** is not representative of a population. One or more parts of the population are favored over others.

**EXAMPLE** 1 **Identifying an Unbiased Sample**

**You want to estimate the number of students in a high school who ride the school bus. Which sample is unbiased?**

Ⓐ   4 students in the hallway

Ⓑ   all students in the marching band

Ⓒ   50 seniors at random

Ⓓ   100 students at random
     during lunch

Choice A is not large enough to provide accurate data.

Choice B is not selected at random.

Choice C is not representative of the population because seniors are more likely to drive to school than other students.

Choice D is representative of the population, selected at random, and large enough to provide accurate data.

∴   So, the correct answer is Ⓓ.

### On Your Own

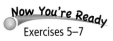
Now You're Ready
Exercises 5–7

1. **WHAT IF?** You want to estimate the number of seniors in a high school who ride the school bus. Which sample is unbiased? Explain.

2. You want to estimate the number of eighth-grade students in your school who consider it relaxing to listen to music. You randomly survey 15 members of the band. Your friend surveys every fifth student whose name appears on an alphabetical list of eighth graders. Which sample is unbiased? Explain.

The results of an unbiased sample are proportional to the results of the population. So, you can use unbiased samples to make predictions about the population.

Biased samples are not representative of the population. So, you should not use them to make predictions about the population because the predictions may not be valid.

EXAMPLE 2 **Determining Whether Conclusions Are Valid**

**You want to know how the residents of your town feel about adding a new stop sign. Determine whether each conclusion is valid.**

   **a.** **You survey the 20 residents who live closest to the new sign. Fifteen support the sign, and five do not. So, you conclude that 75% of the residents of your town support the new sign.**

     The sample is not representative of the population because residents who live close to the sign are more likely to support it.

      ⋮ So, the sample is biased, and the conclusion is not valid.

   **b.** **You survey 100 residents at random. Forty support the new sign, and sixty do not. So, you conclude that 40% of the residents of your town support the new sign.**

     The sample is representative of the population, selected at random, and large enough to provide accurate data.

      ⋮ So, the sample is unbiased, and the conclusion is valid.

EXAMPLE 3 **Making Predictions**

**Movies per Week**

**You ask 75 randomly chosen students how many movies they watch each week. There are 1200 students in the school. Predict the number $n$ of students in the school who watch one movie each week.**

The sample is representative of the population, selected at random, and large enough to provide accurate data. So, the sample is unbiased, and you can use it to make a prediction about the population.

Write and solve a proportion to find $n$.

| Sample | | Population |
|---|---|---|
| $\dfrac{\text{students in survey (one movie)}}{\text{number of students in survey}}$ | $=$ | $\dfrac{\text{students in school (one movie)}}{\text{number of students in school}}$ |

$$\frac{21}{75} = \frac{n}{1200} \qquad \text{Substitute.}$$

$$336 = n \qquad \text{Solve for } n.$$

⋮ So, about 336 students in the school watch one movie each week.

### On Your Own

**Now You're Ready**
Exercises 8, 9, and 12

**3.** In Example 2, each of 25 randomly chosen firefighters supports the new sign. So, you conclude that 100% of the residents of your town support the new sign. Is the conclusion valid? Explain.

**4.** In Example 3, predict the number of students in the school who watch two or more movies each week.

## 10.6 Exercises

Check It Out
Help with Homework
BigIdeasMath.com

### Vocabulary and Concept Check

1. **VOCABULARY** Why would you survey a sample instead of a population?

2. **CRITICAL THINKING** What should you consider when conducting a survey?

### Practice and Problem Solving

**Identify the population and the sample.**

3.

Residents of New Jersey    Residents of Ocean County

4.

4 cards    All cards in a deck

**Determine whether the sample is *biased* or *unbiased*. Explain.**

5. You want to estimate the number of students in your school who play a musical instrument. You survey the first 15 students who arrive at a band class.

6. You want to estimate the number of books students in your school read over the summer. You survey every fourth student who enters the school.

7. You want to estimate the number of people in a town who think that a park needs to be remodeled. You survey every 10th person who enters the park.

**Determine whether the conclusion is valid. Explain.**

8. You want to determine the number of students in your school who have visited a science museum. You survey 50 students at random. Twenty have visited a science museum, and thirty have not. So, you conclude that 40% of the students in your school have visited a science museum.

9. You want to know how the residents of your town feel about building a new baseball stadium. You randomly survey 100 people who enter the current stadium. Eighty support building a new stadium, and twenty do not. So, you conclude that 80% of the residents of your town support building a new baseball stadium.

**Which sample is better for making a prediction? Explain.**

10.

| Predict the number of students in a school who like gym class. | |
| --- | --- |
| Sample A | A random sample of 8 students from the yearbook |
| Sample B | A random sample of 80 students from the yearbook |

11.

| Predict the number of defective pencils produced per day. | |
| --- | --- |
| Sample A | A random sample of 500 pencils from 20 machines |
| Sample B | A random sample of 500 pencils from 1 machine |

**③ 12. FOOD** You ask 125 randomly chosen students to name their favorite food. There are 1500 students in the school. Predict the number of students in the school whose favorite food is pizza.

| Favorite Food | |
|---|---|
| Pizza | 58 |
| Hamburger | 36 |
| Pasta | 14 |
| Other | 17 |

**Determine whether you would survey the population or a sample. Explain.**

**13.** You want to know the average height of seventh graders in the United States.

**14.** You want to know the favorite types of music of students in your homeroom.

**15.** You want to know the number of students in your state who have summer jobs.

| Theater Ticket Sales | |
|---|---|
| **Adults** | **Students** |
| 522 | 210 |

**16. THEATER** You survey 72 randomly chosen students about whether they are going to attend the school play. Twelve say yes. Predict the number of students who attend the school.

**17. CRITICAL THINKING** Explain why 200 people with email addresses may not be a random sample. When might it be a random sample?

**18. LOGIC** A person surveys residents of a town to determine whether a skateboarding ban should be overturned.

   **a.** Describe how the person could conduct the survey so that the sample is biased toward overturning the ban.

   **b.** Describe how the person could conduct the survey so that the sample is biased toward keeping the ban.

**19.** **Reasoning** A guidance counselor surveys a random sample of 60 out of 900 high school students. Using the survey results, the counselor predicts that approximately 720 students plan to attend college. Do you agree with her prediction? Explain.

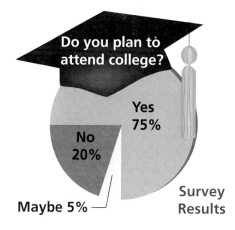

Do you plan to attend college?

Yes 75%
No 20%
Maybe 5%
Survey Results

## Fair Game Review   What you learned in previous grades & lessons

**Write and solve a proportion to answer the question.**   *(Section 6.3)*

**20.** What percent of 60 is 18?

**21.** 70% of what number is 98?

**22.** 30 is 15% of what number?

**23.** What number is 0.6% of 500?

**24. MULTIPLE CHOICE** What is the volume of the pyramid?   *(Section 9.5)*

   **Ⓐ** 40 cm³

   **Ⓑ** 50 cm³

   **Ⓒ** 100 cm³

   **Ⓓ** 120 cm³

5 cm
6 cm
4 cm

You have already used unbiased samples to make inferences about a population. In some cases, making an inference about a population from only one sample is not as precise as using multiple samples.

## 1 ACTIVITY: Using Multiple Random Samples

**Work with a partner. You and a group of friends want to know how many students in your school listen to pop music. There are 840 students in your school. Each person in the group randomly surveys 20 students.**

**Step 1:** The table shows your results. Make an inference about the number of students in your school who prefer pop music.

| Favorite Type of Music | | | |
|---|---|---|---|
| Country | Pop | Rock | Rap |
| 4 | 10 | 5 | 1 |

**Step 2:** The table shows Kevin's results. Use these results to make another inference about the number of students in your school who prefer pop music.

Compare the results of Steps 1 and 2.

| Favorite Type of Music | | | |
|---|---|---|---|
| Country | Pop | Rock | Rap |
| 2 | 13 | 4 | 1 |

**COMMON CORE**

**Probability and Statistics**
In this extension, you will
• use multiple samples to make predictions about populations.
Learning Standard
7.SP.2

**Step 3:** The table shows the results of three other friends. Use these results to make three more inferences about the number of students in your school who prefer pop music.

| Favorite Type of Music | | | | |
|---|---|---|---|---|
| | Country | Pop | Rock | Rap |
| **Steve** | 3 | 8 | 7 | 2 |
| **Laura** | 5 | 10 | 4 | 1 |
| **Ming** | 5 | 9 | 3 | 3 |

**Step 4:** Describe the variation of the five inferences. Which one would you use to describe the number of students in your school who prefer pop music? Explain your reasoning.

**Step 5:** Show how you can use all five samples to make an inference.

## Practice

1. **PACKING PEANUTS** Work with a partner. Mark 24 packing peanuts with either a red or a black marker. Put the peanuts into a paper bag. Trade bags with other students in the class.

   **a.** Generate a sample by choosing a peanut from your bag six times, replacing the peanut each time. Record the number of times you choose each color. Repeat this process to generate four more samples. Organize the results in a table.

   **b.** Use each sample to make an inference about the number of red peanuts in the bag. Then describe the variation of the five inferences. Make inferences about the numbers of red and black peanuts in the bag based on all the samples.

   **c.** Take the peanuts out of the bag. How do your inferences compare to the population? Do you think you can make a more accurate prediction? If so, explain how.

**Hours Worked Each Week**

**1:** 6, 8, 6, 6, 7, 4, 10, 8, 7, 8
**2:** 10, 4, 4, 6, 8, 6, 7, 12, 8, 8
**3:** 10, 9, 8, 6, 5, 8, 6, 6, 9, 10
**4:** 4, 8, 4, 4, 5, 4, 4, 6, 5, 6
**5:** 6, 8, 8, 6, 12, 4, 10, 8, 6, 12
**6:** 10, 10, 8, 9, 16, 8, 7, 12, 16, 14
**7:** 4, 5, 6, 6, 4, 5, 6, 6, 4, 4
**8:** 16, 20, 8, 12, 10, 8, 8, 14, 16, 8

**Work with a partner. You want to know the mean number of hours students with part-time jobs work each week. You go to 8 different schools. At each school, you randomly survey 10 students with part-time jobs. Your results are shown at the left.**

**Step 1:** Find the mean of each sample.

**Step 2:** Make a box-and-whisker plot of the sample means.

**Step 3:** Use the box-and-whisker plot to estimate the actual mean number of hours students with part-time jobs work each week.

How does your estimate compare to the mean of the entire data set?

③ **ACTIVITY: Using a Simulation**

**Work with a partner. Another way to generate multiple samples of data is to use a simulation. Suppose 70% of all seventh graders watch reality shows on television.**

**Step 1:** Design a simulation involving 50 packing peanuts by marking 70% of the peanuts with a certain color. Put the peanuts into a paper bag.

**Step 2:** Simulate choosing a sample of 30 students by choosing peanuts from the bag, replacing the peanut each time. Record the results. Repeat this process to generate eight more samples. How much variation do you expect among the samples? Explain.

**Step 3:** Display your results.

## Practice

2. **SPORTS DRINKS** You want to know whether student-athletes prefer water or sports drinks during games. You go to 10 different schools. At each school, you randomly survey 10 student-athletes. The percents of student-athletes who prefer water are shown.

   60%   70%   60%   50%   80%   70%   30%   70%   80%   40%

   a. Make a box-and-whisker plot of the data.

   b. Use the box-and-whisker plot to estimate the actual percent of student-athletes who prefer water. How does your estimate compare to the mean of the data?

3. **PART-TIME JOBS** Repeat Activity 2 using the medians of the samples.

4. **TELEVISION** In Activity 3, how do the percents in your samples compare to the given percent of seventh graders who watch reality shows on television?

5. **REASONING** Why is it better to make inferences about a population based on multiple samples instead of only one sample? What additional information do you gain by taking multiple random samples? Explain.

# 10.7 Comparing Populations

**Essential Question** How can you compare data sets that represent two populations?

## 1 ACTIVITY: Comparing Two Data Distributions

**Work with a partner. You want to compare the shoe sizes of male students in two classes. You collect the data shown in the table.**

| Male Students in Eighth-Grade Class | | | | | | | | | | | | | | |
|---|---|---|---|---|---|---|---|---|---|---|---|---|---|---|
| 7 | 9 | 8 | $7\frac{1}{2}$ | $8\frac{1}{2}$ | 10 | 6 | $6\frac{1}{2}$ | 8 | 8 | $8\frac{1}{2}$ | 9 | 11 | $7\frac{1}{2}$ | $8\frac{1}{2}$ |

| Male Students in Sixth-Grade Class | | | | | | | | | | | | | | |
|---|---|---|---|---|---|---|---|---|---|---|---|---|---|---|
| 6 | $5\frac{1}{2}$ | 6 | $6\frac{1}{2}$ | $7\frac{1}{2}$ | $8\frac{1}{2}$ | 7 | $5\frac{1}{2}$ | 5 | $5\frac{1}{2}$ | $6\frac{1}{2}$ | 7 | $4\frac{1}{2}$ | 6 | 6 |

**a.** How can you display both data sets so that you can visually compare the measures of center and variation? Make the data display you chose.

**b.** Describe the shape of each distribution.

**c.** Complete the table.

| | Mean | Median | Mode | Range | Interquartile Range (IQR) | Mean Absolute Deviation (MAD) |
|---|---|---|---|---|---|---|
| **Male Students in Eighth-Grade Class** | | | | | | |
| **Male Students in Sixth-Grade Class** | | | | | | |

**d.** Compare the measures of center for the data sets.

**e.** Compare the measures of variation for the data sets. Does one data set show more variation than the other? Explain.

**f.** Do the distributions overlap? How can you tell using the data display you chose in part (a)?

**g.** The double box-and-whisker plot below shows the shoe sizes of the members of two girls basketball teams. Can you conclude that at least one girl from each team has the same shoe size? Can you conclude that at least one girl from the Bobcats has a larger shoe size than one of the girls from the Tigers? Explain your reasoning.

**COMMON CORE**

**Probability and Statistics**

In this lesson, you will
- use measures of center and variation to compare populations.
- use random samples to compare populations.

Learning Standards
7.SP.3
7.SP.4

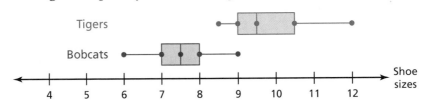

## 2

**ACTIVITY: Comparing Two Data Distributions**

Work with a partner. Compare the shapes of the distributions. Do the two data sets overlap? Explain. If so, use measures of center and the least and the greatest values to describe the overlap between the two data sets.

**a.**

**b.** **Heights (inches)**

**Recognize Usefulness of Tools**

How is each type of data display useful? Which do you prefer? Explain.

**c.** **Ages of People in Two Exercise Classes**

| 10:00 A.M. Class | | 8:00 P.M. Class |
|---|---|---|
| | 1 | 8  9 |
| | 2 | 1  2  2  7  9  9 |
| | 3 | 0  3  4  5  7 |
| 9  7  3  2  2  2 | 4 | 0 |
| 7  5  4  3  1 | 5 | |
| 7  0  0 | 6 | |
| 0 | 7 | |

**Key: 1 | 8 = 18**

## What Is Your Answer?

**3. IN YOUR OWN WORDS** How can you compare data sets that represent two populations?

**Practice** Use what you learned about comparing data sets to complete Exercise 3 on page 452.

Recall that you use the mean and the mean absolute deviation (MAD) to describe symmetric distributions of data. You use the median and the interquartile range (IQR) to describe skewed distributions of data.

To compare two populations, use the mean and the MAD when both distributions are symmetric. Use the median and the IQR when either one or both distributions are skewed.

**EXAMPLE** **1** **Comparing Populations**

**The double dot plot shows the time that each candidate in a debate spent answering each of 15 questions.**

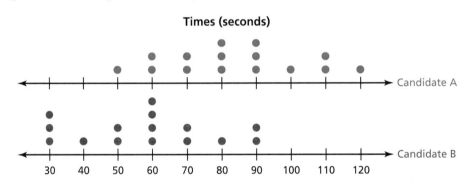

Times (seconds)

> **Study Tip**
>
> You can more easily see the visual overlap of dot plots that are aligned vertically.

a. **Compare the populations using measures of center and variation.**

Both distributions are approximately symmetric, so use the mean and the MAD.

| **Candidate A** | **Candidate B** |
|---|---|
| Mean $= \dfrac{1260}{15} = 84$ | Mean $= \dfrac{870}{15} = 58$ |
| MAD $= \dfrac{244}{15} \approx 16$ | MAD $= \dfrac{236}{15} \approx 16$ |

> **Study Tip**
>
> When two populations have similar variabilities, the value in part (b) describes the visual overlap between the data. In general, the greater the value, the less the overlap.

So, the variation in the times was about the same, but Candidate A had a greater mean time.

b. **Express the difference in the measures of center as a multiple of the measure of variation.**

$$\frac{\text{mean for Candidate A} - \text{mean for Candidate B}}{\text{MAD}} = \frac{26}{16} \approx 1.6$$

So, the difference in the means is about 1.6 times the MAD.

**On Your Own**

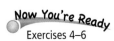
Now You're Ready
Exercises 4–6

1. **WHAT IF?** Each value in the dot plot for Candidate A increases by 30 seconds. How does this affect the answers in Example 1? Explain.

You do not need to have all the data from two populations to make comparisons. You can use random samples to make comparisons.

EXAMPLE 2 Using Random Samples to Compare Populations

**You want to compare the costs of speeding tickets in two states.**

a. **The double box-and-whisker plot shows a random sample of 10 speeding tickets issued in two states. Compare the samples using measures of center and variation. Can you use this to make a valid comparison about speeding tickets in the two states? Explain.**

Both distributions are skewed right, so use the median and the IQR.

⋮• The median and the IQR for State A, 70 and 20, are less than the median and the IQR for State B, 80 and 30. However, the sample size is too small and the variability is too great to conclude that speeding tickets generally cost more in State B.

b. **The double box-and-whisker plot shows the medians of 100 random samples of 10 speeding tickets for each state. Compare the variability of the sample medians to the variability of the sample costs in part (a).**

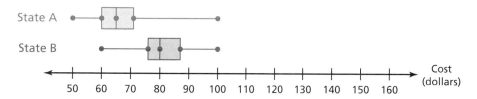

The IQR of the sample medians for each state is about 10.

⋮• So, the sample medians vary much less than the sample costs.

c. **Make a conclusion about the costs of speeding tickets in the two states.**

The sample medians show less variability. Most of the sample medians for State B are greater than the sample medians for State A.

⋮• So, speeding tickets generally cost more in State B than in State A.

● **On Your Own**

Now You're Ready
Exercise 8

2. **WHAT IF?** A random sample of 8 speeding tickets issued in State C has a median of $120. Can you conclude that a speeding ticket in State C costs more than in States A and B? Explain.

Check It Out
Help with Homework
BigIdeasMath ✓com

## Vocabulary and Concept Check

1. **REASONING** When comparing two populations, when should you use the mean and the MAD? the median and the IQR?

2. **WRITING** Two data sets have similar variabilities. Suppose the measures of center of the data sets differ by 4 times the measure of variation. Describe the visual overlap of the data.

## Practice and Problem Solving

3. **SNAKES** The tables show the lengths of two types of snakes at an animal store.

| Garter Snake Lengths (inches) | | | | | |
|---|---|---|---|---|---|
| 26 | 30 | 22 | 15 | 21 | 24 |
| 28 | 32 | 24 | 25 | 18 | 35 |

| Water Snake Lengths (inches) | | | | | |
|---|---|---|---|---|---|
| 34 | 25 | 24 | 35 | 40 | 32 |
| 41 | 27 | 37 | 32 | 21 | 30 |

   a. Find the mean, median, mode, range, interquartile range, and mean absolute deviation for each data set.

   b. Compare the data sets.

① 4. **HOCKEY** The double box-and-whisker plot shows the goals scored per game by two hockey teams during a 20-game season.

   a. Compare the populations using measures of center and variation.

   b. Express the difference in the measures of center as a multiple of the measure of variation.

5. **TEST SCORES** The dot plots show the test scores for two classes taught by the same teacher.

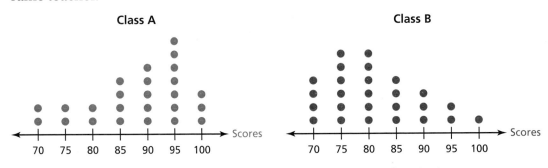

   a. Compare the populations using measures of center and variation.

   b. Express the difference in the measures of center as a multiple of each measure of variation.

**6. ATTENDANCE** The tables show the attendances at volleyball games and basketball games at a school during the year.

| Volleyball Game Attendance | | | | | | |
|---|---|---|---|---|---|---|
| 112 | 95 | 84 | 106 | 62 | 68 | 53 |
| 75 | 88 | 93 | 127 | 98 | 117 | 60 |
| 49 | 54 | 85 | 74 | 88 | 132 | |

| Basketball Game Attendance | | | | | | |
|---|---|---|---|---|---|---|
| 202 | 190 | 173 | 155 | 169 | 188 | 195 |
| 176 | 141 | 152 | 181 | 198 | 214 | 179 |
| 163 | 186 | 184 | 207 | 219 | 228 | |

  **a.** Compare the populations using measures of center and variation.

  **b.** Express the difference in the measures of center as a multiple of each measure of variation.

**7. NUMBER SENSE** Compare the answers to Exercises 4(b), 5(b), and 6(b). Which value is the greatest? What does this mean?

**8. MAGAZINES** You want to compare the number of words per sentence in a sports magazine to the number of words per sentence in a political magazine.

  **a.** The data represent random samples of 10 sentences in each magazine. Compare the samples using measures of center and variation. Can you use this to make a valid comparison about the magazines? Explain.

      Sports magazine: 9, 21, 15, 14, 25, 26, 9, 19, 22, 30

      Political magazine: 31, 22, 17, 5, 23, 15, 10, 20, 20, 17

  **b.** The double box-and-whisker plot shows the means of 200 random samples of 20 sentences. Compare the variability of the sample means to the variability of the sample numbers of words in part (a).

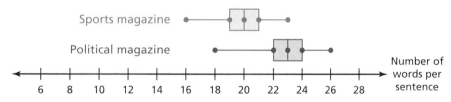

  **c.** Make a conclusion about the numbers of words per sentence in the magazines.

**9.** **Project** You want to compare the average amounts of time students in sixth, seventh, and eighth grade spend on homework each week.

  **a.** Design an experiment involving random sampling that can help you make a comparison.

  **b.** Perform the experiment. Can you make a conclusion about which students spend the most time on homework? Explain your reasoning.

 **Fair Game Review** What you learned in previous grades & lessons

**Graph the inequality on a number line.** *(Section 4.1)*

**10.** $x > 5$      **11.** $b \leq -3$      **12.** $n < -1.6$      **13.** $p \geq 2.5$

**14. MULTIPLE CHOICE** The number of students in the marching band increased from 100 to 125. What is the percent of increase? *(Section 6.5)*

    **A** 20%         **B** 25%         **C** 80%         **D** 500%

1. Which sample is better for making a prediction? Explain. *(Section 10.6)*

| Predict the number of students in your school who play at least one sport. | |
|---|---|
| Sample A | A random sample of 10 students from the school student roster |
| Sample B | A random sample of 80 students from the school student roster |

2. **GYMNASIUM** You want to estimate the number of students in your school who think the gymnasium should be remodeled. You survey 12 students on the basketball team. Determine whether the sample is *biased* or *unbiased*. Explain. *(Section 10.6)*

3. **TOWN COUNCIL** You want to know how the residents of your town feel about a recent town council decision. You survey 100 residents at random. Sixty-five support the decision, and thirty-five do not. So, you conclude that 65% of the residents of your town support the decision. Determine whether the conclusion is valid. Explain. *(Section 10.6)*

4. **FIELD TRIP** Of 60 randomly chosen students surveyed, 16 chose the aquarium as their favorite field trip. There are 720 students in the school. Predict the number of students in the school who would choose the aquarium as their favorite field trip. *(Section 10.6)*

5. **FOOTBALL** The double box-and-whisker plot shows the points scored per game by two football teams during the regular season. *(Section 10.7)*

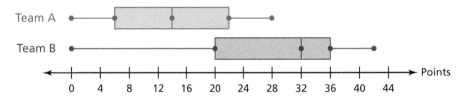

   a. Compare the populations using measures of center and variation.

   b. Express the difference in the measures of center as a multiple of the measure of variation.

6. **SUMMER CAMP** The dot plots show the ages of campers at two summer camps. *(Section 10.7)*

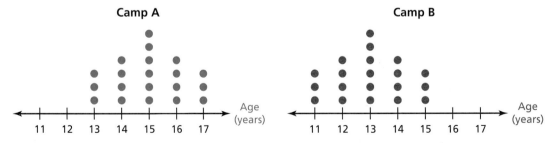

   a. Compare the populations using measures of center and variation.

   b. Express the difference in the measures of center as a multiple of the measure of variation.

# 10 Chapter Review

Check It Out
Vocabulary Help
BigIdeasMath ✓com

## Review Key Vocabulary

experiment, *p. 402*
outcomes, *p. 402*
event, *p. 402*
favorable outcomes, *p. 402*
probability, *p. 408*
relative frequency, *p. 412*

experimental probability,
  *p. 414*
theoretical probability, *p. 415*
sample space, *p. 422*
Fundamental Counting
  Principle, *p. 422*
compound event, *p. 424*

independent events, *p. 430*
dependent events, *p. 431*
simulation, *p. 436*
population, *p. 440*
sample, *p. 440*
unbiased sample, *p. 442*
biased sample, *p. 442*

## Review Examples and Exercises

 **Outcomes and Events** *(pp. 400–405)*

You randomly choose one toy race car.

**a.** In how many ways can choosing a green car occur?

**b.** In how many ways can choosing a car that is *not* green occur? What are the favorable outcomes of choosing a car that is *not* green?

**a.** There are 5 green cars. So, choosing a green car can occur in 5 ways.

**b.** There are 2 cars that are *not* green. So, choosing a car that is *not* green can occur in 2 ways.

| green | *not* green |
|---|---|
| green, green, green, green, green | blue, red |

∴ The favorable outcomes of the event are blue and red.

### Exercises

You spin the spinner. (a) Find the number of ways the event can occur. (b) Find the favorable outcomes of the event.

**1.** Spinning a 1

**2.** Spinning a 3

**3.** Spinning an odd number

**4.** Spinning an even number

**5.** Spinning a number greater than 0

**6.** Spinning a number less than 3

**10.2** **Probability** *(pp. 406–411)*

**You flip a coin. What is the probability of flipping tails?**

$$P(\text{event}) = \frac{\text{number of favorable outcomes}}{\text{number of possible outcomes}}$$

$$P(\text{tails}) = \frac{1}{2} \longleftarrow \boxed{\text{There is 1 tails.}}$$
$$\longleftarrow \boxed{\text{There is a total of 2 sides.}}$$

⋰ The probability of flipping tails is $\frac{1}{2}$, or 50%.

## Exercises

**7.** You roll a number cube. Find the probability of rolling an even number.

**10.3** **Experimental and Theoretical Probability** *(pp. 412–419)*

**a. The bar graph shows the results of spinning the spinner 70 times. What is the experimental probability of spinning a 2?**

The bar graph shows 12 twos. So, the spinner landed on two 12 times in a total of 70 spins.

$$P(\text{event}) = \frac{\text{number of times the event occurs}}{\text{total number of trials}}$$

Spinning a Spinner

$\boxed{\text{Two was landed on 12 times.}}$

$$P(2) = \frac{12}{70} = \frac{6}{35}$$

$\boxed{\text{There was a total of 70 spins.}}$

⋰ The experimental probability is $\frac{6}{35}$, or about 17%.

**b. The theoretical probability of choosing a purple grape from a bag is $\frac{2}{9}$. There are 8 purple grapes in the bag. How many grapes are in the bag?**

$$P(\text{purple}) = \frac{\text{number of purple grapes}}{\text{total number of grapes}}$$

$$\frac{2}{9} = \frac{8}{g} \qquad \text{Substitute. Let } g \text{ be the total number of grapes.}$$

$$g = 36 \qquad \text{Solve for } g.$$

⋰ So, there are 36 grapes in the bag.

## Exercises

**Use the bar graph on page 456 to find the experimental probability of the event.**

**8.** Spinning a 3

**9.** Spinning an odd number

**10.** *Not* spinning a 5

**11.** Spinning a number greater than 3

**Use the spinner to find the theoretical probability of the event.**

**12.** Spinning blue

**13.** Spinning a 1

**14.** Spinning an even number

**15.** Spinning a 4

**16.** The theoretical probability of spinning an even number on a spinner is $\frac{2}{3}$. The spinner has 8 even-numbered sections. How many sections are on the spinner?

---

**10.4** **Compound Events** *(pp. 420–427)*

**a.** **How many different home theater systems can you make from 6 DVD players, 8 TVs, and 3 brands of speakers?**

$$6 \times 8 \times 3 = 144 \qquad \text{Fundamental Counting Principle}$$

∴ So, you can make 144 different home theater systems.

**b.** **You flip two pennies. What is the probability of flipping two heads?**

Use a tree diagram to find the probability. Let H = heads and T = tails.

There is one favorable outcome in the sample space for flipping two heads: HH.

$$P(\text{event}) = \frac{\text{number of favorable outcomes}}{\text{number of possible outcomes}}$$

$$P(2 \text{ heads}) = \frac{1}{4} \qquad \text{Substitute.}$$

∴ The probability is $\frac{1}{4}$, or 25%.

## Exercises

**17.** You have 6 bracelets and 15 necklaces. Find the number of ways you can wear one bracelet and one necklace.

**18.** You flip two coins and roll a number cube. What is the probability of flipping two tails and rolling an even number?

### 10.5 Independent and Dependent Events  *(pp. 428–437)*

**You randomly choose one of the tiles and flip the coin. What is the probability of choosing a vowel and flipping heads?**

Choosing one of the tiles does not affect the outcome of flipping the coin. So, the events are independent.

$P(\text{vowel}) = \dfrac{2}{7}$ ← There are 2 vowels (A and E).
← There is a total of 7 tiles.

$P(\text{tails}) = \dfrac{1}{2}$ ← There is 1 tails side.
← There is a total of 2 sides.

Use the formula for the probability of independent events.

$$P(A \text{ and } B) = P(A) \cdot P(B)$$

$$= \dfrac{2}{7} \cdot \dfrac{1}{2} = \dfrac{1}{7}$$

∴ The probability of choosing a vowel and flipping heads is $\dfrac{1}{7}$, or about 14%.

### Exercises

**You randomly choose one of the tiles above and flip the coin. Find the probability of the compound event.**

**19.** Choosing a blue tile and flipping tails

**20.** Choosing the letter G and flipping tails

**You randomly choose one of the tiles above. Without replacing the first tile, you randomly choose a second tile. Find the probability of the compound event.**

**21.** Choosing a green tile and then a blue tile

**22.** Choosing a red tile and then a vowel

### 10.6 Samples and Populations  *(pp. 440–447)*

**You want to estimate the number of students in your school whose favorite subject is math. You survey every third student who leaves the school. Determine whether the sample is *biased* or *unbiased*.**

The sample is representative of the population, selected at random, and large enough to provide accurate data.

∴ So, the sample is unbiased.

**458**     Chapter 10     Probability and Statistics

23. You want to estimate the number of students in your school whose favorite subject is biology. You survey the first 10 students who arrive at biology club. Determine whether the sample is *biased* or *unbiased*. Explain.

**10.7** **Comparing Populations** *(pp. 448–453)*

The double box-and-whisker plot shows the test scores for two French classes taught by the same teacher.

a. **Compare the populations using measures of center and variation.**

Both distributions are skewed left, so use the median and the IQR.

⋮· The median for Class A, 92, is greater than the median for Class B, 88. The IQR for Class B, 12, is greater than the IQR for Class A, 8. The scores in Class A are generally greater and have less variability than the scores in Class B.

b. **Express the difference in the measures of center as a multiple of each measure of variation.**

$$\frac{\text{median for Class A} - \text{median for Class B}}{\text{IQR for Class A}} = \frac{4}{8} = 0.5$$

$$\frac{\text{median for Class A} - \text{median for Class B}}{\text{IQR for Class B}} = \frac{4}{12} = 0.3$$

⋮· So, the difference in the medians is about 0.3 to 0.5 times the IQR.

Exercises

24. **SPANISH TEST** The double box-and-whisker plot shows the test scores of two Spanish classes taught by the same teacher.

a. Compare the populations using measures of center and variation.

b. Express the difference in the measures of center as a multiple of each measure of variation.

**You randomly choose one game piece. (a) Find the number of ways the event can occur. (b) Find the favorable outcomes of the event.**

**1.** Choosing green

**2.** Choosing *not* yellow

**3.** Use the Fundamental Counting Principle to find the total number of different sunscreens possible.

| Sunscreen | |
|---|---|
| SPF | 10, 15, 30, 45, 50 |
| Type | Lotion, Spray, Gel |

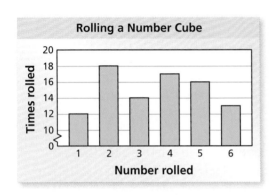

**Use the bar graph to find the experimental probability of the event.**

**4.** Rolling a 1 or a 2

**5.** Rolling an odd number

**6.** *Not* rolling a 5

**Use the spinner to find the theoretical probability of the event(s).**

**7.** Spinning an even number

**8.** Spinning a 1 and then a 2

**You randomly choose one chess piece. Without replacing the first piece, you randomly choose a second piece. Find the probability of choosing the first piece, then the second piece.**

**9.** Bishop and bishop   **10.** King and queen

**11. LUNCH** You want to estimate the number of students in your school who prefer to bring a lunch from home rather than buy one at school. You survey five students who are standing in the lunch line. Determine whether the sample is *biased* or *unbiased*. Explain.

**12. AGES** The double box-and-whisker plot shows the ages of the viewers of two television shows in a small town.

   **a.** Compare the populations using measures of center and variation.

   **b.** Express the difference in the measures of center as a multiple of each measure of variation.

1. A school athletic director asked each athletic team member to name his or her favorite professional sports team. The results are below:

- D.C. United: 3
- Florida Panthers: 8
- Jacksonville Jaguars: 26
- Jacksonville Sharks: 7
- Miami Dolphins: 22
- Miami Heat: 15
- Miami Marlins: 20
- Minnesota Lynx: 4
- New York Knicks: 5
- Orlando Magic: 18
- Tampa Bay Buccaneers: 17
- Tampa Bay Lightning: 12
- Tampa Bay Rays: 28
- Other: 6

One athletic team member is picked at random. What is the likelihood that this team member's favorite professional sports team is *not* located in Florida? *(7.SP.5)*

A. certain

B. likely, but not certain

C. unlikely, but not impossible

D. impossible

2. Each student in your class voted for his or her favorite day of the week. Their votes are shown below:

**Favorite Day of the Week**

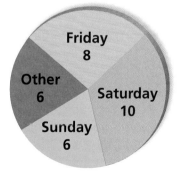

A student from your class is picked at random. What is the probability that this student's favorite day of the week is Sunday? *(7.SP.7b)*

**3.** How far, in millimeters, will the tip of the hour hand of the clock travel in 2 hours? (Use $\frac{22}{7}$ for $\pi$.)  *(7.G.4)*

84 mm

**F.** 44 mm

**G.** 88 mm

**H.** 264 mm

**I.** 528 mm

**4.** Nathaniel solved the proportion in the box below.

$$\frac{16}{40} = \frac{p}{27}$$

$$16 \cdot p = 40 \cdot 27$$

$$16p = 1080$$

$$\frac{16p}{16} = \frac{1080}{16}$$

$$p = 67.5$$

What should Nathaniel do to correct the error that he made?  *(7.RP.2c)*

**A.** Add 40 to 16 and 27 to $p$.

**B.** Subtract 16 from 40 and 27 from $p$.

**C.** Multiply 16 by 27 and $p$ by 40.

**D.** Divide 16 by 27 and $p$ by 40.

**5.** A North American hockey rink contains 5 face-off circles. Each of these circles has a radius of 15 feet. What is the total area, in square feet, of all the face-off circles? (Use 3.14 for $\pi$.)  *(7.G.4)*

**F.** 706.5 ft$^2$

**G.** 2826 ft$^2$

**H.** 3532.5 ft$^2$

**I.** 14,130 ft$^2$

**6.** A spinner is divided into eight congruent sections, as shown below.

You spin the spinner twice. What is the probability that the arrow will stop in a yellow section both times? *(7.SP.8a)*

**7.** What is the surface area, in square inches, of the square pyramid? *(7.G.6)*

8 in.

6 in.

**A.** 24 in.²

**B.** 96 in.²

**C.** 132 in.²

**D.** 228 in.²

**8.** The value of one of Kevin's baseball cards was $6.00 when he first got it. The value of this card is now $15.00. What is the percent increase in the value of the card? *(7.RP.3)*

**F.** 40%

**G.** 90%

**H.** 150%

**I.** 250%

**9.** You roll a number cube twice. You want to roll two even numbers. *(7.SP.8a)*

*Part A* Determine whether the events are independent or dependent.

*Part B* Find the number of favorable outcomes and the number of possible outcomes of each roll.

*Part C* Find the probability of rolling two even numbers. Explain your reasoning.

# Appendix A
# My Big Ideas Projects

# My Big Ideas Projects

# A.1 Literature Project

## The Mathematics of Jules Verne

### 1 Project Overview

Jules Verne (1828–1905) was a famous French science fiction writer. He wrote about space, air, and underwater travel before aircraft and submarines were commonplace, and before any means of space travel had been devised.

For example, in his 1865 novel *From the Earth to the Moon*, he wrote about three astronauts who were launched from Florida and recovered through a splash landing. The first actual moon landing wasn't until 1969.

**Essential Question** How does the knowledge of mathematics influence science fiction writing?

Read one of Jules Verne's science fiction novels. Then write a book report about some of the mathematics used in the novel.

**Sample:** A league is an old measure of distance. It is approximately equal to 4 kilometers. You can convert 20,000 leagues to miles as follows.

$$20{,}000 \text{ leagues} \cdot \frac{4 \text{ km}}{1 \text{ league}} \cdot \frac{1 \text{ mile}}{1.61 \text{ km}} \approx 50{,}000 \text{ miles}$$

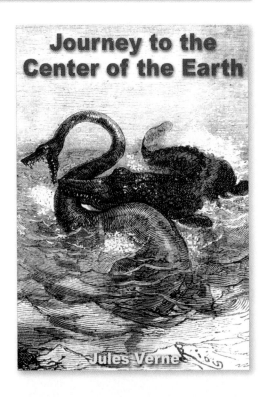

## 2 Things to Include

- Describe the major events in the plot.

- Write a brief paragraph describing the setting of the story.

- List and identify the main characters. Explain the contribution of each character to the story.

- Explain the major conflict in the story.

- Describe at least four examples of mathematics used in the story.

- Which of Jules Verne's scientific predictions have come true since he wrote the novel?

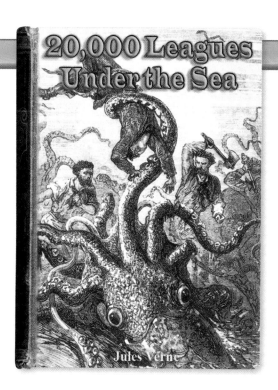

## 3 Things to Remember

- You can download one of Jules Verne's novels at *BigIdeasMath.com*.

- Add your own illustrations to your project.

- Organize your report in a folder, and think of a title for your report.

Jules Verne (1828–1905)

## Mathematics in Ancient Greece

**1** **Getting Started**

Corinthian
Helmet

The ancient Greek period began around 1100 B.C. and lasted until the Roman conquest of Greece in 146 B.C.

The civilization of the ancient Greeks influenced the languages, politics, educational systems, philosophy, science, mathematics, and arts of Western Civilization. It was a primary force in the birth of the Renaissance in Europe between the 14th and 17th centuries.

**Essential Question** How do you use mathematical knowledge that was originally discovered by the Greeks?

**Sample:** Ancient Greek symbols for the numbers from 1 through 10 are shown in the table.

| I | II | III | IIII | Γ | ΓI | ΓII | ΓIII | ΓIIII | △ |
|---|----|-----|------|---|----|-----|------|-------|---|
| 1 | 2 | 3 | 4 | 5 | 6 | 7 | 8 | 9 | 10 |

These same symbols were used to write the numbers between 11 and 39. Here are some examples.

$$\triangle\,\Gamma\,III = 18 \qquad \triangle\triangle\triangle\,\Gamma = 35 \qquad \triangle\triangle\,IIII = 24$$

Alexander
the Great

Parthenon

## 2 Things to Include

- Describe at least one contribution that each of the following people made to mathematics.

  Pythagoras (c. 570 B.C.–c. 490 B.C.)

  Aristotle (c. 384 B.C.–c. 322 B.C.)

  Euclid (c. 300 B.C.)

  Archimedes (c. 287 B.C.–c. 212 B.C.)

  Eratosthenes (c. 276 B.C.–c. 194 B.C.)

- Which of the people listed above was the teacher of Alexander the Great? What subjects did Alexander the Great study when he was in school?

- How did the ancient Greeks represent fractions?

- Describe how the ancient Greeks used mathematics. How does this compare with the ways in which mathematics is used today?

| | | | |
|---|---|---|---|
| A α | alpha | N ν | nu |
| B β | beta | Ξ ξ | xi |
| Γ γ | gamma | O o | omicron |
| Δ δ | delta | Π π | pi |
| E ε | epsilon | P ρ | rho |
| Z ζ | zeta | Σ σ | sigma |
| H η | eta | T τ | tau |
| Θ θ | theta | Y υ | upsilon |
| I ι | iota | Φ φ | phi |
| K κ | kappa | X χ | chi |
| Λ λ | lambda | Ψ ψ | psi |
| M μ | mu | Ω ω | omega |

## 3 Things to Remember

- Add your own illustrations to your project.

- Try to include as many different math concepts as possible. Your goal is to include at least one concept from each of the chapters you studied this year.

- Organize your report in a folder, and think of a title for your report.

Greek Pottery

Trireme Greek Warship

# A.3 Art Project

## Circle Art

### 1 Getting Started

Circles have been used in art for thousands of years.

**Essential Question** How have circles influenced ancient and modern art?

Find examples of art in which circles were used. Describe how the artist might have used properties of circles to make each piece of art.

**Sample:** Here is a technique for making a pattern that uses circles in a stained glass window.

1. Use a compass to draw a circle.

2. Without changing the radius, draw a second circle whose center lies on the first circle.

3. Without changing the radius, draw a third circle whose center is one of the points of intersection.

4. Without changing the radius, continue to draw new circles at the points of intersection.

5. Color the design to make a pattern.

Mexican Tile

Ancient Roman Mosaic

## 2  Things to Include

- Describe how ancient artists drew circles.

- Describe the symbolism of circles in ancient art.

- Find examples of how circles are used to create mosaic tile patterns.

- Measure the angles that are formed by the patterns in the circle art you find. For instance, you might describe the angles formed by the netting in the Native American dreamcatcher.

- Find examples of the use of circles in modern art.

- Use circles to create your own art. Describe how you used mathematics and the properties of circles to make your art.

Native American
Dreamcatcher

## 3  Things to Remember

- Add your own illustrations to your project.

- Try to include as many different math concepts as possible. Your goal is to include at least one concept from each of the chapters you studied this year.

- Organize your report in a folder, and think of a title for your report.

## Classifying Animals

### 1 Getting Started

Biologists classify animals by placing them in phylums, or groups, with similar characteristics. Latin names, such as Chordata (having a spinal cord) or Arthropoda (having jointed limbs and rigid bodies) are used to describe these groups.

Biological classification is difficult, and scientists are still developing a complete system. There are seven main ranks of life on Earth; kingdom, phylum, class, order, family, genus, and species. However, scientists usually use more than these seven ranks to classify organisms.

**Essential Question** How does the classification of living organisms help you understand the similarities and differences of animals?

Write a report about how animals are classified. Choose several different animals and list the phylum, class, and order of each animal.

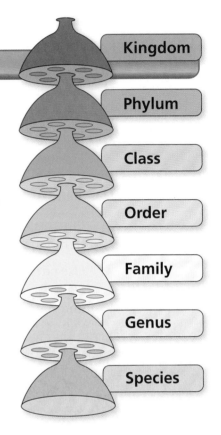

Kingdom
Phylum
Class
Order
Family
Genus
Species

**Wasp**
Phylum: Arthropoda
Class: Insecta
Order: Hymenoptera
(membranous wing)

**Sample:** A bat is classified as an animal in the phylum Chordata, class Mammalia, and order Chiroptera. *Chiroptera* is a Greek word meaning "hand-wing."

**Bat**
Phylum: Chordata
Class: Mammalia
Order: Chiroptera
(hand-wing)

**Monkey**
Phylum: Chordata
Class: Mammalia
Order: Primate (large brain)

**Kangaroo**
Phylum: Chordata
Class: Mammalia
Order: Diprotodontia
(two front teeth)

**2** **Things to Include**

- List the different classes of phylum Chordata. Have you seen a member of each class?

- List the different classes of phylum Arthropoda. Have you seen a member of each class?

- Show how you can use graphic organizers to help classify animals. Which types of graphic organizers seem to be most helpful? Explain your reasoning.

- Summarize the number of species in each phlyum in an organized way. Be sure to include fractions, decimals, and percents.

**Parrot**
Phylum: Chordata
Class: Aves
Order: Psittaciformes
    (strong, curved bill)

**Frog**
Phylum: Chordata
Class: Amphibia
Order: Anura (no tail)

**Spider**
Phylum: Arthropoda
Class: Arachnida
Order: Araneae (spider)

**3** **Things to Remember**

- Organize your report in a folder, and think of a title for your report.

**Lobster**
Phylum: Arthropoda
Class: Malacostraca
Order: Decopada (ten footed)

**Crocodile**
Phylum: Chordata
Class: Reptilia
Order: Crocodilia
    (pebble-worm)

**Cougar**
Phylum: Chordata
Class: Mammalia
Order: Carnivora (meat eater)

# Selected Answers

## Section 1.1

### Integers and Absolute Value
*(pages 6 and 7)*

**1.** 9, −1, 15

**3.** −6; All of the other expressions are equal to 6.

**5.** 6    **7.** 10    **9.** 13    **11.** 12    **13.** 8    **15.** 18

**17.** 45    **19.** 125    **21.** $|-4| < 7$    **23.** $|-4| > -6$    **25.** $|5| = |-5|$

**27.** Because $|-5| = 5$, the statement is incorrect; $|-5| > 4$

**29.** −8, 5    **31.** −7, −6, $|5|$, $|-6|$, 8    **33.** −17, $|-11|$, $|20|$, 21, $|-34|$

**35.** −4

**37. a.** MATE;     **b.** TEAM;

**39.** $n \geq 0$    **41.** The number closer to 0 is the greater integer.

**43. a.** Player 3    **b.** Player 2    **c.** Player 1

**45.** false; The absolute value of zero is zero, which is neither positive nor negative.

**47.** 144    **49.** 3170

## Section 1.2

### Adding Integers
*(pages 12 and 13)*

**1.** Change the sign of the integer.

**3.** positive; 20 has the greater absolute value and is positive.

**5.** negative; The common sign is a negative sign.

**7.** false; A positive integer and its absolute value are equal, not opposites.

**9.** −10    **11.** 7    **13.** 0    **15.** 10

**17.** −4    **19.** −11    **21.** −4    **23.** −34

**25.** −10 and −10 are not opposites; $-10 + (-10) = -20$    **27.** $48

**29.** Use the Associative Property to add 13 and −13 first; −8

**31.** *Sample answer:* Use the Commutative Property to switch the last two terms; −12

**33.** *Sample answer:* Use the Commutative Property to switch the last two terms; 11

**35.** −27    **37.** 21    **39.** −85

**41.** *Sample answer:* $-26 + 1$; $-12 + (-13)$    **43.** −3

**45.** $d = -10$    **47.** $m = -7$

**49.** Find the number in each row or column that already has two numbers in it before guessing.

**51.** 8    **53.** 183

**1.** You add the integer's opposite.

**3.** What is 3 less than $-2$?; $-5$; 5

**5.** C

**7.** B

**9.** 13

**11.** $-5$

**13.** $-10$

**15.** 3

**17.** 17

**19.** 1

**21.** $-22$

**23.** $-20$

**25.** $-3 - 9$

**27.** 6

**29.** 9

**31.** 8

**33.** $m = 14$

**35.** $c = 15$

**37.** 2

**39.** 3

**41.** *Sample answer:* $x = -2, y = -1$; $x = -3, y = -2$

**43.** sometimes; It's positive only if the first integer is greater.

**45.** always; It's always positive because the first integer is always greater.

**47.** all values of $a$ and $b$

**49.** when $a$ and $b$ have the same sign and $|a| \geq |b|$ or $b = 0$

**51.** $-45$

**53.** 468

**55.** 2378

**1.** **a.** They are the same.   **b.** They are different.

**3.** negative; different signs

**5.** negative; different signs

**7.** false; The product of the first two negative integers is positive. The product of the positive result and the third negative integer is negative.

**9.** $-21$

**11.** 12

**13.** 27

**15.** 12

**17.** 0

**19.** $-30$

**21.** 78

**23.** 121

**25.** $-240,000$

**27.** 54

**29.** $-700$

**31.** 0

**33.** $-1$

**35.** $-36$

**37.** 54

**39.** The answer should be negative; $-10^2 = -(10 \cdot 10) = -100$

**41.** 32

**43.** $-7500$, 37,500

**45.** $-12$

**47.** **a.**

| Month | Price of Skates | |
|---|---|---|
| June | 165 | $= \$165$ |
| July | $165 + \ (-12)$ | $= \$153$ |
| August | $165 + 2(-12)$ | $= \$141$ |
| September | $165 + 3(-12)$ | $= \$129$ |

**b.** The price drops $12 every month.

**c.** no; yes; In August you have $135 but the cost is $141. In September you have $153 and the cost is only $129.

**49.** 3

**51.** 14

**53.** D

## Section 1.5 — Dividing Integers
(pages 32 and 33)

1. They have the same sign; They have different signs; The dividend is zero.

3. *Sample answer:* $-4, 2$
5. negative
7. negative

9. $-3$
11. 6
13. 0
15. $-6$
17. 7
19. $-10$

21. undefined
23. 12

25. The quotient should be 0; $0 \div (-5) = 0$
27. 15 pages

29. $-8$
31. 65
33. 5

35. 4
37. $-400$ ft/min
39. 5

41. *Sample answer:* $-20, -15, -10, -5, 0$; Start with $-10$, then pair $-15$ with $-5$ and $-20$ with 0.

43.
45. B

## Section 2.1 — Rational Numbers
(pages 48 and 49)

1. no; The denominator cannot be 0.

3. rational numbers, integers
5. rational numbers, integers, whole numbers

7. repeating
9. terminating

11. 0.875
13. $-0.\overline{7}$
15. $1.8\overline{3}$
17. $-5.58\overline{3}$

19. The bar should be over both digits to the right of the decimal point; $-\dfrac{7}{11} = -0.\overline{63}$

21. $\dfrac{9}{20}$
23. $-\dfrac{39}{125}$
25. $-1\dfrac{16}{25}$
27. $-12\dfrac{81}{200}$

29. $-2.5, -1.1, -\dfrac{4}{5}, 0.8, \dfrac{9}{5}$
31. $-\dfrac{9}{4}, -0.75, -\dfrac{6}{10}, \dfrac{5}{3}, 2.1$

33. $-2.4, -2.25, -\dfrac{11}{5}, \dfrac{15}{10}, 1.6$
35. spotted turtle

37. $-1.82 < -1.81$
39. $-4\dfrac{6}{10} > -4.65$

41. $-2\dfrac{13}{16} < -2\dfrac{11}{14}$
43. Michelle

45. no; The base of the skating pool is at $-10$ feet, which is deeper than $-9\dfrac{5}{6}$ feet.

47. **a.** when $a$ is negative
    **b.** when $a$ and $b$ have the same sign, $a \neq 0 \neq b$

49. $\dfrac{7}{30}$
51. 21.15

## Section 2.2

### Adding Rational Numbers
*(pages 54 and 55)*

1. Because $|-8.46| > |5.31|$, subtract $|5.31|$ from $|-8.46|$ and the sign is negative.

3. What is the distance between $-4.5$ and $3.5$?; 8; $-1$

5. $-1\frac{4}{5}$ 　　　　 7. $-\frac{5}{14}$ 　　　　 9. $-\frac{7}{12}$ 　　　　 11. 1.844

13. The decimals are not lined up correctly; Line up the decimals; $-3.95$

15. $-1\frac{5}{12}$ 　　　 17. $1\frac{5}{12}$ 　　　 19. $\frac{1}{2}$ 　　　 21. $-9\frac{3}{4}$

23. The sum is an integer when the sum of the fractional parts of the numbers adds up to an integer.

25. less than; The water level for the three-month period compared to the normal level is $-1\frac{7}{16}$.

27. no; This is only true when $a$ and $b$ have the same sign.

29. Commutative Property of Addition; 7 　　　　 31. Associative Property of Addition; $1\frac{1}{8}$

33. A

## Section 2.3

### Subtracting Rational Numbers
*(pages 62 and 63)*

1. Instead of subtracting, add the opposite of $\frac{3}{5}$, $-\frac{3}{5}$. Then, add $\left|-\frac{4}{5}\right|$ and $\left|-\frac{3}{5}\right|$, and the sign is negative.

3. $1\frac{1}{2}$ 　　　 5. $-3.5$ 　　　 7. $-18\frac{13}{24}$ 　　　 9. $-2.6$

11. 14.963 　　　 13. $3\frac{1}{4}$ 　　　 15. $3\frac{1}{3}$

17. **a.** 410.7 feet 　　　 **b.** 136.9 feet per hour 　　　 19. 1.2

21. The difference is an integer when (1) the decimals have the same sign and the digits to the right of the decimal point are the same, or (2) the decimals have different signs and the sum of the decimal parts of the numbers add up to 1.

23. $-1\frac{7}{8}$ miles 　　　 25. Subtract the least number from the greatest number.

27. *Sample answer:* $x = -1.8$ and $y = -2.4$; $x = -5.5$ and $y = -6.1$

29. always; It's always positive because the first decimal is always greater.

31. 35.88 　　　 33. $8\frac{2}{3}$ 　　　 35. C

**Multiplying and Dividing Rational Numbers**
*(pages 68 and 69)*

**1.** The same rules for signs of integers are applied to rational numbers.

**3.** positive  **5.** negative  **7.** $-\dfrac{4}{5}$  **9.** 0.25

**11.** $-\dfrac{2}{3}$  **13.** $-\dfrac{1}{100}$  **15.** $2\dfrac{5}{14}$  **17.** $3.\overline{63}$

**19.** $-6$  **21.** $-2.5875$  **23.** $-\dfrac{4}{9}$  **25.** 9

**27.** 0.025  **29.** $8\dfrac{1}{4}$

**31.** The answer should be negative; $-2.2 \times 3.7 = -8.14$

**33.** $-66°$  **35.** $-19.59$  **37.** $-22.667$  **39.** $-5\dfrac{11}{24}$

**41.** *Sample answer:* $-\dfrac{9}{10}, \dfrac{2}{3}$  **43.** $3\dfrac{5}{8}$ gal  **45.** $-1.28$ sec

**47.** $-1.5$  **49.** $4\dfrac{1}{2}$  **51.** D

**Algebraic Expressions**
*(pages 84 and 85)*

**1.** Terms of an expression are separated by addition. Rewrite the expression as $3y + (-4) + (-5y)$. The terms in the expression are $3y$, $-4$, and $-5y$.

**3.** no; The like terms $3x$ and $2x$ should be combined.
$$3x + 2x - 4 = (3 + 2)x - 4$$
$$= 5x - 4$$

**5.** Terms: $t$, 8, $3t$; Like terms: $t$ and $3t$  **7.** Terms: $2n$, $-n$, $-4$, $7n$; Like terms: $2n$, $-n$, and $7n$

**9.** Terms: $1.4y$, 5, $-4.2$, $-5y^2$, $z$; Like terms: 5 and $-4.2$

**11.** $2x^2$ is not a like term because $x$ is squared. The like terms are $3x$ and $9x$.

**13.** $11x + 2$  **15.** $-2.3v - 5$  **17.** $3 - \dfrac{1}{2}y$  **19.** $-p - 30$

**21.** $10.2x$; The weight carried by each hiker is 10.2 pounds.

**23.** yes; Both expressions simplify to $11x^2 + 3y$.

**25.** $(9 + 3x)\,\text{ft}^2$

**27.** *Sample answer:*

$5x + 25$

**29.** When you subtract the two red strips, you subtract their intersection twice. So, you need to add it back into the expression once.

**31.** 0.52 m, 0.545 m, 0.55 m, 0.6 m, 0.65 m

## Adding and Subtracting Linear Expressions
*(pages 90 and 91)*

**1.** not linear; An exponent of a variable is not equal to 1.

**3.** not linear; An exponent of a variable is not equal to 1.

**5.** What is $x$ more than $3x - 1$?; $4x - 1$; $2x - 1$

**7.** *Sample answer:* $(2x + 7) - (2x - 4) = 11$

**9.** $2b + 9$

**11.** $6x - 18$

**13.** 17

**15.** $m + 1$

**17.** $55w + 145$

**19.** $-3g - 4$

**21.** $-12y + 20$

**23.** $-2c$

**25.** The $-3$ was not distributed to both terms inside the parentheses.

$$(4m + 9) - 3(2m - 5) = 4m + 9 - 6m + 15$$
$$= 4m - 6m + 9 + 15$$
$$= -2m + 24$$

**27.** no; If the variable terms are opposites, the sum is a numerical expression.

**29.** $0.25x + 0.15$

**31.** $\left| x - 3 \right|$, or equivalently $\left| -x + 3 \right|$; 0; 6

**33.** $\dfrac{2}{5}$

**35.** D

## Factoring Expressions
*(page 93)*

**1.** $3(3 + 7)$

**3.** $2(4x + 1)$

**5.** $4(5z - 2)$

**7.** $4(9a + 4b)$

**9.** $\dfrac{1}{3}(b - 1)$

**11.** $2.2(x + 2)$

**13.** $-\dfrac{1}{2}(x - 12)$

**15.** $(3x - 8)$ ft

**17.** *Sample answer:* $2x - 1$ and $x$, $2x$ and $x - 1$

## Solving Equations Using Addition or Subtraction *(pages 100 and 101)*

**1.** Subtraction Property of Equality

**3.** No, $m = -8$ not $-2$ in the first equation.

**5.** $a = 19$

**7.** $k = -20$

**9.** $c = 3.6$

**11.** $q = -\dfrac{1}{6}$

**13.** $g = -10$

**15.** $y = -2.08$

**17.** $q = -\dfrac{7}{18}$

**19.** $w = -1\dfrac{13}{24}$

**21.** The 8 should have been subtracted rather than added.

$$\begin{array}{r} x + 8 = 10 \\ \underline{-8 \quad -8} \\ x = \quad 2 \end{array}$$

**23.** $c + 10 = 3$; $c = -7$

**25.** $p - 6 = -14$; $p = -8$

**27.** $p + 2.54 = 1.38$; $-\$1.16$ million

**29.** $x + 8 = 12$; 4 cm

## Section 3.3    Solving Equations Using Addition or Subtraction *(continued) (pages 100 and 101)*

**31.** $x + 22.7 = 34.6$; 11.9 ft

**33.** Because your first jump is higher, your second jump went a farther distance than your first jump.

**35.** $m + 30.3 + 40.8 = 180$; 108.9°

**37.** $-9$      **39.** 6, $-6$      **41.** $-56$

**43.** $-9$      **45.** B

## Section 3.4    Solving Equations Using Multiplication or Division *(pages 106 and 107)*

**1.** Multiplication is the inverse operation of division, so it can undo division.

**3.** dividing by 5

**5.** multiplying by $-8$

**7.** $h = 5$      **9.** $n = -14$      **11.** $m = -2$      **13.** $x = -8$

**15.** $p = -8$      **17.** $n = 8$      **19.** $g = -16$      **21.** $f = 6\frac{3}{4}$

**23.** They should divide by $-4.2$.

$$-4.2x = 21$$
$$\frac{-4.2x}{-4.2} = \frac{21}{-4.2}$$
$$x = -5$$

**25.** $\frac{2}{5}x = \frac{3}{20}$; $x = \frac{3}{8}$

**27.** $\frac{x}{-1.5} = 21$; $x = -31.5$

**29.** $\frac{x}{30} = 12\frac{3}{5}$; 378 ft

**31 and 33.** Sample answers are given.

**31. a.** $-2x = 4.4$    **b.** $\frac{x}{1.1} = -2$

**33. a.** $4x = -5$    **b.** $\frac{x}{5} = -\frac{1}{4}$

**35.** $-1.26n = -10.08$; 8 days      **37.** $-50$ ft      **39.** $-5, 5$

**41.** $-7$      **43.** 12      **45.** B

## Section 3.5    Solving Two-Step Equations *(pages 112 and 113)*

**1.** Eliminate the constants on the side with the variable. Then solve for the variable using either division or multiplication.

**3.** D      **5.** A      **7.** $b = -3$      **9.** $t = -4$

**11.** $g = 4.22$      **13.** $p = 3\frac{1}{2}$      **15.** $h = -3.5$      **17.** $y = -6.4$

**19.** Each side should be divided by $-3$, not 3.

$$-3x + 2 = -7$$
$$-3x = -9$$
$$\frac{-3x}{-3} = \frac{-9}{-3}$$
$$x = 3$$

**21.** $a = 1\dfrac{1}{3}$

**23.** $b = 13\dfrac{1}{2}$

**25.** $v = -\dfrac{1}{30}$

**27.** $2.5 + 2.25x = 9.25$;  3 games

**29.** $v = -5$

**31.** $d = -12$

**33.** $m = -9$

**35.** *Sample answer:*  You travel halfway up a ladder. Then you climb down two feet and are 8 feet above the ground. How long is the ladder? $x = 20$

**37.** the initial fee

**39.** Find the number of insects remaining and then find the number of insects you caught.

**41.** decrease the length by 10 cm;  $2(25 + x) + 2(12) = 54$

**43.** $-6\dfrac{2}{3}$

**45.** 6.2

## Section 4.1    Writing and Graphing Inequalities
*(pages 128 and 129)*

**1.** A closed circle would be used because $-42$ is a solution.

**3.** no; $x < 5$ is all values of $x$ less than 5.  $5 < x$ is all values of $x$ greater than 5.

**5.** $x \le -4$; all values of $x$ less than or equal to $-4$

**7.** $w + 2.3 > 18$

**9.** $b - 4.2 < -7.5$

**11.** yes

**13.** no

**15.** yes

**17.**

**19.**

**21.** $p \ge 53$

**23.** yes

**25.** yes

**27.** **a.** any value that is greater than $-2$

**b.** any value that is less than or equal to $-2$; $b \le -2$

**c.** They represent the entire set of real numbers; yes

**29.** $p = 11$

**31.** $x = -7$

**1.** Yes, because of the Subtraction Property of Inequality.

**3.** $x \geq 11$;

**5.** $-4 \leq g$;

**7.** $-6 < y$;

**9.** $t \leq -2$;

**11.** $-\frac{3}{7} > b$;

**13.** $-2.8 < d$;

**15.** $\frac{3}{4} \geq m$;

**17.** $h \leq -2.4$;

**19.** The wrong side of the number line is shaded;

**21.** $7 + 7 + x < 28$; $x < 14$ ft

**23.** $8 + 8 + 10 + 10 + x \leq 51$; $x \leq 15$ m

**25.** $x - 3 \geq 5$; $x \geq 8$ ft

**27.** The three items must use less than 2400 watts total.

**29.** $x = 9$        **31.** $b = -22$        **33.** A

Hint

## Section 4.3

### Solving Inequalities Using Multiplication or Division *(pages 143–145)*

**1.** Multiply each side by 3.

**3.** *Sample answer: $-4x < 16$*

**5.** $x \geq -1$

**7.** $x \leq -35$

**9.** $x \leq \frac{3}{2}$

**11.** $c \leq -36$;

**13.** $x < -32$;

**15.** $k > 2$;

**17.** $y \leq -3$;

**19.** The inequality sign should not have been reversed.

$$\frac{x}{3} < -9$$
$$3 \cdot \frac{x}{3} < 3 \cdot (-9)$$
$$x < -27$$

**21.** $\frac{x}{7} < -3$; $x < -21$

**23.** $-2x > 30$; $x < -15$

**25. a.** $2.40x \leq 9.60$; $x \leq 4$ avocados

    **b.** no; You must buy a whole number of avocados.

**27.** $n \geq -3$;

**29.** $h \leq -24$;

**31.** $y > \dfrac{14}{3}$;

**33.** $m > -27$;

**35.** $b > 6$;

**37.** $-2.5x < -20$; $x > 8$ h

**39.** $10x \geq 120$; $x \geq 12$ cm

**41.** $\dfrac{x}{5} < 100$; $x < \$500$

**43.** *Answer should include, but is not limited to:* Use the correct number of months that the novel has been out.

**45.** $n \geq -12$ and $n \leq -5$;

**47.** $s < 14$;

**49.** $v = 45$

**51.** $m = 4$

## Section 4.4

# Solving Two-Step Inequalities
*(pages 150 and 151)*

**1.** *Sample answer:* They use the same techniques, but when solving an inequality, you must be careful to reverse the inequality symbol when you multiply or divide by a negative number.

**3.** C

**5.** $y < 1$;

**7.** $h > \dfrac{9}{2}$;

**9.** $b \leq -6$;

**11.** They did not perform the operations in the proper order.
$$\dfrac{x}{3} + 4 < 6$$
$$\dfrac{x}{3} < 2$$
$$x < 6$$

**13.** $w \leq 3$;

**15.** $d > -9$;

**17.** $c \geq -1.95$;

**19.** $x > 4$;

**21.** $-12x - 38 < -200$; $x > 13.5$ min

**23. a.** $9.5(70 + x) \geq 1000$; $x \geq 35\dfrac{5}{19}$, which means that at least 36 more tickets must be sold.

**b.** Because each ticket costs $1 more, fewer tickets will be needed for the theater to earn $1000.

**25.**

| Flutes | 7 | 21 | 28 |
|---|---|---|---|
| Clarinets | 4 | 12 | 16 |

$7 : 4$, $21 : 12$, and $28 : 16$

**27.** A

## Section 5.1

**Ratios and Rates**
*(pages 167–169)*

1. It has a denominator of 1.

3. *Sample answer:* A basketball player runs 10 feet down the court in 2 seconds.

5. $0.10 per fluid ounce

7. $72

9. 870 MB

11. $\dfrac{5}{9}$

13. $\dfrac{7}{3}$

15. $\dfrac{4}{3}$

17. 60 miles per hour

19. $2.40 per pound

21. 54 words per minute

23. 4.5 servings per package

25. 4.8 MB per minute

27. 280 square feet per hour

29. no; Although the relative number of boys and girls are the same, the two ratios are inverses.

31. $51

33. 2 cups of juice concentrate, 16 cups of water

35. **a.** rest: 72 beats per minute
   running: 150 beats per minute
   **b.** 234 beats

37. Try searching for "fire hydrant colors."

39. **a.** you; $\dfrac{1}{3}$ mile per hour faster

   **b.** $3\dfrac{1}{2}$ hours

   **c.** you; $1\dfrac{1}{6}$ miles

41. <

43. B

## Section 5.2

**Proportions**
*(pages 174 and 175)*

1. Both ratios are equal.

3. *Sample answer:* $\dfrac{6}{10}, \dfrac{12}{20}$

5. yes

7. no

9. yes

11. no

13. no

15. yes

17. no

19. yes

21. yes; Both can do 45 sit-ups per minute.

23. yes

25. yes

27. yes; The ratio of height to base for both triangles is $\dfrac{4}{5}$.

29. Organize the information by using a table.

31. no; The ratios are not equivalent; $\dfrac{13}{19} \neq \dfrac{14}{20} \neq \dfrac{15}{21}$ etc.

33. $-13$

35. $-18$

37. D

## Extension 5.2 — Graphing Proportional Relationships
*(pages 176 and 177)*

**1.** 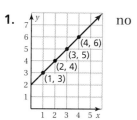 no

**3.** (0, 0): You earn $0 for working 0 hours.

(1, 15): You earn $15 for working 1 hour; unit rate: $\dfrac{\$15}{1\,h}$

(4, 60): You earn $60 for working 4 hours; unit rate: $\dfrac{\$60}{4\,h} = \dfrac{\$15}{1\,h}$

**5.** yes; 5 ft/h

**7.** $y = \dfrac{4}{3}$

## Section 5.3 — Writing Proportions
*(pages 182 and 183)*

**1.** You can use the columns or the rows of the table to write a proportion.

**3.** *Sample answer:* $\dfrac{x}{12} = \dfrac{5}{6}$; $x = 10$

**5.** $\dfrac{x}{50} = \dfrac{78}{100}$

**7.** $\dfrac{x}{150} = \dfrac{96}{100}$

**9.** $\dfrac{n \text{ winners}}{85 \text{ entries}} = \dfrac{34 \text{ winners}}{170 \text{ entries}}$

**11.** $\dfrac{100 \text{ meters}}{x \text{ seconds}} = \dfrac{200 \text{ meters}}{22.4 \text{ seconds}}$

**13.** $\dfrac{\$24}{3 \text{ shirts}} = \dfrac{c}{7 \text{ shirts}}$

**15.** $\dfrac{5 \text{ 7th grade swimmers}}{16 \text{ swimmers}} = \dfrac{s \text{ 7th grade swimmers}}{80 \text{ swimmers}}$

**17.** $y = 16$

**19.** $c = 24$

**21.** $g = 14$

**23.** $\dfrac{1}{200} = \dfrac{19.5}{x}$; Dimensions for the model are in the numerators and the corresponding dimensions for the actual space shuttle are in the denominators.

**25.** Draw a diagram of the given information.  Hint

**27.** $x = 9$

**29.** $x = 140$

## Section 5.4 — Solving Proportions
*(pages 190 and 191)*

**1.** mental math; Multiplication Property of Equality; Cross Products Property

**3.** yes; Both cross products give the equation $3x = 60$.

**5.** $h = 80$

**7.** $n = 15$

**9.** $y = 7\dfrac{1}{3}$

**11.** $k = 5.6$

**13.** $n = 10$

**15.** $d = 5.76$

**17.** $m = 20$

**19.** $d = 15$

**21.** $k = 5.4$

**23.** 108 pens

**25.** $x = 1.5$

**27.** $k = 4$

## Section 5.4 — Solving Proportions (continued)
(pages 190 and 191)

**29.** $\dfrac{2.5}{x} = \dfrac{1}{0.26}$; about 0.65

**31.** true; Both cross products give the equation $3a = 2b$.

**33.** 15.5 lb

**35.** no; The relationship is not proportional. It should take more people less time to build the swing set.

**37.** 4 bags

**39 and 41.**

**43.** D

## Section 5.5 — Slope
(pages 196 and 197)

**1.** yes; Slope is the rate of change of a line.

**3.** 5; A ramp with a slope of 5 increases 5 units vertically for every 1 unit horizontally. A ramp with a slope of $\dfrac{1}{5}$ increases 1 unit vertically for every 5 units horizontally.

**5.** $\dfrac{3}{2}$

**7.** 1

**9.** $\dfrac{4}{5}$

**11.**

slope = 32.5;
32.5 miles per gallon

**13.**

slope = 7

**15.**

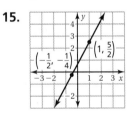

slope = $\dfrac{11}{6}$

**17. a.**

**b.** 2.5; Every millimeter represents 2.5 miles.

**c.** 120 mi

**d.** 90 mm

**19.** $y = 6$

**21.** $-\dfrac{3}{5}$

**23.** C

## Section 5.6 — Direct Variation
*(pages 202 and 203)*

**1.** $y = kx$, where $k$ is a number and $k \neq 0$.

**3.** Is the graph of the relationship a line?; yes; no

**5.**

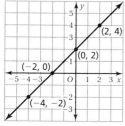

no; The line does not pass through the origin.

**7.** no; The line does not pass through the origin.

**9.** yes; The line passes through the origin; $k = \dfrac{2}{3}$

**11.** yes; The equation can be written as $y = kx$; $k = \dfrac{5}{2}$

**13.** no; The equation cannot be written as $y = kx$.

**15.** yes; The equation can be written as $y = kx$; $k = \dfrac{1}{2}$

**17.** no; The equation cannot be written as $y = kx$.

**19.**

yes; $y = 0.45x$

**21.** $k = \dfrac{5}{3}$; $y = \dfrac{5}{3}x$

**23.** $y = 2.54x$

**25.** When $x = 0$, $y = 0$. So, the graph of a proportional relationship always passes through the origin.

**27.** no

**29.** Every graph of direct variation is a line; however, not all lines show direct variation because the line must pass through the origin.

**31.** 0.5625

**33.** 0.96

## Section 6.1 — Percents and Decimals
*(pages 218 and 219)*

**1.** B

**3.** C

**5.** *Sample answer:* 0.11, 0.13, 0.19

**7.** 0.78

**9.** 0.185

**11.** 0.33

**13.** 0.4763

**15.** 1.66

**17.** 0.0006

**19.** 74%

**21.** 89%

**23.** 99%

**25.** 48.7%

**27.** 368%

**29.** 3.71%

**31.** The decimal point was moved in the wrong direction. $0.86 = 0.86 = 86\%$

**33.** 34%

**35.** $\dfrac{9}{25} = 0.36$

**37.** $\dfrac{203}{1250} = 0.1624$

**39.** 40%

**41. a.** $16.\overline{6}\%$ or $16\dfrac{2}{3}\%$   **b.** $\dfrac{5}{6}$

**43.** $\dfrac{31}{100}$

**45.** $4\dfrac{8}{25}$

**47.** $-1.6n - 1$

**49.** $-b - \dfrac{3}{2}$

**1.**

| Fraction | Decimal | Percent |
|---|---|---|
| $\frac{18}{25}$ | 0.72 | 72% |
| $\frac{17}{20}$ | 0.85 | 85% |
| $\frac{13}{50}$ | 0.26 | 26% |
| $\frac{31}{50}$ | 0.62 | 62% |
| $\frac{9}{20}$ | 0.45 | 45% |

**3.** $0.04$; $0.04 = 4\%$, but $40\%$, $\frac{2}{5}$, and $0.4$ are all equal to $40\%$.

**5.** 20%  **7.** $\frac{13}{25}$  **9.** 76%

**11.** 0.12  **13.** 140%  **15.** 80%

**17.**

**19.**

**21.**

**23.** Russia, Brazil, United States, India

**25.** $21\%, 0.2\overline{1}, \frac{11}{50}, \frac{2}{9}$

**27.** *D*

**29.** *C*

**31.** Write the numbers as percents or decimals to make the ordering easier.

**33.** yes  **35.** yes

---

**1.** The percent proportion is $\frac{a}{w} = \frac{p}{100}$ where $a$ is part of the whole $w$, and $p\%$, or $\frac{p}{100}$, is the percent.

**3.** $\frac{a}{40} = \frac{60}{100}$; $a = 24$  **5.** 19.2  **7.** 50  **9.** about 38.5%

**11.** $\frac{12}{25} = \frac{p}{100}$; $p = 48$  **13.** $\frac{9}{w} = \frac{25}{100}$; $w = 36$

**15.** $\frac{a}{124} = \frac{75}{100}$; $a = 93$  **17.** $\frac{a}{40} = \frac{0.4}{100}$; $a = 0.16$

**19.** 34 represents the part, not the whole.  **21.** $6000

$$\frac{a}{w} = \frac{p}{100}$$

$$\frac{34}{w} = \frac{40}{100}$$

$$w = 85$$

**23.** $\frac{14.2}{w} = \frac{35.5}{100}$; $w = 40$  **25.** $\frac{a}{\frac{7}{8}} = \frac{25}{100}$; $a = \frac{7}{32}$  **27.** $8.40

**29. a.** a scale along the vertical axis

   **b.** 6.25%; *Sample answer:* Although you do not know the actual number of votes, you can visualize each bar as a model with the horizontal lines breaking the data into equal parts. The sum of all the parts is 16. Greg has the least parts with 1, which is 100% ÷ 16 = 6.25%.

   **c.** 31 votes

**31. a.** 62.5%    **b.** $52x$       **33.** $-0.6$                **35.** B

---

## Section 6.4 — The Percent Equation
*(pages 236 and 237)*

**1.** A part of the whole is equal to a percent times the whole.

**3.** 55 is 20% of what number?;  275;  11

**5.** 37.5%                          **7.** 84                                **9.** 64

**11.** $45 = p \cdot 60$;  75%                    **13.** $0.008 \cdot 150$;  1.2

**15.** $12 = 0.005 \cdot w$;  2400                **17.** $102 = 1.2 \cdot w$;  85

**19.** 30 represents the part of the whole.

   $30 = 0.6 \cdot w$

   $50 = w$

**21.** $5400                       **23.** 26 years old            **25.** 56 signers

**27.** If the percent is less than 100%, the percent of a number is less than the number; 50% of 80 is 40; If the percent is equal to 100%, the percent of a number will equal the number; 100% of 80 is 80; If the percent is greater than 100%, the percent of a number is greater than the number; 150% of 80 is 120.

**29.** Remember when writing a proportion that either the units are the same on each side of the proportion, or the numerators have the same units and the denominators have the same units.

**31.** 92%                **33.** 0.88                **35.** 0.36

---

## Section 6.5 — Percents of Increase and Decrease
*(pages 244 and 245)*

**1.** If the original amount decreases, the percent of change is a percent of decrease. If the original amount increases, the percent of change is a percent of increase.

**3.** The new amount is now 0.                       **5.** 24 L

**7.** 17 penalties                **9.** decrease; 66.7%            **11.** increase; 225%

**13.** decrease; 12.5%            **15.** decrease; 37.5%            **17.** 12.5% decrease

**19. a.** about 16.7%

   **b.** 280 people; To get the same percent error, the amount of error needs to be the same. Because your estimate was 40 people below the actual attendance, an estimate of 40 people above the actual attendance will give the same percent error.

### Percents of Increase and Decrease (continued)
*(pages 244 and 245)*

**21.** decrease; 25%

**23.** decrease; 70%

**25. a.** about 16.95% increase

  **b.** 161,391 people

**27.** 15.6 ounces; 16.4 ounces

**29.** less than; *Sample answer:* Let $x$ represent the number. A 10% increase is equal to $x + 0.1x$, or $1.1x$. A 10% decrease of this new number is equal to $1.1x - 0.1(1.1x)$, or $0.99x$. Because $0.99x < x$, the result is less than the original number.

**31.** 10 girls

**33.** $39.2 = p \cdot 112$; 35%

**35.** $18 = 0.32 \cdot w$; 56.25

## Section 6.6

### Discounts and Markups
*(pages 250 and 251)*

**1.** *Sample answer:* Multiply the original price by $100\% - 25\% = 75\%$ to find the sale price.

**3. a.** 6% tax on a discounted price; The discounted price is less, so the tax is less.

  **b.** 30% markup on a $30 shirt; 30% of $30 is less than $30.

**5.** $35.70

**7.** $76.16

**9.** $53.33

**11.** $450

**13.** $172.40

**15.** 20%

**17.** $55

**19.** $175

**21.** "Multiply $45.85 by 0.1" and "Multiply $45.85 by 0.9, then subtract from $45.85." Both will give the sale price of $4.59. The first method is easier because it is only one step.

**23.** no; $31.08

**25.** $30

**27.** 180

**29.** C

## Section 6.7

### Simple Interest
*(pages 256 and 257)*

**1.** $I$ = simple interest, $P$ = principal, $r$ = annual interest rate (in decimal form), $t$ = time (in years)

**3.** You have to change 6% to a decimal and 8 months to a fraction of a year.

**5. a.** $300 **b.** $1800

**7. a.** $292.50 **b.** $2092.50

**9. a.** $308.20 **b.** $1983.20

**11. a.** $1722.24 **b.** $6922.24

**13.** 3%

**15.** 4%

**17.** 2 yr

**19.** 1.5 yr

**21.** $1440

**23.** 2 yr

**25.** $2720

**27.** $6700.80

**29.** $8500

**31.** 5.25%

**33.** 4 yr

**35.** 12.5 yr; Substitute $2000 for $P$ and $I$, 0.08 for $r$, and solve for $t$.

**37.** Year 1 = $520; Year 2 = $540.80; Year 3 = $562.43

**39.** $b \geq 1$;

**41.** A

## Section 7.1
### Adjacent and Vertical Angles
*(pages 274 and 275)*

1. two pairs; four pairs

3. $\angle ABC = 120°$, $\angle CBD = 60°$, $\angle DBE = 120°$, $\angle ABE = 60°$

5. *Sample answer:* adjacent: $\angle FGH$ and $\angle HGJ$, $\angle FGK$ and $\angle KGJ$;
   vertical: $\angle FGH$ and $\angle JGK$, $\angle FGK$ and $\angle JGH$

7. $\angle ACB$ and $\angle BCD$ are adjacent angles, not vertical angles.

9. vertical; 128          11. vertical; 25          13. adjacent; 20

15.           17.

19. **a.** *Sample answer:*          **b.** *Sample answer:*          **c.** *Sample answer:*

21. never          23. sometimes

25.           27. $n < -9$

29. $m < 4$

## Section 7.2
### Complementary and Supplementary Angles
*(pages 280 and 281)*

1. The sum of the measures of two complementary angles is 90°. The sum of the measures of two supplementary angles is 180°.

3. sometimes; Either $x$ or $y$ may be obtuse.

5. never; Because $x$ and $y$ must both be less than 90° and greater than 0°.

7. complementary          9. supplementary          11. neither

13. complementary; 55          15. $\angle 1 = 130°$, $\angle 2 = 50°$, $\angle 3 = 130°$

17. [figure: 20°]          19. [figure: 80°]

**21.** *Sample answer:* 1) Draw one angle, then draw the other using a side of the first angle;
2) Draw a right angle, then draw the shared side.

**23. a.** 25°    **b.** 65°       **25.** 54°

**27.** $x = 10$; $y = 20$       **29.** $n = -\dfrac{5}{12}$       **31.** B

### Section 7.3

## Triangles
*(pages 286 and 287)*

**1.** *Angles:* When a triangle has 3 acute angles, it is an acute triangle. When a triangle has 1 obtuse angle, it is an obtuse triangle. When a triangle has 1 right angle, it is a right triangle. When a triangle has 3 congruent angles, it is an equiangular triangle.

*Sides:* When a triangle has no congruent sides, it is a scalene triangle. When a triangle has 2 congruent sides, it is an isosceles triangle. When a triangle has 3 congruent sides, it is an equilateral triangle.

**3.** *Sample answer:*

**5.** *Sample answer:*

**7.** equilateral equiangular

**9.** right scalene

**11.** obtuse scalene

**13.** acute isosceles

**15.**

obtuse scalene triangle

**17.**

**19.**

**21.** no; The sum of the angle measures must be 180°.

**23.** many; You can change the angle formed by the two given sides to create many triangles.

**25.** no; The sum of any two side lengths must be greater than the remaining length.

**27. a.** green: 65; purple: 25; red: 45

   **b.** The angles opposite the congruent sides are congruent.

   **c.** An isosceles triangle has at least two angles that are congruent.

**29.** no; The equation cannot be written as $y = kx$.       **31.** B

## Extension 7.3   Angle Measures of Triangles
*(pages 288 and 289)*

**1.** 91; obtuse scalene triangle              **3.** 90; right scalene triangle

**5.** 48; acute isosceles triangle             **7.** yes

**9.** no; $28\dfrac{2}{3}$                            **11.** 67.5; acute isosceles triangle

**13.** 24; obtuse isosceles triangle            **15.** 35; obtuse scalene triangle

**17. a.** 72

   **b.** You can change the distance between the bottoms of the two upright cards; yes; $x$ must be greater than 60 and less than 90; If $x$ were less than or equal to 60, the two upright cards would have to be exactly on the edges of the base card or off the base card. It is not possible to stack cards at these angles. If $x$ were equal to 90, then the two upright cards would be vertical, which is not possible. The card structure would not be stable. In practice, the limits on $x$ are probably closer to $70 < x < 80$.

## Section 7.4   Quadrilaterals
*(pages 296 and 297)*

**1.** all of them

**3.** kite; It is the only type of quadrilateral listed that does not have opposite sides that are parallel and congruent.

**5.** trapezoid            **7.** kite                 **9.** rectangle

**11.** 110                 **13.** 58°

**15.**           **17.**

**19.** always            **21.** never           **23.** sometimes

**Quadrilaterals** *(continued)*
*(pages 296 and 297)*

**25.**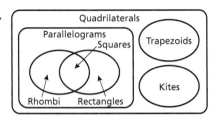

**27.** $\dfrac{1}{4}$

**29.** $\dfrac{6}{5}$

**Scale Drawings**
*(pages 303–305)*

**1.** A scale is the ratio that compares the measurements of the drawing or model with the actual measurements. A scale factor is a scale without any units.

**3.** Convert one of the lengths into the same units as the other length. Then, form the scale and simplify.

**5.** 10 ft by 10 ft

**7.** 112.5%

**9.** 50 mi

**11.** 110 mi

**13.** 15 in.

**15.** 21.6 yd

**17.** The 5 cm should be in the numerator.

$$\frac{1\ \text{cm}}{20\ \text{m}} = \frac{5\ \text{cm}}{x\ \text{m}}$$

$$x = 100\ \text{m}$$

**19.** 2.4 cm; 1 cm : 10 mm

**21.** **a.** *Answer should include, but is not limited to:* Make sure words and picture match the product.

**b.** Answers will vary.

**23.** **a.** 16 cm; 16 cm$^2$    **b.** 40 mm; 100 mm$^2$

**25.**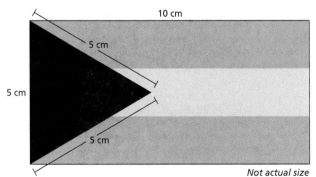
10 cm
5 cm
5 cm
5 cm
*Not actual size*

**27.** 15 ft$^2$

**29.** 3 ft$^2$

**31.** Find the size of the object that would represent the model of the Sun.

**33 and 35.**

## Section 8.1

### Circles and Circumference
*(pages 321–323)*

1. The radius is one-half the diameter.

3. 2.5 cm

5. $1\frac{3}{4}$ in.

7. 4 in.

9. about 31.4 in.

11. about 56.52 in.

13. **a.** about 25 m; about 50 m

    **b.** about 2 times greater

15. about 7.71 ft

17. about 31.4 cm; about 62.8 cm

19. about 69.08 m; about 138.16 m

21. about 200.96 cm

23. Draw a diagram of the given information.

25. **a.** about 254.34 mm; First find the length of the minute hand. Then find $\frac{3}{4}$ of the circumference of a circle whose radius is the length of the minute hand.

    **b.** about 320.28 mm; Subtract $\frac{1}{12}$ of the circumference of a circle whose radius is the length of the hour hand from the circumference of a circle whose radius is the length of the minute hand.

27. 20 m

29. D

## Section 8.2

### Perimeters of Composite Figures
*(pages 328 and 329)*

1. no; The perimeter of the composite figure does not include the measure of the shared side.

3. 19.5 units

5. 25.5 units

7. 19 units

9. 56 m

11. 30 cm

13. about 26.85 in.

15. about 36.84 ft

17. First find the total distance you run.

19. *Sample answer:* By adding the triangle shown by the dashed line to the L-shaped figure, you *reduce* the perimeter.

21. 279.68

23. 205

1. Divide the diameter by 2 to get the radius. Then use the formula $A = \pi r^2$ to find the area.

3. about 254.34 mm$^2$

5. about 314 in.$^2$

7. about 3.14 cm$^2$

9. about 2461.76 mm$^2$

11. about 113.04 in.$^2$

13. about 628 cm$^2$

15. about 1.57 ft$^2$

17. What fraction of the circle is the dog's running area?

19. about 9.8125 in.$^2$; The two regions are identical, so find one-half the area of the circle.

21. about 4.56 ft$^2$; Find the area of the shaded regions by subtracting the areas of both unshaded regions from the area of the quarter-circle containing them. The area of each unshaded region can be found by subtracting the area of the smaller shaded region from the semicircle. The area of the smaller shaded region can be found by drawing a square about the region.

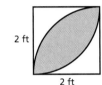

2 ft

2 ft

Subtract the area of a quarter-circle from the area of the square to find an unshaded area. Then subtract both unshaded areas from the square's area to find the shaded region's area.

23. 53

25. A

1. *Sample answer:* You could add the areas of an 8-inch × 4-inch rectangle and a triangle with a base of 6 inches and a height of 6 inches. Also you could add the area of a 2-inch × 4-inch rectangle to the area of a trapezoid with a height of 6 inches, and base lengths of 4 inches and 10 inches.

3. 28.5 units$^2$

5. 25 units$^2$

7. 25 units$^2$

9. 132 cm$^2$

11. *Answer will include but is not limited to:* Tracings of a hand and foot on grid paper, estimates of the areas, and a statement of which is greater.

13. 23.5 in.$^2$

15. 24 m$^2$

17. Each envelope can be broken up into 5 smaller figures to find the area.

19. $y \div 6$

21. $7w$

## Section 9.1  Surface Areas of Prisms
*(pages 359–361)*

1. *Sample answer:* 1) Use a net. 2) Use the formula $S = 2\ell w + 2\ell h + 2wh$.

3. Find the area of the bases of the prism; 24 in.$^2$; 122 in.$^2$

5.

   38 in.$^2$

7. 324 m$^2$

9. 49.2 yd$^2$

11. 136 m$^2$

13. 294 yd$^2$

15. $2\frac{2}{3}$ ft$^2$

17. 177 in.$^2$

19. yes; Because you do not need to frost the bottom of the cake, you only need 249 square inches of frosting.

21. 68 m$^2$

23. $x = 4$ in.

25. The dimensions of the red prism are three times the dimensions of the blue prism. The surface area of the red prism is 9 times greater than the surface area of the blue prism.

27. **a.** 0.125 pint  **b.** 1.125 pints

   **c.** red and green: The ratio of the paint amounts (red to green) is 4 : 1 and the ratio of the side lengths is 2 : 1.

     green and blue: The ratio of the paint amounts (blue to green) is 9 : 1 and the ratio of the side lengths is 3 : 1.

     The ratio of the paint amounts is the square of the ratio of the side lengths.

29. 160 ft$^2$

31. 28 ft$^2$

## Section 9.2  Surface Areas of Pyramids
*(pages 366 and 367)*

1. no; The lateral faces of a pyramid are triangles.

3. triangular pyramid; The other three are names for the pyramid.

5. 178.3 mm$^2$

7. 144 ft$^2$

9. 170.1 yd$^2$

11. 1240.4 mm$^2$

13. 6 m

Hint

15. 283.5 cm$^2$

17. Determine how long the fabric needs to be so you can cut the fabric most efficiently.

19. 124 cm$^2$

21. $A \approx 452.16$ units$^2$; $C \approx 75.36$ units

23. $A \approx 572.265$ units$^2$; $C \approx 84.78$ units

1. $2\pi rh$

3.

   $30\pi \approx 94.2 \text{ ft}^2$

5.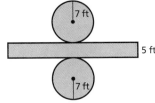

   $168\pi \approx 527.5 \text{ ft}^2$

7. $156\pi \approx 489.8 \text{ ft}^2$

9. $120\pi \approx 376.8 \text{ ft}^2$

11. $28\pi \approx 87.9 \text{ m}^2$

13. $432\pi \approx 1356.48 \text{ ft}^2$

Hmmm.

15. The surface area of the cylinder with the height of 8.5 inches is greater than the surface area of the cylinder with the height of 11 inches.

17. After removing the wedge, is there any new surface area added?

19. $10 \text{ ft}^2$

21. $47.5 \text{ in.}^2$

1. cubic units

3. The volume of an object is the amount of space it occupies. The surface area of an object is the sum of the areas of all its faces.

5. $288 \text{ cm}^3$

7. $210 \text{ yd}^3$

9. $420 \text{ mm}^3$

11. $645 \text{ mm}^3$

13. The area of the base is wrong.

$$V = \frac{1}{2}(7)(5) \cdot 10$$
$$= 175 \text{ cm}^3$$

15. $225 \text{ in.}^3$

17. $7200 \text{ ft}^3$

19. $1728 \text{ in.}^3$

$1 \times 1 \times 1 = 1 \text{ ft}^3$    $12 \times 12 \times 12 = 1728 \text{ in.}^3$

21. 20 cm

Hint

23. You can write the volume in cubic inches and use prime factorization to find the dimensions.

25. $90

27. $240.50

# Volumes of Pyramids
*(pages 386 and 387)*

1. The volume of a pyramid is $\frac{1}{3}$ times the area of the base times the height. The volume of a prism is the area of the base times the height.

3. 3 times

5. 20 mm$^3$

7. 80 in.$^3$

9. 252 mm$^3$

11. 700 mm$^3$

13. 156 ft$^3$

15. 340.4 in.$^3$

17. 12,000 in.$^3$; The volume of one paperweight is 12 cubic inches. So, 12 cubic inches of glass is needed to make one paperweight. So, it takes $12 \times 1000 = 12,000$ cubic inches to make 1000 paperweights.

19. *Sample answer:* 5 ft by 4 ft

21. 153°; 63°

23. 60°; none

# Cross Sections of Three-Dimensional Figures
*(pages 388 and 389)*

1. triangle

3. rectangle

5. triangle

7. The intersection is the shape of the base.

9. circle

11. circle

13. rectangle

15. The intersection occurs at the vertex of the cone.

# Outcomes and Events
*(pages 404 and 405)*

1. event; It is a collection of several outcomes.

3. 8

5. 1, 2, 3, 4, 5, 6, 7, 8, 9

7. 1, 3, 5, 7, 9

9. 1, 3

11. 3, 6, 9

13. **a.** 1 way   **b.** green

15. **a.** 1 way   **b.** yellow

17. **a.** 7 ways

   **b.** red, red, red, purple, purple, green, yellow

19. 7 ways

21. true

23. true

25. 30 rock CDs

27. $x = 2$

29. $w = 12$

31. C

# Probability
*(pages 410 and 411)*

1. The probability of an event is the ratio of the number of favorable outcomes to the number of possible outcomes.

3. *Sample answer:* You will not have any homework this week.; You will fall asleep tonight.

5. either; Both spinners have the same number of chances to land on "Forward."

7. impossible

9. unlikely

11. $\dfrac{1}{10}$

13. $\dfrac{9}{10}$

15. 0

17. 20

19. **a.** $\dfrac{2}{3}$; likely   **b.** $\dfrac{1}{3}$; unlikely

    **c.** $\dfrac{1}{2}$; equally likely to happen or not happen

21. There are 2 combinations for each.

23. $x < 4$;

25. $w > -3$;

27. C

# Experimental and Theoretical Probability
*(pages 417–419)*

1. Perform an experiment several times. Count how often the event occurs and divide by the number of trials.

3. There is a 50% chance you will get a favorable outcome.

5. experimental probability; The population is too large to survey every person, so a sample will be used to predict the outcome.

7. $\dfrac{12}{25}$, or 48%

9. $\dfrac{7}{25}$, or 28%

11. 0, or 0%

13. 45 tiles

15. $\dfrac{1}{3}$, or about 33.3%

17. $\dfrac{1}{2}$, or 50%

19. 1, or 100%

21. $\dfrac{25}{26}$, or about 96.2%

23. 36 songs

25. theoretical: $\dfrac{1}{5}$, or 20%;

    experimental: $\dfrac{37}{200}$, or 18.5%

    The experimental probability is close to the theoretical probability.

27. theoretical: $\dfrac{1}{5}$, or 20%;

    experimental: $\dfrac{1}{5}$, or 20%

    The probabilities are equal.

29. **a.** $\dfrac{1}{12}$; 50 times   **b.** $\dfrac{11}{50}$; 132 times

    **c.** A larger number of trials should result in a more accurate probability, which gives a more accurate prediction.

**31.** Make a list of all possible ways to get each sum.

**35.** 4%

**37.** D

**33. a.** As a number of trials increases, the most likely sum will change from 6 to 7.

**b.** As an experiment is repeated over and over, the experimental probability of an event approaches the theoretical probability of the event.

## Section 10.4 Compound Events
*(pages 425–427)*

**1.** A sample space is the set of all possible outcomes of an event. Use a table or tree diagram to list all the possible outcomes.

**3.** You could use a tree diagram or the Fundamental Counting Principle. Either way, the total number of possible outcomes is 30.

**5.** 125,000

**7.** Sample space: Realistic Lion, Realistic Bear, Realistic Hawk, Realistic Dragon, Cartoon Lion, Cartoon Bear, Cartoon Hawk, Cartoon Dragon; 8 possible outcomes

**9.** 20

**11.** 60

**13.** The possible outcomes of each question should be multiplied, not added. The correct answer is $2 \times 2 \times 2 \times 2 \times 2 = 32$.

**15.** $\frac{1}{10}$, or 10%

**17.** $\frac{1}{5}$, or 20%

**19.** $\frac{2}{5}$, or 40%

**21.** $\frac{1}{18}$, or $5\frac{5}{9}$%

**23.** $\frac{1}{9}$, or $11\frac{1}{9}$%

**25. a.** $\frac{1}{9}$, or about 11.1%

**b.** It increases the probability that your guesses are correct to $\frac{1}{4}$, or 25%, because you are only choosing between 2 choices for each question.

**27. a.** $\frac{1}{1000}$, or 0.1%

**b.** There are 1000 possible combinations. With 5 tries, someone would guess 5 out of the 1000 possibilities. So, the probability of getting the correct combination is $\frac{5}{1000}$, or 0.5%.

**29. a.** The Fundamental Counting Principle is more efficient. A tree diagram would be too large.

**b.** 1,000,000,000 or one billion

**c.** *Sample answer:* Not all possible number combinations are used for Social Security Numbers (SSN). SSNs are coded into geographical, group, and serial numbers. Some SSNs are reserved for commercial use and some are forbidden for various reasons.

**31.** *Sample answer:* adjacent: $\angle XWY$ and $\angle ZWY$, $\angle XWY$ and $\angle XWV$; vertical: $\angle VWX$ and $\angle YWZ$, $\angle YWX$ and $\angle VWZ$

**33.** B

**1.** What is the probability of choosing a 1 and then a blue chip?; $\dfrac{1}{15}$; $\dfrac{1}{10}$

**3.** independent; The outcome of the first roll does not affect the outcome of the second roll.

**5.** $\dfrac{1}{8}$

**7.** $\dfrac{3}{8}$

**9.** $\dfrac{1}{42}$

**11.** $\dfrac{2}{21}$

**13.** The two events are dependent, so the probability of the second event is $\dfrac{1}{3}$.

$$P(\text{red and green}) = \dfrac{1}{4} \cdot \dfrac{1}{3} = \dfrac{1}{12}$$

**15.** $\dfrac{1}{6}$, or about 16.7%

**17.** $\dfrac{2}{35}$

**19.** $\dfrac{5}{162}$, or about 3.1%

**21.** $\dfrac{4}{81}$, or about 4.9%

**23.** $\dfrac{3}{4}$

**25. a.** Because the probability that both you and your best friend are chosen is $\dfrac{1}{132}$, you and your best friend are not in the same group. The probability that you both are chosen would be 0 because only one leader is chosen from each group.

    **b.** $\dfrac{1}{11}$   **c.** 23

**27.**

right scalene

**29.**

acute isosceles

**1. a.** *Sample answer:* Roll four number cubes. Let an odd number represent a correct answer and an even number represent an incorrect answer. Run 40 trials.

    **b.** Check students' work. The probability should be "close" to 6.25% (depending on the number of trials, because that is the theoretical probability).

**3.** *Sample answer:* Using the spreadsheet in Example 3 and using digits 1–4 as successes, the experimental probability is 16%.

1. Samples are easier to obtain.

3. Population: Residents of New Jersey
   Sample: Residents of Ocean County

5. biased; The sample is not selected at random and is not representative of the population because students in a band class play a musical instrument.

7. biased; The sample is not representative of the population because people who go to a park are more likely to think that the park needs to be remodeled.

9. no; the sample is not representative of the population because people going to the baseball stadium are more likely to support building a new baseball stadium. So, the sample is biased and the conclusion is not valid.

11. Sample A; It is representative of the population.

13. a sample; It is much easier to collect sample data in this situation.

15. a sample; It is much easier to collect sample data in this situation.

17. Not everyone has an email address, so the sample may not be representative of the entire population. *Sample answer:* When the survey question is about technology or which email service you use, the sample may be representative of the entire population.

19. Use the survey results to find the number of students in the school who plan to attend college.

21. 140

23. 3

1. When comparing two populations, use the mean and the MAD when each distribution is symmetric. Use the median and the IQR when either one or both distributions are skewed.

3. **a.** garter snake: mean = 25, median = 24.5, mode = 24, range = 20, IQR = 7.5, MAD ≈ 4.33
   water snake: mean = 31.5, median = 32, mode = 32, range = 20, IQR = 10, MAD ≈ 5.08

   **b.** The water snakes have greater measures of center because the mean, median, and mode are greater. The water snakes also have greater measures of variation because the interquartile range and mean absolute deviation are greater.

**5. a.** Class A: median = 90, IQR = 12.5
Class B: median = 80, IQR = 10

The variation in the test scores is about the same,
but Class A has greater test scores.

**b.** The difference in the medians is 0.8 to 1 times the IQR.

**7.** Arrange the dot plots in Exercise 5 vertically and construct a double
box-and-whisker plot in Exercise 6 to help you visualize the distributions.

**9. a.** Check students' work. Experiments should include taking many samples
of a manageable size from each grade level. This will be more doable if the
work of sampling is divided among the whole class, and the results are
pooled together.

**b.** Check students' work. The data may or may not support a conclusion.

**11.**

**13.**

# Key Vocabulary Index

Mathematical terms are best understood when you see them used and defined *in context*. This index lists where you will find key vocabulary. A full glossary is available in your Record and Practice Journal and at *BigIdeasMath.com*.

# Student Index

This student-friendly index will help you find vocabulary, key ideas, and concepts. It is easily accessible and designed to be a reference for you whether you are looking for a definition, real-life application, or help with avoiding common errors.

## A

**Absolute value,** 2–7
  defined, 4
  error analysis, 6
  real-life application, 5
Addition
  of expressions
    linear, 86–91
    modeling, 86–87
  of integers, 8–15
    with different signs, 8–11
    error analysis, 12
    with the same sign, 8–11
  Property
    of Equality, 98
    of Inequality, 132
    real-life application, 99
  of rational numbers, 50–55
    error analysis, 54
    real-life application, 53
    writing, 54, 55, 65
  to solve inequalities, 130–135
Addition Property of Equality, 98
  real-life application, 99
Addition Property of Inequality, 132
**Additive inverse**
  defined, 10
  writing, 12
Additive Inverse Property, 10
**Adjacent angles**
  constructions, 270–275
  defined, 272
Algebra
  equations
    equivalent, 98
    modeling, 96–97, 102, 108–109
    solving, 96–113
    two-step, 108–113
    writing, 100, 106, 112
  expressions
    linear, 86–91
    modeling, 86–87, 91
    simplifying, 80–85
    writing, 84, 90
  formulas, *See* Formulas
Algebra tiles
  equations, 96–97, 102, 108–109
  expressions, 86–87

Angle(s)
  adjacent
    constructions, 270–275
    defined, 272
  classifying, 277–278
    triangles by, 284
  complementary
    constructions, 276–281
    defined, 278
  congruent
    defined, 272
    reading, 284
  measures
    of a quadrilateral, 295
    of a triangle, 288–289
  naming, 272
    error analysis, 274
  sums
    for a quadrilateral, 295
    for a triangle, 288–289
  supplementary
    constructions, 276–281
    defined, 278
  vertical
    constructions, 270–275
    defined, 272
Area, *See also* Surface area
  of a circle, 332–337
    formula, 334
  of a composite figure, 338–343

## B

**Biased sample(s),** defined, 442

## C

**Center,** defined, 318
Choose Tools, *Throughout. For example, see:*
  angles, 271
  circles and circumference, 321
  probability, 406, 426
**Circle(s)**
  area of, 332–337
    formula, 334
  center of, 318
  circumference and, 316–323
    defined, 319
    formula, 319
    research, 323
  defined, 318

  diameter of, 318
  pi, 316, 319
  semicircle, 320
**Circumference**
  and circles, 316–323
  defined, 319
  formula, 319
  research, 323
Common Error
  inequalities, 141
  percent equations, 234
**Complementary angles**
  constructing, 276–281
  defined, 278
**Complex fraction,** defined, 165
**Composite figure(s)**
  area of, 338–343
  defined, 326
  perimeters of, 324–329
    error analysis, 328
**Compound event(s),** *See also* Events, Probability
  defined, 424
  error analysis, 426
  writing, 425
**Congruent angles**
  defined, 272
  reading, 284
**Congruent sides**
  defined, 284
  reading, 284
Connections to math strands
  Algebra, 13, 19, 27, 32
  Geometry, 91, 101, 113, 175, 245, 366
**Constant of proportionality,**
  defined, 200
Constructions
  angles
    adjacent, 270–275
    vertical, 270–275
  quadrilaterals, 292–297
  triangles, 282–289
    error analysis, 286
Critical Thinking, *Throughout. For example, see:*
  circles, area of, 337
  composite figures
    area of, 343
    perimeter of, 329
  cylinders, 372
  discounts, 251

Rectangle(s)
    area of, formula, 341
    defined, 294
**Regular pyramid,** defined, 364
**Relative frequency,** *See also*
        Probability
    defined, 412
    formula, 412
Repeated Reasoning, *Throughout.*
        *For example, see:*
    adding rational numbers, 55
    circles
        area, 337
        circumference, 337
    cylinders, 373
    probability, 427
    slope, 193
    solving equations, 109
    surface area, 373
**Repeating decimal(s)**
    defined, 46
    error analysis, 48
    writing, 48
Rhombus, defined, 294

Ⓢ

**Sample(s),** 440–447
    biased, 442
    defined, 440
    research, 441
    unbiased, 442
**Sample space,** defined, 422
**Scale**
    defined, 300
    research, 299
**Scale drawing,** 298–305
    defined, 300
    error analysis, 303
    research, 299
    scale, 300
    scale factor, 301
**Scale factor**
    defined, 301
    finding, 301
**Scale model**
    defined, 300
    finding distance in, 301
    modeling, 305
**Semicircle(s)**
    area of, 341
    defined, 320
Side(s)
    classifying triangles by, 284
    congruent, 284
        defined, 284
        reading, 284

**Simple interest,** 252–257
    defined, 254
    error analysis, 256
    writing, 256
**Simplest form,** defined, 82
**Simulation(s),** 436–437
    defined, 436
**Slant height,** defined, 364
**Slope,** 192–197
    defined, 194
    error analysis, 196
Solids, *See also:* Surface area,
        Three-dimensional figures,
        Volume
    cross section, 388–389
    cube, 358
    cylinder, 368–373
    prism, 354–361, 376–381
    pyramid, 362–367, 382–387
**Solution of an inequality,** defined,
        126
**Solution set,** defined, 126
Square, defined, 294
Statistics, *See also* Probability
    populations, 440–453
        comparing, 448–453
        defined, 440
        project, 453
        research, 441
        writing, 452
    samples, 440–447
        biased, 442
        defined, 440
        research, 441
        unbiased, 442
Structure, *Throughout. For*
        *example, see:*
    angles, 275
    composite figures, 343
    direct variation, 199
    expressions
        factoring, 93
        linear, 91
        simplifying, 85
    integers
        adding, 9
        dividing, 29
        multiplying, 23
        subtracting, 15
    parallelograms, 297
    percent proportions, 231
    probability, 435
    solving equations, 97
    subtracting rational numbers, 63
    surface area of a prism, 361
    volume
        of a prism, 377
        of a pyramid, 383

Study Tip
    additive inverse, 11
    amount of error, 243
    circumference, 319
    comparing populations, 450
    constructing triangles, 285
    cross products, 173
    decimal points, 216
    direct variation, 200
    discounts, 248
    exponents, 25
    expressions
        linear, 89
        simplifying, 82
    inequalities
        graphing, 127
        solving, 132
    opposites, 11
    ordering decimals and percents,
        222
    percent proportions, 228
    powers, 25
    prisms, 378
    probability, 408
        Fundamental Counting
            Principle, 422
        simulating outcomes, 436,
            437
    proportional relationships, 177
    pyramids, 384
        volume of, 384
    quadrilaterals, 294
    radius, 319
    rational numbers, 47, 52
    ratios, 302
    scale, 300
    scale drawings, 302
    simplifying equations, 110
    slope, 194
    squares, 294
    volume
        of a prism, 378
        of a pyramid, 384
Subtraction
    of expressions
        error analysis, 91
        linear, 86–91
        modeling, 87
        real-life application, 89
    of integers, 14–19
        error analysis, 18
        real-life application, 17
        writing, 18
    Property
        of Equality, 98
        of Inequality, 132

# Common Core State Standards

## Domain: Ratios and Proportional Relationships

**Analyze proportional relationships and use them to solve real-world and mathematical problems.**

**7.RP.1**   Compute unit rates associated with ratios of fractions, including ratios of lengths, areas, and other quantities measured in like or different units.

**7.RP.2**   Recognize and represent proportional relationships between quantities.
   **a.**   Decide whether two quantities are in a proportional relationship.
   **b.**   Identify the constant of proportionality (unit rate) in tables, graphs, equations, and diagrams, and verbal descriptions of proportional relationships.
   **c.**   Represent proportional relationships by equations.
   **d.**   Explain what a point $(x, y)$ on the graph of a proportional relationship means in terms of the situation, with special attention to the points $(0, 0)$ and $(1, r)$ where $r$ is the unit rate.

**7.RP.3**   Use proportional relationships to solve multistep ratio and percent problems.

## Domain: The Number System

**Apply and extend previous understandings of operations with fractions to add, subtract, multiply, and divide rational numbers.**

**7.NS.1**   Apply and extend previous understandings of addition and subtraction to add and subtract rational numbers; represent addition and subtraction on a horizontal or vertical number line diagram.
   **a.**   Describe situations in which opposite quantities combine to make 0.
   **b.**   Understand $p + q$ as the number located a distance $|q|$ from $p$, in the positive or negative direction depending on whether $q$ is positive or negative. Show that a number and its opposite have the sum of 0 (are additive inverses). Interpret sums of rational numbers by describing real-world contexts.
   **c.**   Understand subtraction of rational numbers as adding the additive inverse, $p - q = p + (-q)$. Show that the distance between two rational numbers on the number line is the absolute value of their difference, and apply this principle in real-world contexts.
   **d.**   Apply properties of operations as strategies to add and subtract rational numbers.

**7.NS.2**   Apply and extend previous understandings of multiplication and division and of fractions to multiply and divide rational numbers.
   **a.**   Understand that multiplication is extended from fractions to rational numbers by requiring that operations continue to satisfy the properties of operations, particularly the distributive property, leading to products such as $(-1)(-1) = 1$ and the rules for multiplying signed numbers. Interpret products of rational numbers by describing real-world contexts.
   **b.**   Understand that integers can be divided, provided that the divisor is not zero, and every quotient of integers (with non-zero divisor) is a rational number. If $p$ and $q$ are integers, then $-(p/q) = (-p)/q = p/(-q)$. Interpret quotients of rational numbers by describing real-world contexts.
   **c.**   Apply properties of operations as strategies to multiply and divide rational numbers.
   **d.**   Convert a rational number to a decimal using long division; know that the decimal form of a rational number terminates in 0s or eventually repeats.

**7.NS.3**    Solve real-world and mathematical problems involving the four operations with rational numbers.

## Domain: Expressions and Equations

### Use properties of operations to generate equivalent expressions.

**7.EE.1**    Apply properties of operations as strategies to add, subtract, factor, and expand linear expressions with rational coefficients.

**7.EE.2**    Understand that rewriting an expression in different forms in a problem context can shed light on the problem and how the quantities in it are related.

### Solve real-life and mathematical problems using numerical and algebraic expressions and equations.

**7.EE.3**    Solve multi-step real-life and mathematical problems posed with positive and negative rational numbers in any form (whole numbers, fractions, and decimal), using tools strategically. Apply properties of operations to calculate with numbers in any form; convert between forms as appropriate; and assess the reasonableness of answers using mental computation and estimation strategies.

**7.EE.4**    Use variables to represent quantities in a real-world or mathematical problem, and construct simple equations and inequalities to solve problems by reasoning about the quantities.

    **a.**   Solve word problems leading to equations of the form $px + q = r$ and $p(x + q) = r$, where $p$, $q$, and $r$ are specific rational numbers. Solve equations of these forms fluently. Compare an algebraic solution to an arithmetic solution, identifying the sequence of the operations used in each approach.

    **b.**   Solve word problems leading to inequalities of the form $px + q > r$ or $px + q < r$, where $p$, $q$, and $r$ are specific rational numbers. Graph the solution set of the inequality and interpret it in the context of the problem.

## Domain: Geometry

### Draw, construct, and describe geometrical figures and describe the relationships between them.

**7.G.1**    Solve problems involving scale drawings of geometric figures, including computing actual lengths and areas from a scale drawing and reproducing a scale drawing at a different scale.

**7.G.2**    Draw (freehand, with ruler and protractor, and with technology) geometric shapes with given conditions. Focus on constructing triangles from three measures of angles or sides, noticing when the conditions determine a unique triangle, more than one triangle, or no triangle.

**7.G.3**    Describe the two-dimensional figures that result from slicing three-dimensional figures, as in plane sections of right rectangular prisms and right rectangle pyramids.

### Solve real-life and mathematical problems involving angle measure, area, surface area, and volume.

**7.G.4**    Know the formulas for the area and circumference of a circle and use them to solve problems; give an informal derivation of the relationship between the circumference and area of a circle.

**7.G.5**    Use facts about supplementary, complementary, vertical, and adjacent angles in a multi-step problem to write and solve simple equations for an unknown angle in a figure.

**7.G.6**    Solve real-world and mathematical problems involving area, volume and surface area of two- and three-dimensional objects composed of triangles, quadrilaterals, polygons, cubes, and right prisms.

# Domain: Statistics and Probability

## Use random sampling to draw inferences about a population.

**7.SP.1** Understand that statistics can be used to gain information about a population by examining a sample of the population; generalizations about a population from a sample are valid only if the sample is representative of that population. Understand that random sampling tends to produce representative samples and support valid inferences.

**7.SP.2** Use data from a random sample to draw inferences about a population with an unknown characteristic of interest. Generate multiple samples (or simulated samples) of the same size to gauge the variation in estimates or predictions.

## Draw informal comparative inferences about two populations.

**7.SP.3** Informally assess the degree of visual overlap of two numerical data distributions with similar variabilities, measuring the difference between the centers by expressing it as a multiple of a measure of variability.

**7.SP.4** Use measures of center and measures of variability for numerical data from random samples to draw informal comparative inferences about two populations.

## Investigate chance processes and develop, use, and evaluate probability models.

**7.SP.5** Understand that the probability of a chance event is a number between 0 and 1 that expresses the likelihood of the event occurring. Larger numbers indicate greater likelihood. A probability near 0 indicates an unlikely event, a probability around 1/2 indicates an event that is neither unlikely nor likely, and a probability near 1 indicates a likely event.

**7.SP.6** Approximate the probability of a chance event by collecting data on the chance process that produces it and observing its long-run relative frequency, and predict the approximate relative frequency given the probability.

**7.SP.7** Develop a probability model and use it to find probabilities of events. Compare probabilities from a model to observed frequencies; if the agreement is not good, explain possible sources of the discrepancy.
   **a.** Develop a uniform probability model by assigning equal probability to all outcomes, and use the model to determine probabilities of events.
   **b.** Develop a probability model (which may not be uniform) by observing frequencies in data generated from a chance process.

**7.SP.8** Find probabilities of compound events using organized lists, tables, tree diagrams, and simulation.
   **a.** Understand that, just as with simple events, the probability of a compound event is the fraction of outcomes in the sample space for which the compound event occurs.
   **b.** Represent sample spaces for compound events using methods such as organized lists, tables and tree diagrams. For an event described in everyday language, identify the outcomes in the sample space which compose the event.
   **c.** Design and use a simulation to generate frequencies for compound events.

# Photo Credits

## Chapter 9

**352** stephan kerkhofs/Shutterstock.com, Cigdem Sean Cooper/Shutterstock.com, ©iStockphoto.com/Andreas Gradin; **355** ©iStockphoto.com/Remigiusz Załucki; **358** *center left* Bob the Wikipedian / CC-BY-SA-3.0; *center right* ©iStockphoto.com/Sherwin McGehee; **360** Yasonya/Shutterstock.com; **361** Stankevich/Shutterstock.com; **362** *top left* ©iStockphoto.com/Luke Daniek; *top right* ©iStockphoto.com/Jeff Whyte; *bottom left* ©Michael Mattox. Image from BigStockPhoto.com; *bottom right* ©iStockphoto.com/Hedda Gjerpen; **363** ©iStockphoto.com/josh webb; **372** ©iStockphoto.com/Tomasz Pietryszek; **373** *center left* Newcastle Drum Centre; *center right* ©iStockphoto.com/scol22; **381** *top left* ©iStockphoto.com/david franklin; *top right* ©Ruslan Kokarev. Image from BigStockPhoto.com; *center right* ©iStockphoto.com/Lev Mel, ©iStockphoto.com/Ebru Baraz; **382** *bottom left* ©iStockphoto.com/Jiri Vatka; *bottom right* Patryk Kosmider/Shutterstock.com; **386** ©iStockphoto.com/ranplett, Image © Courtesy of Museum of Science, Boston; **387** *center right* ©iStockphoto.com/James Kingman; *center left* ©iStockphoto.com/Yails; **389** *Exercise 12* ©iStockphoto.com/AlexStar; *Exercise 13* Voznikevich Konstantin/Shutterstock.com; *Exercise 14* SOMMAI/Shutterstock.com

## Chapter 10

**398** infografick/Shutterstock.com, ©iStockphoto.com/Ann Marie Kurtz; **401** ryasick photography/Shutterstock.com; **402** *center* ©iStockphoto.com/sweetym; *bottom left* Big Ideas Learning, LLC; **403** ©iStockphoto.com/sweetym; **404** ©iStockphoto.com/Joe Potato Photo, ©iStockphoto.com/sweetym; **405** *top left* ©iStockphoto.com/Jennifer Morgan; *top center* United States coin image from the United States Mint; **409** *top left* design56/Shutterstock.com; *center right* ©iStockphoto.com/spxChrome; **410** *center right* Daniel Skorodyelov/Shutterstock.com; *bottom left* Mitrofanova/Shutterstock.com; **411** Jamie Wilson/Shutterstock.com; **412** *top right* James Steidl/Shutterstock.com; *bottom right* traudl/Shutterstock.com; **413** Warren Goldswain/Shutterstock.com; **414** ©iStockphoto.com/Frank van de Bergh; **415** ©iStockphoto.com/Eric Ferguson; **419** Feng Yu/Shutterstock.com; **420** *top right* John McLaird/Shutterstock.com; *center right* Robert Asento/Shutterstock.com; *bottom right* Mark Aplet/Shutterstock.com; **425** ©iStockphoto.com/Justin Horrocks; **426** ©iStockphoto.com/sweetym; **427** *top right* John McLaird/Shutterstock.com; *center* tele52/Shutterstock.com; **428 and 429** ©iStockphoto.com/Joe Potato Photo, ©iStockphoto.com/sweetym; **431** Univega/Shutterstock.com; **432** James Steidl/Shutterstock.com; **434** *top left* ©iStockphoto.com/sweetym; *center right* ©iStockphoto.com/Andy Cook; **435** -Albachiaraa-/Shutterstock.com; **439** ©iStockphoto.com/Doug Cannell; **440** *Activity 1a left* ©iStockphoto.com/Shannon Keegan; *Activity 1a right* ©iStockphoto.com/Lorelyn Medina; *Activity 1b left* Joel Sartore/joelsartore.com; *Activity 1b right* Feng Yu/Shutterstock.com; *Activity 1c left* ©iStockphoto.com/kledge; *Activity 1c right* ©iStockphoto.com/spxChrome; *Activity 1d* ©iStockphoto.com/Alex Slobodkin; **442** ©iStockphoto.com/Philip Lange; **444** ©iStockphoto.com/blackwaterimages, ©iStockphoto.com/Rodrigo Blanco, ©iStockphoto.com/7nuit; **447** Feng Yu/Shutterstock.com; **451** artis777/Shutterstock.com; **457** GoodMood Photo/Shutterstock.com; **458** Asaf Eliason/Shutterstock.com; **460** *top right* ©iStockphoto.com/Frank Ramspott; *bottom left* ©iStockphoto.com/7nuit; **462** mylisa/Shutterstock.com; **463** ra-design/Shutterstock.com

## Appendix A

**A0** *background* ©iStockphoto.com/Björn Kindler; *top left* ©iStockphoto.com/MichaelMattner; *top right* AKaiser/Shutterstock.com; **A1** *top right* ©iStockphoto.com/daver2002ua; *bottom left* ©iStockphoto.com/Eric Isselée; *bottom right* ©iStockphoto.com/Victor Paez; **A2** sgame/Shutterstock.com; **A3** *top right* AKaiser/Shutterstock.com; *bottom left* Photononstop/SuperStock, ©iStockphoto.com/Adam Radosavljevic; *bottom right* ©iStockphoto.com/MichaelMattner; **A4** ©iStockphoto.com/Duncan Walker; **A5** *top right* ©iStockphoto.com/DNY59; *bottom left* ©iStockphoto.com/Boris Katsman; *bottom right* © User:MatthiasKabel/Wikimedia Commons / CC-BY-SA-3.0 / GFDL; **A6** *top right* ©iStockphoto.com/Keith Webber Jr.; *bottom left* ©iStockphoto.com/Eliza Snow; *bottom right* ©iStockphoto.com/David H. Seymour; **A7** *top right* ©iStockphoto.com/daver2002ua; *bottom left* ©iStockphoto.com/Victor Paez; **A8** *wasp* ©iStockphoto.com/arlindo71; *bat* ©iStockphoto.com/Alexei Zaycev; *monkey and kangaroo* ©iStockphoto.com/Eric Isselée; **A9** *macaw* ©iStockphoto.com/Eric Isselée; *frog* ©iStockphoto.com/Brandon Alms; *spider* ©iStockphoto.com/arlindo71; *puma* ©iStockphoto.com/Eric Isselée; *lobster* ©iStockphoto.com/Joan Vicent Cantó Roig; *crocodile* ©iStockphoto.com/Mehmet Salih Guler

**Cartoon illustrations** Tyler Stout

# Common Core State Standards

## Kindergarten

| | |
|---|---|
| Counting and Cardinality | — Count to 100 by Ones and Tens; Compare Numbers |
| Operations and Algebraic Thinking | — Understand and Model Addition and Subtraction |
| Number and Operations in Base Ten | — Work with Numbers 11–19 to Gain Foundations for Place Value |
| Measurement and Data | — Describe and Compare Measurable Attributes; Classify Objects into Categories |
| Geometry | — Identify and Describe Shapes |

## Grade 1

| | |
|---|---|
| Operations and Algebraic Thinking | — Represent and Solve Addition and Subtraction Problems |
| Number and Operations in Base Ten | — Understand Place Value for Two-Digit Numbers; Use Place Value and Properties to Add and Subtract |
| Measurement and Data | — Measure Lengths Indirectly; Write and Tell Time; Represent and Interpret Data |
| Geometry | — Draw Shapes; Partition Circles and Rectangles into Two and Four Equal Shares |

## Grade 2

| | |
|---|---|
| Operations and Algebraic Thinking | — Solve One- and Two-Step Problems Involving Addition and Subtraction; Build a Foundation for Multiplication |
| Number and Operations in Base Ten | — Understand Place Value for Three-Digit Numbers; Use Place Value and Properties to Add and Subtract |
| Measurement and Data | — Measure and Estimate Lengths in Standard Units; Work with Time and Money |
| Geometry | — Draw and Identify Shapes; Partition Circles and Rectangles into Two, Three, and Four Equal Shares |

# Grade 3

| | |
|---|---|
| **Operations and Algebraic Thinking** | – Represent and Solve Problems Involving Multiplication and Division; Solve Two-Step Problems Involving Four Operations |
| **Number and Operations in Base Ten** | – Round Whole Numbers; Add, Subtract, and Multiply Multi-Digit Whole Numbers |
| **Number and Operations—Fractions** | – Understand Fractions as Numbers |
| **Measurement and Data** | – Solve Time, Liquid Volume, and Mass Problems; Understand Perimeter and Area |
| **Geometry** | – Reason with Shapes and Their Attributes |

# Grade 4

| | |
|---|---|
| **Operations and Algebraic Thinking** | – Use the Four Operations with Whole Numbers to Solve Problems; Understand Factors and Multiples |
| **Number and Operations in Base Ten** | – Generalize Place Value Understanding; Perform Multi-Digit Arithmetic |
| **Number and Operations—Fractions** | – Build Fractions from Unit Fractions; Understand Decimal Notation for Fractions |
| **Measurement and Data** | – Convert Measurements; Understand and Measure Angles |
| **Geometry** | – Draw and Identify Lines and Angles; Classify Shapes |

# Grade 5

| | |
|---|---|
| **Operations and Algebraic Thinking** | – Write and Interpret Numerical Expressions |
| **Number and Operations in Base Ten** | – Perform Operations with Multi-Digit Numbers and Decimals to Hundredths |
| **Number and Operations—Fractions** | – Add, Subtract, Multiply, and Divide Fractions |
| **Measurement and Data** | – Convert Measurements within a Measurement System; Understand Volume |
| **Geometry** | – Graph Points in the First Quadrant of the Coordinate Plane; Classify Two-Dimensional Figures |

# Mathematics Reference Sheet

## Conversions

### U.S. Customary
1 foot = 12 inches
1 yard = 3 feet
1 mile = 5280 feet
1 acre ≈ 43,560 square feet
1 cup = 8 fluid ounces
1 pint = 2 cups
1 quart = 2 pints
1 gallon = 4 quarts
1 gallon = 231 cubic inches
1 pound = 16 ounces
1 ton = 2000 pounds
1 cubic foot ≈ 7.5 gallons

### U.S. Customary to Metric
1 inch = 2.54 centimeters
1 foot ≈ 0.3 meter
1 mile ≈ 1.61 kilometers
1 quart ≈ 0.95 liter
1 gallon ≈ 3.79 liters
1 cup ≈ 237 milliliters
1 pound ≈ 0.45 kilogram
1 ounce ≈ 28.3 grams
1 gallon ≈ 3785 cubic centimeters

### Time
1 minute = 60 seconds
1 hour = 60 minutes
1 hour = 3600 seconds
1 year = 52 weeks

### Temperature
$$C = \frac{5}{9}(F - 32)$$

$$F = \frac{9}{5}C + 32$$

### Metric
1 centimeter = 10 millimeters
1 meter = 100 centimeters
1 kilometer = 1000 meters
1 liter = 1000 milliliters
1 kiloliter = 1000 liters
1 milliliter = 1 cubic centimeter
1 liter = 1000 cubic centimeters
1 cubic millimeter = 0.001 milliliter
1 gram = 1000 milligrams
1 kilogram = 1000 grams

### Metric to U.S. Customary
1 centimeter ≈ 0.39 inch
1 meter ≈ 3.28 feet
1 kilometer ≈ 0.62 mile
1 liter ≈ 1.06 quarts
1 liter ≈ 0.26 gallon
1 kilogram ≈ 2.2 pounds
1 gram ≈ 0.035 ounce
1 cubic meter ≈ 264 gallons

## Number Properties

Commutative Properties of Addition and Multiplication
$$a + b = b + a$$
$$a \cdot b = b \cdot a$$

Associative Properties of Addition and Multiplication
$$(a + b) + c = a + (b + c)$$
$$(a \cdot b) \cdot c = a \cdot (b \cdot c)$$

Addition Property of Zero
$$a + 0 = a$$

Multiplication Properties of Zero and One
$$a \cdot 0 = 0$$
$$a \cdot 1 = a$$

Distributive Property:
$$a(b + c) = ab + ac$$
$$a(b - c) = ab - ac$$

## Properties of Equality

Addition Property of Equality
If $a = b$, then $a + c = b + c$.

Subtraction Property of Equality
If $a = b$, then $a - c = b - c$.

Multiplication Property of Equality
If $a = b$, then $a \cdot c = b \cdot c$.

Multiplicative Inverse Property
$$n \cdot \frac{1}{n} = \frac{1}{n} \cdot n = 1, n \neq 0$$

Division Property of Equality
If $a = b$, then $a \div c = b \div c, c \neq 0$.

## Properties of Inequality

**Addition Property of Inequality**
    If $a > b$, then $a + c > b + c$.

**Subtraction Property of Inequality**
    If $a > b$, then $a - c > b - c$.

**Multiplication Property of Inequality**
    If $a > b$ and $c$ is positive, then $a \cdot c > b \cdot c$.
    If $a > b$ and $c$ is negative, then $a \cdot c < b \cdot c$.

**Division Property of Inequality**
    If $a > b$ and $c$ is positive, then $a \div c > b \div c$.
    If $a > b$ and $c$ is negative, then $a \div c < b \div c$.

## Circumference and Area of a Circle

$C = \pi d$ or $C = 2\pi r$

$A = \pi r^2$

$\pi \approx \dfrac{22}{7}$, or 3.14

## Angles of Polygons

**Sum of the Angle Measures of a Triangle**

$x + y + z = 180$

**Sum of the Angle Measures of a Quadrilateral**

$w + x + y + z = 360$

## Surface Area

**Prism**

$S = 2\ell w + 2\ell h + 2wh$

$S$ = areas of bases
    + areas of lateral faces

**Pyramid**

$S$ = area of base
    + areas of lateral faces

**Cylinder**

$S = 2\pi r^2 + 2\pi rh$

## Volume

**Prism**

$V = Bh$

$V = Bh$

**Pyramid**

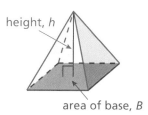

$V = \dfrac{1}{3}Bh$

## Simple Interest

Simple interest formula

$I = Prt$